GREAT DAMS IN SOUTHERN AFRICA

Painting by Waaldo Dingemans · Photograph by Eric Hayne (Pty) Ltd

THE AUTHOR WEARING THE GOWN AND HOOD OF THE HONORARY DOCTORATE OF SCIENCE FROM THE UNIVERSITY OF CAPE TOWN AND THE ORDER OF SAINT MICHAEL AND SAINT GEORGE

Dr H. OLIVIER
C.M.G., PH.D., D.ENG., D.SC. (HON.), F.I.C.E.,
F.S.A.I.C.E., F.ASCE, F.R.S.A.

GREAT DAMS

IN SOUTHERN AFRICA

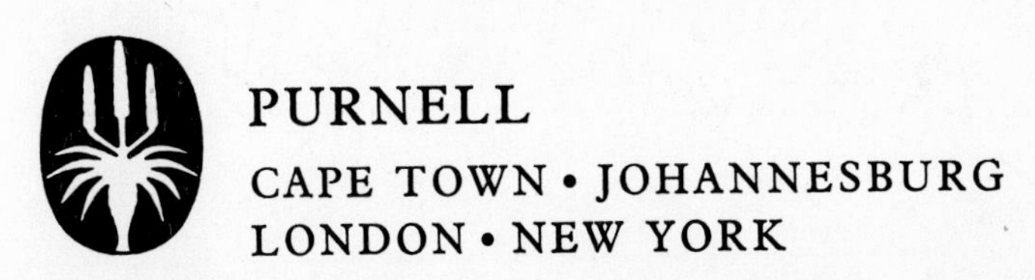

PURNELL
CAPE TOWN • JOHANNESBURG
LONDON • NEW YORK

Published by
Purnell & Sons S.A. Pty Ltd
97 Keerom Street Cape Town

ISBN 0 86843 004 8

Colour separations: Photo Sepro Pty Ltd

Layout Nick Hobbelman

Set in Plantin: Printed and Bound
in South Africa by
Printpak (Cape) Ltd

To: All the countries of Africa

Salus populi suprema lex

(THE PEOPLE'S WELFARE IS THE SUPREME LAW)

Baron Verulam (Francis Bacon)

Verw.
Ref. I.24/2

REPUBLIEK VAN SUID-AFRIKA
REPUBLIC OF SOUTH AFRICA

Kantoor van die Eerste Minister
Prime Minister's Office

FOREWORD BY THE HONOURABLE B.J. VORSTER, PRIME MINISTER OF SOUTH AFRICA

The first dam known to have been built by man was constructed in Northern Africa some 4 000 years ago and remains of it can still be seen in a Wadi in Egypt. In Southern Africa some of the most modern dams in the world have been and still are being constructed. It is very fitting that a book describing some of these achievements has now been compiled by a South African engineer who has devoted his life to water development projects in many parts of the world.

Water is the elixir of life. It is a catalyst which is essential for all development: industrial and agricultural. The earliest extensive civilisations started in river valleys, along the Nile in Egypt and along the Tigris and Euphrates in Mesopotamia, when man learnt how to make use of water to improve his living conditions. Modern civili= sation needs more water than ever to maintain and improve its standards.

In Southern Africa the rainfall in general decreases from the Equator southwards. We have, therefore, large perennial rivers with a reliable flow in the North and more erratic rivers in the South.

By making use of imaginative concepts for the construction of dams and interlinking water projects, a reliable water supply even during drought periods can be made available for irrigation, urban and industrial use and power develop= ment. Thereby great opportunities can be created for inter= national co-operation to the mutual benefit of all concerned.

This reference book describes 46 of the major dams already constructed in Southern Africa. Other such projects of a regional and interregional character are on the drawing boards. There can be no turning back from the husbanding of precious water resources to ensure that enough will be available and to maintain its quality. This will require the building of more dams and I am sure that men of vision and initiative will be found to do this in such a way that optimum benefits will be obtained.

Contents

Foreword

Preface 11

Prologue 13

Acknowledgements 17

Introduction 1 19

Introduction 2 28

Introduction 3 35

CHAPTER 1 Uganda 43

CHAPTER 2 Zaire 54

CHAPTER 3 Tanzania 61

CHAPTER 4 Malawi 63

CHAPTER 5 Zambia 65

CHAPTER 6 Moçambique 75

CHAPTER 7 Rhodesia 89

CHAPTER 8 Angola/S.W. Africa 123

CHAPTER 9 South West Africa 130

CHAPTER 10 Swaziland 138

CHAPTER 11 Botswana 143

CHAPTER 12 Lesotho 145

CHAPTER 13 Transkei 148

CHAPTER 14 South Africa 153

General Index 229

Consulting Engineers and Contractors Index 231

Preface

H. LYRA

.. T. H. LYRA DA SILVA
President International Commission on Large Dams (1976–1979)

The International Commission on Large Dams (ICOLD) *has a current membership of 74 nations, all concerned with the design, construction, maintenance and operation of large dams.*
If Southern Africa is taken to be the region to the south of the equator, of the many countries existing in this area only South Africa, Rhodesia and Zambia are at present members of ICOLD. *Moçambique and Angola, until recently, were represented before their independence by Portugal.*
There are many dams in Southern Africa which fall within the definition of a Large Dam as laid down by ICOLD. *This definition reads:*

> *"All dams above 15 m in height measured from the lowest portion of the general foundation area to the crest.*
> *Dams between 10 m and 15 m can be included provided they comply with at least one of the following conditions:*
> *(a) length of crest not less than 500 m.*
> *(b) capacity of reservoir formed by the dam not less than 1 x* $10^6 m^3$.
> *(c) maximum flood discharge dealt with by the dam not less than 2 000* m^3/s.
> *(d) dam has specially difficult foundation problems.*
> *(e) dam is of unusual design."*

The publishers are of the opinion that, apart from individual papers published in various journals from time to time describing some major projects in this vast region, little is known internationally about the functions and dimensions of such projects in the different territories.
In order to remedy this situation they decided to compile what may be regarded as an illustrated reference book listing the principal or pertinent data of 46 major projects in fourteen countries selected in so far as was possible by the authorities concerned or from published data.
It is considered that such a concentration of vital statistics in one volume will be of great value to people in all parts of the world interested in the water resources and in the development achieved and to be achieved in this part of the world which, in general, must be regarded as developing territories.
It must be pointed out that it would not be practicable to describe in a book of this nature all the dams which qualify under the ICOLD *definition. For instance the total number of dams in the Republic of South Africa, including farm dams, is more than 500 000, of which 306 are classified as large in the* ICOLD *Register of 1973–74.*
It was therefore decided to select a list of major dam or multi-purpose projects which will record in one volume for the various interested persons and institutions in the world historical and vital statistical data of key water projects in Southern Africa which have or will have a bearing on future development potential in the sub-continent with special reference to inter-basin linkage.

It is known that many other projects are being planned in the countries covered, but it was felt that future planning is liable to change and hence the work is restricted to existing projects of special interest. The work will no doubt be brought up to date as and when considered necessary in the light of other major developments in this field.
It is hoped that the work will be regarded as a reference book by member countries of ICOLD *and even those who are not yet members, and particularly by organisations that are interested in water resource development in Southern Africa.*

c/o Comite Brasileiro do Grandes Barragens
Rua Real Grandeza 219
RIO DE JANEIRO R.J.
BRAZIL

Prologue

HENRY OLÍVIER

It was for me an honour to be approached by the Publishers to compile an illustrated volume of Great Dams in Southern Africa and to write three chapters for the book dealing with aspects of future interdependence in Southern Africa as regards development of water resources to the mutual economic and social benefit of all the nations involved and indeed to the possible benefit of all Africa.

It became a pleasant task to collect and collate the relevant reference data in close collaboration with the many authorities and individuals concerned. In the process I also learnt much about the dimensions and functions of the various dams in the regions other than those with which I had been intimately concerned in the past.

The intrinsic value of the concept by the Publishers was brought home to me as the work neared completion. One would have to seek far and wide, as indeed I had to do, in order to obtain the principal statistics for each of the 46 projects selected. Now they are concentrated in one volume for easy reference by anyone anywhere in the world who might be interested in Southern Africa. The rest of the world is becoming very interested in Southern Africa for diverse reasons.

As regards electricity the concept of a Southern African Electricity Transmission Grid is slowly becoming a fact. A few more links using the most up-to-date transmission technology and Southern Africa will have an inter-connected electricity grid such as links most European countries and which will permit harnessing to best economic use first the water and power resources, secondly the creation of infrastructure which will, in turn, spark the development of large mineral resources in this vast region.

As regards food, the Third World Countries face rapidly declining nutritional standards. Southern Africa has the potential to look after itself and with peaceful co-operation meaningful progress can be made in this direction, first to become self-sufficient in staple foods and next to export surpluses to other nations.

Some of the territories involved are landlocked and hence a look is taken at recent technological developments in the sphere of river transport. In the long run it is difficult to improve on the economics of water transport and these territories will have to look increasingly to such possible use of their own or international rivers, natural lakes and artificial lakes created by dams.

I am now nearing the end of my professional career, but I look forward to reviewing the situation in five to ten years time and to gauge the extent to which these forward looking plans have matured.

'LINDFIELD',
6 JAMESON AVENUE
MELROSE
JOHANNESBURG
2196

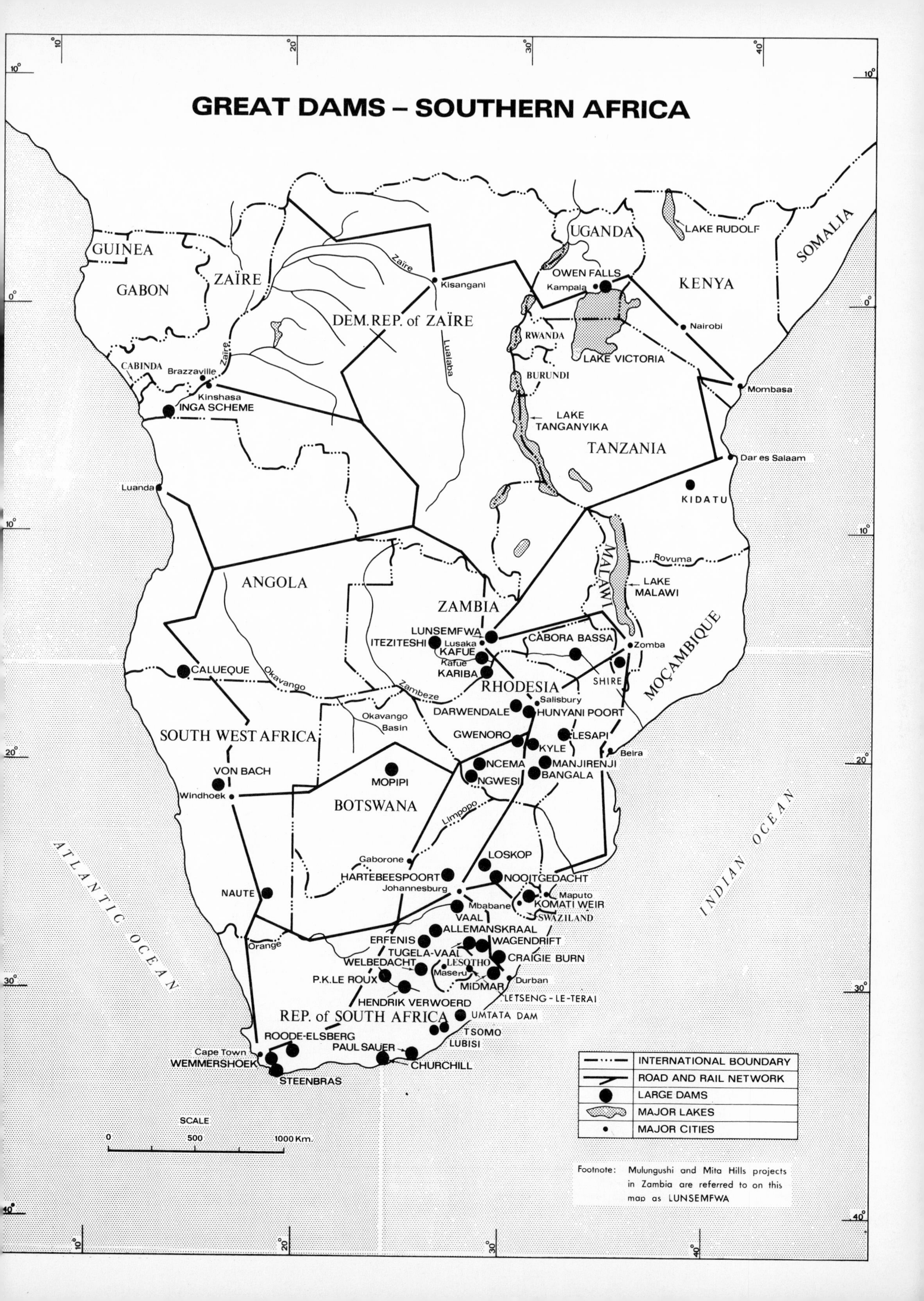

GREAT DAMS – SOUTHERN AFRICA
GUINEA
GABON
ZAÏRE
DEM. REP. of ZAÏRE
UGANDA
KENYA
SOMALIA
RWANDA
BURUNDI
TANZANIA
ANGOLA
ZAMBIA
MALAWI
MOÇAMBIQUE
RHODESIA
SOUTH WEST AFRICA
BOTSWANA
SWAZILAND
LESOTHO
REP. of SOUTH AFRICA
CABINDA
LAKE RUDOLF
LAKE VICTORIA
LAKE TANGANYIKA
LAKE MALAWI
ATLANTIC OCEAN
INDIAN OCEAN
Zaïre
Lualaba
Rovuma
Okavango
Okavango Basin
Zambeze
Kafue
Limpopo
Orange
Kisangani
Kampala
Nairobi
Mombasa
Dar es Salaam
Brazzaville
Kinshasa
Luanda
Lusaka
Zomba
Salisbury
Beira
Windhoek
Gaborone
Johannesburg
Mbabane
Maputo
Maseru
Durban
Cape Town
OWEN FALLS
INGA SCHEME
KIDATU
LUNSEMFWA
ITEZITESHI
KAFUE
KARIBA
CABORA BASSA
SHIRE
CALUEQUE
DARWENDALE
HUNYANI POORT
GWENORO
KYLE
LESAPI
NCEMA
NGWESI
MANJIRENJI
BANGALA
VON BACH
MOPIPI
LOSKOP
HARTEBEESPOORT
NOOITGEDACHT
KOMATI WEIR
NAUTE
VAAL
ALLEMANSKRAAL
ERFENIS
WAGENDRIFT
TUGELA-VAAL
CRAIGIE BURN
WELBEDACHT
P.K.LE ROUX
MIDMAR
LETSENG - LE-TERAI
HENDRIK VERWOERD
UMTATA DAM
TSOMO
LUBISI
ROODE-ELSBERG
PAUL SAUER
CHURCHILL
WEMMERSHOEK
STEENBRAS
INTERNATIONAL BOUNDARY
ROAD AND RAIL NETWORK
LARGE DAMS
MAJOR LAKES
MAJOR CITIES
SCALE
0
500
1000 Km.
Footnote: Mulungushi and Mita Hills projects in Zambia are referred to on this map as LUNSEMFWA

Acknowledgements

Acknowledgements by the Publishers and the Author are due to the following persons and organisations for assistance in providing relevant data, photographs and other material for compiling the reference section covering the projects in the various zones:

Chapter 1 **Uganda**
The Uganda Electricity Board, Sir Alexander Gibb and Partners and the Institution of Civil Engineers for permission to reproduce data relating to the Owen Falls Hydro-electric Scheme.

Chapter 2 **Zaire**
Mr. W. E. Wentges, Chairman of Siemens (Pty) Ltd. of South Africa; Messrs. SICAI of Rome and Electro-consult of Milan for providing details of the INGA project, and Italafrica/Kinshaha for permission to reproduce two photographs of the Inga projects which appeared in the 1977 Calendar of Société Nationale d' Electricité Kinshaha.

Chapter 3 **Tanzania**
Reinius E. and Steby B. *Kidatu Powerplant, Tanzania.* International Water Power & Dam Construction. Vol. 28 No. 11, November 1976.

Chapter 4 **Malawi**
The Electricity Supply Commission of Malawi and Messrs. Kennedy & Donkin (Africa) for permission to print details of the hydro-electric potential of the Shire River.

Chapter 5 **Zambia**
Mr. A. G. V. Pearce, Consulting Mechanical and Electrical Engineer, Nchanga Consolidated Copper Mines and Messrs. Watermeyer, Legge, Piesold and Uhlmann, Consulting Engineers, for details of the Mulungushi and Mita Hills dams.
Mr. H. Kaunda, Chairman, Nchanga Consolidated Copper Mines and Mr. A. Mkandawire, Zambian Electricity Supply Commission for permission to include details of the Kariba underground power station (Zambian bank), the Itezhitezhi and Kafue projects.
Nchanga Consolidated Copper Mines for permission to reproduce three photographs of the Mita Hills dam and Lunsemfwa power station.

Chapter 6 **Mocambique**
The Anglo American Corporation for permission to reproduce a panoramic view of the Cabora Bassa dam and lake. Mr. Enno Vocke of Hochtief, West Germany, for permission to reproduce photographs of Cabora Bassa. The operating company Hidrotecnica Cabora Bassa (H.C.B.) at Songo; Members of the Consortium Zamco, LTA Limited of Johannesburg for permission to reproduce details of the project.

Chapter 7 **Rhodesia**
Mr. Cecil Wetmore, Secretary for Transport and Power, Mr. J. Ward, C.B.E., Chairman of CAPCO, and Mr. Gordon Allison, Chief Civil Engineer of CAPCO, for permission to include details and to reproduce photographs of the Kariba project;
Eng. S. W. Loewenson, C.Eng., Director of the Department of Water Development and his Chief Designs Engineer, Eng. G. P. Carmichael, C.Eng., for selection of dams and providing principal data and photographs relating to these;
Mr. Michael V. Gardner, Director of Tourism, for permission to reproduce a photograph from their brochures; Brig. M. O. Collins, C.B.E., (Consultant) and Mr. Roger Sleigh, Department of the Surveyor General, for details of seismic surveys and crustal behaviour connected with Lake Kariba;
The Rhodesian Ministry of Information for photographs of some of the Rhodesian dams included.

Chapter 8 **Angola - South West Africa**
The South African Institution of Civil Engineers' journal *The Civil Engineer in South Africa* for permission to reproduce details published in their issue of 12th December 1975 relating to the Ruacana - Calueque projects; Dr. D. Stephenson, Chairman of the Division of Hydraulic and Water Engineering (SAICE) and Messrs. R. Immelman, W. C. S. Legge and J. B. Richard for permission to reproduce the photographs taken by them of the Ruacana-Calueque projects; The South West Africa Water and Electricity Corporation and Mr. E. P. Chunnett (Consulting Engineer) for permission to reproduce details from his article on these projects which appeared in the July 1976 issue of "South African Tunnelling";
The South West Africa Water and Electricity Corporation and Mr. A. D. W. Wolmarans for using extracts from his Paper entitled *Power and Water and South West Africa* which appeared in *The Professional Engineer* of October, 1975.

Chapter 9 **South West Africa**
The South West Africa Water Electricity Corporation, Mr. E. P. Chunnett and LTA Limited for details of the Von Bach dam and Photo Pfohl of Windhoek for reproduction of a photograph taken by them.

Chapter 10 **Swaziland**
The Commonwealth Development Corporation; Mr. M. F. Ollerenshaw, its Regional Manager; Mr. J. R. Tuckett, General Manager of the Swaziland Irrigation Scheme and his project engineer Mr. J. F. T. Haine, for details of the Mhlume Canal and the irrigation project and for vetting the chapter on this project.

Chapter 11 **Botswana**
The Anglo American Corporation and Messrs. B. G. A. Lund and Partner for details of the Putimolonwane Pan (Mopipi Dam) project, and for permission to reproduce a photograph of the pan.

Chapter 12 **Lesotho**
The Anglo American Corporation and Mr. D. Standish White for permission to publish a photograph of the Letseng-le-Terai overflow rockfill dam and to Messrs. Gibb Hawkins and Partners and Sir Alexander Gibb and Partners for details of the design of this dam.

Chapter 13 **Transkei**
The Secretary for Water Affairs (RSA) for details of the Lubisi and Tsomo dams and the Secretary for Agriculture and Forestry (Transkei) for details of the Umtata dam.

Chapter 14 **R.S.A.**
Dr. J. P. Kriel, Pr.Eng., Secretary for Water Affairs and members of the Department of Water Affairs, for selection of dams and projects in the Republic of South Africa and for providing the principal data and photographs. Dr. R. Straszacker, Chairman of ESCOM for permission to include details of Hendrik Verwoerd and Vanderkloof (P. K. le Roux) power stations, (Orange River Projects), and the Apollo converter station at Irene for re-conversion of Direct into Alternating Current, received from the Cabora Bassa Hydro-electric project.
Mr. P. J. Treurnich, Public Relations Officer of ESCOM, for making available selected photographs and Mr. E. S. Smook of ESCOM, for providing the principal Mechanical and Electrical data for the Hendrik Verwoerd and Vanderkloof Power Stations.
Mr. Gordon Douglas, F.R.P.S., F.P.S.(SA), for permission to reproduce his photographs of the following dams: Wemmershoek, Vaaldam, Steenbras, Wagendrift and Welbedacht;
Water 1975 published by Erudita Publications (Pty.) Ltd., Ferreirasdorp, Johannesburg, for permission to reproduce parts of the text and sketches from this book relating to the Orange River and Tugela Vaal projects; Mr. Otto Adendorff of the Department of Information for permission to reproduce some of his photographs and slides.
Mr. Theo Owen of SATOUR for permission to reproduce the photograph of the Paul Sauer dam taken by Mr. G. Thompson, A.I.I.P.;
The Union Corporation and Professors Jennings and Midgley for providing details of the Bafokeng overflow slimes dam.

Special acknowledgements are due to the following for assistance rendered in compiling the book:

The Ministry of Information for supplying a photograph of the Prime Minister, the Hon. B. J. Vorster, for inclusion with his Foreword;
Mr. R. G. Allnutt for preparing all the art work and drawings;
Mr. Gordon Douglas, F.R.P.S., FPS(SA), for permission to reproduce his photograph of the Hendrik Verwoerd dam on the cover of this book;
Mr. John Faber, London, for details about grain silos;
Mr. P. S. Chennell and Mr. C. Palmer, Marketing Managers for Hovermarine U.K., for details of Hovercraft and Hoverbarges;
MacMillan S.A. (Publishers) (Pty.) Ltd. for permission to reproduce some data previously published by them in *Damit* (1975);
Mr. Ken Anderson for collating text in relation to photographs, drawings and art work;
Messrs. Eric Hayne (Pty.) Ltd. for permission to reproduce their colour photograph of the painting of the Author by Waalko Dingemans;
Last but not least to Mrs. M. M. Hyde, my Secretary, for typing the manuscript which required many drafts.

1 Introduction

INTERDEPENDENCE IN ENERGY AND WATER BETWEEN COUNTRIES IN SOUTHERN AFRICA.

Until 1973 the word "energy" was a commonplace word in the dull sense. Like the water and the air, electricity was considered to be there for all to enjoy – for all time.

Then came the bombshell: the oil and energy crisis sparked by the quadrupling of oil prices by the OPEC countries. Overnight a serious situation arose with world wide ramifications. Suddenly it was realised how dependent civilisation is on energy – in every day life and for the continued growth of the economy and the improvement in standards of living. It was Lilienthal of Tennessee Valley fame who said some years ago that one kilowatt is worth 10 men extra to the economy. Thus in effect a nation could multiply its effective population – its productivity by such factors through the application and use of available energy.

It is significant that the higher a nation is placed up the scale of development the higher is its demand for units of energy per head of population. In countries most advanced as regards industrial and agricultural development, such as the United States of America, the annual consumption of electrical energy in 1973 was of the order of 9 254 units (kWh) per head of population. The consumption in Sweden was the same. In the Republic of South Africa the consumption in that year was some 2 667 units per head per annum with the growth rate in demand redoubling every ten years. These figures compare with 1 537, the world average, 151 for Algeria and 336 for Ghana.

In one sense the Republic, which is the industrial giant of the continent, is not in danger of an energy crisis linked with oil – as in the Americas and Europe. About 99 per cent of the electrical energy comes from coal-fired thermal stations. Coal of the type needed is plentiful. The same cannot unfortunately be said for most of the neighbours to the north.

Whereas coal may be plentiful in the Republic, water is not. Water is life. Without it there is nothing. These large thermal stations of the size of Kriel at 3 000 megawatts consume large quantities of water in evaporation for cooling purposes. The Kriel station on the basis of Escom figures for 1973 would consume 35 million imperial gallons (160 million litres) per day, or about 65 cubic feet per second. (1,8 cubic metres per second).

It is for this reason that Escom has been experimenting with a "dry" cooling system at their Grootvlei power station in an effort to reduce water consumption.

Not only is water valuable and becoming more so, but in many cases the power stations are located for overall economic reasons on or near the coal fields. These are not always conveniently located near water sources, which means that water has to be brought over great distances with corresponding cost increases.

The nuclear stations will be in the same position but some of these can be located near the sea and salt water used for cooling.

There is yet another angle. Dr. Baghar Mostofi, the Chief Executive of the Iranian National Petroleum Company and a member of the Iranian National Oil Board, delivered a lecture to the National War College of the U.S.A. in Tehran on 7th May 1975 in which he stated *inter alia:*

'While conscious of its responsibility towards the consumers as a producer of oil, Iran is also fully aware that its reserves of crude oil and gas, no matter how large, are none the less exhaustible. Therefore, in future, as other sources of energy come more and more into the forefront, for supplying the world's energy needs Iran plans, in accordance with a co-ordinated and realistic phasing, to change its role as supplier of crude oil to that of a supplier of petrochemicals. The emphasis in our programme, however, is on the manufacture of bulk petrochemicals which will provide for many decades the necessary material for further processing in Iran and abroad.

In this way also by upgrading its hydrocarbon resources and benefiting from their added values Iran will be in a position to obtain, in spite of a reduced oil production, the necessary foreign exchange and sources of funds for development of her economy in the 21st century. Thus, any petrochemical venture in this country is assured of longest possible life.

Dedicated to this proposition, the government of Iran has given priority to petrochemical feedstock over the energy requirement within the country and is planning to install nuclear power stations to take care of our own extensive energy needs.'

Thus, availability as well as price consideration will dominate future thinking and planning.

If one takes into account such a possible trend coupled with the needs of the steel and other industries and a sharp increase in coal exports to nations which will progressively be starved of oil as the producing nations cut back in supplies, it may be necessary to look again at coal reserves in Southern Africa both as regards quantity and possible price trends. The coal consumed by Escom in 1973 was 30 million metric tons. On past trends this consumption will be double this figure by 1983.

In assessing a nation's energy requirements one must not overlook nutritional viability. The available energy for labour and effort in a nation is as important, if not more so, than electrical energy, and nutritional viability is also very closely related to water.

Diet deficiency has a pronounced impact on national economy as regards output per man hour, expectancy of life, health requirements, import of foodstuffs, hence foreign currency problems and, therefore, political alignments.

As far as the capacity for work is concerned, the main results of a poor diet are that the body avoids effort, there is a lowering of resistance to disease, and the accident rate rises. The symptoms of the avoidance of effort are lethargy and a lack of initiative and drive.

Alarming trends are revealed in recent surveys carried out by the U.S. Department of Agriculture in the sixties which showed that whereas the total and *per capita* food production for the more developed countries show a steady increase over the period 1954–66, the *per capita* food production for the less developed countries during the same period showed an alarming drop back to 1954 figures, starting in 1963. Africa, with the exception of South Africa and Rhodesia, is in the latter category.

The world is bulging from accelerating increases in population, and is shrinking as regards all forms of communication. In combination, these twin dynamic trends generate pressures to equalize standards of living and render it increasingly difficult to isolate hunger geographically, with subsequent great potential danger for all mankind.

Distinct trends of thought in water resources engineering are emerging as states-

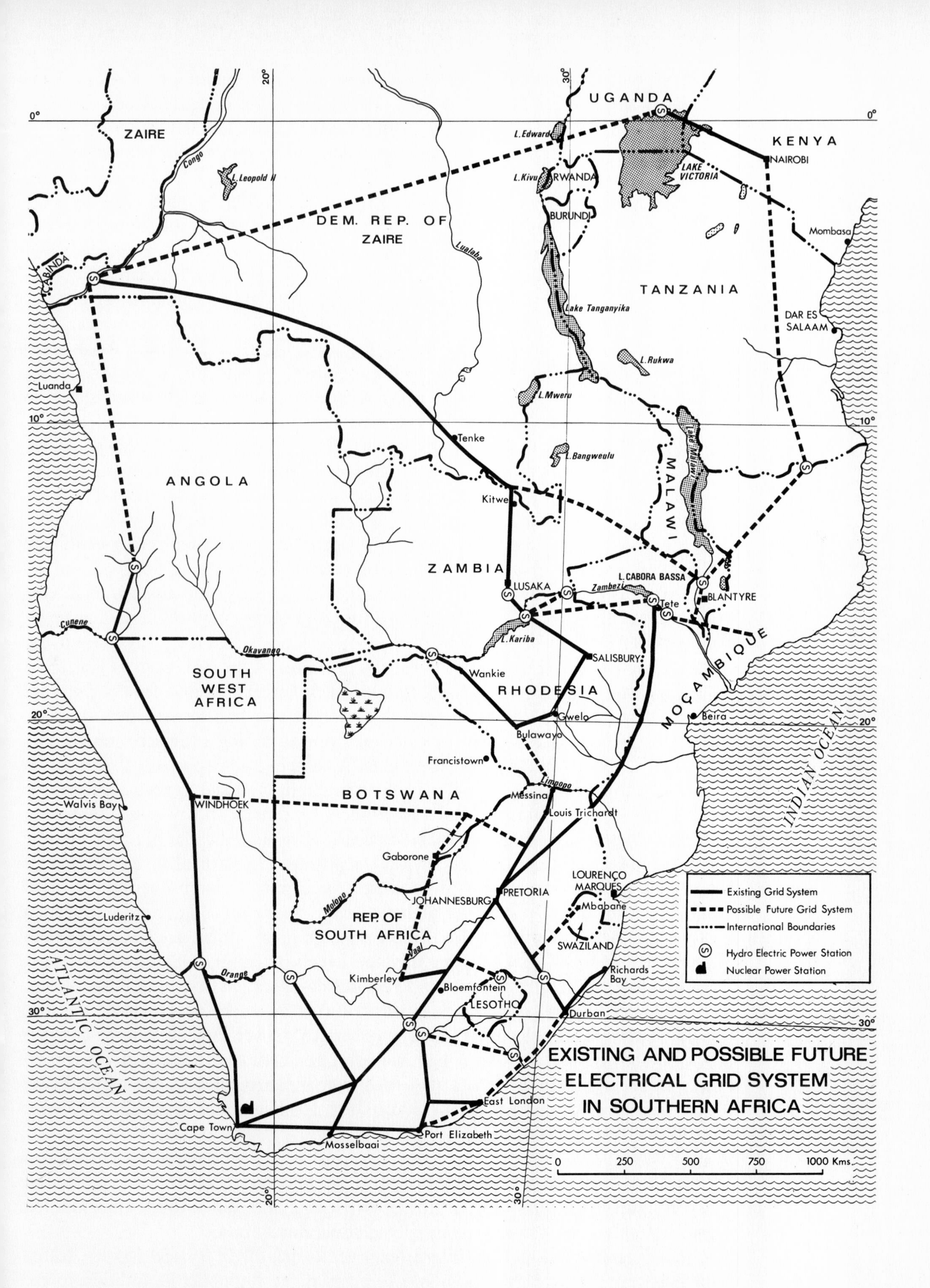
EXISTING AND POSSIBLE FUTURE ELECTRICAL GRID SYSTEM IN SOUTHERN AFRICA
Existing Grid System
Possible Future Grid System
International Boundaries
Hydro Electric Power Station
Nuclear Power Station
0 250 500 750 1000 Kms.
ZAIRE
UGANDA
KENYA
DEM. REP. OF ZAIRE
TANZANIA
ANGOLA
ZAMBIA
MALAWI
SOUTH WEST AFRICA
RHODESIA
MOÇAMBIQUE
BOTSWANA
REP. OF SOUTH AFRICA
SWAZILAND
LESOTHO
CABINDA
RWANDA
BURUNDI
ATLANTIC OCEAN
INDIAN OCEAN
L. Edward
L. Kivu
L. Leopold II
LAKE VICTORIA
Lake Tanganyika
L. Rukwa
L. Mweru
L. Bangweulu
Lake Malawi
L. CABORA BASSA
L. Kariba
Congo
Lualaba
Zambezi
Cunene
Okavango
Limpopo
Molopo
Vaal
Orange
NAIROBI
Mombasa
DAR ES SALAAM
Luanda
Tenke
Kitwe
LUSAKA
Tete
BLANTYRE
SALISBURY
Wankie
Gwelo
Beira
Bulawayo
Francistown
Messina
Louis Trichardt
WINDHOEK
Walvis Bay
Gaborone
LOURENÇO MARQUES
PRETORIA
JOHANNESBURG
Mbabane
Luderitz
Kimberley
Bloemfontein
Richards Bay
Durban
East London
Cape Town
Mosselbaai
Port Elizabeth
0°
10°
20°
30°

men or policy-makers become increasingly aware, firstly, of the potential, through irrigated agriculture, for generating economic activity at regional and national levels, and secondly of the consequential opportunity, through such economic activity, for promoting peaceful inter-regional relationships. One example of the extent of such global interest was the "Water for Peace" conference held in Washington, D.C., during May 1967, attended at ministerial, expert and observer levels by representatives from some 75 countries resulting in the presentation of approximately 700 papers on this subject.

It has become clear that, except in the least developed territories, irrigation cannot be separated from overall water resources planning. Economic (as distinct from purely financial) evaluation of storage, distribution and infrastructure projects in relation to resultant benefits automatically involves public funds and, therefore, national considerations, having regard to such factors as tax structures, economic defence aspects and the criterion that the greatest benefit must accrue to the greatest number.

The recent terrible droughts experienced in the north-western areas of Africa and in Ethiopia have focussed international attention on this whole question of nutritional standards and the political and other consequences which can flow from neglecting this aspect of energy.

One has to stand back and survey the scene, from an engineering-economic viewpoint, and look first at Southern Africa and maybe one day Africa.

Water and energy are the keys. Water is the key to food production for maintaining national energy; it is required to cool the plant for the mighty thermal generating sets for electrical energy and it can generate electrical energy by itself if the waters of a river catchment are concentrated behind a large dam such as Kariba or Cabora Bassa and dropped through the available height to transform the pressure from static head to kinetic energy which in turn is transformed into units of electrical energy. But it is not necessary to have such single big schemes. In the North of Scotland and Snowy Mountain (Australia) hydro-projects the power is produced by transferring one river to the other and so on and to produce the same energy by using a number of dams, tunnels or canals, and power stations. In essence, hydro-electric energy is the quantity of water available per unit of time multiplied by the height of drop, or head, available. Thus a small river with a high fall could produce the same power and energy output as a large river with a small fall. The ideal situation is when you have a large river with a high head or fall such as the Zambezi at Kariba and Cabora Bassa. Of course there are not many such sites left, but this does not mean that the hydro potential of Southern Africa is limited, particularly if harnessed on a regional co-operative basis, as will be shown.

Until recently the great problem was the vast distances of Africa. It was not possible to transmit electrical energy economically over long distances. In 1960 Kariba showed the way with high voltage transmission distances of up to 600 kilometres, and in 1976 Cabora Bassa really broke through with the transmission of a large block of power over a distance of 1 400 kilometres to Apollo converter station at Irene near Pretoria.

To disseminate, barter, sell or buy electrical energy one needs a transmission network. If there is a glut in agricultural output one can change the crop pattern or export to another region. In the electrical field one is confined to the transmission grid available or which can be constructed at reasonable cost.

Quietly and unsung, Escom have expanded their high voltage 400 kV transmission network over the length and breadth of the Republic, in addition to the lower voltage feeders. By 1973 the length of 400 kV national grid transmission lines had reached 4 500 kilometres and the expansion planned budgets for 8 000 kilometres by 1978 apart from the 1 400 kilometres of line which now links the national grid with Cabora Bassa.

It is the existence of this national grid, situated in the heart of an industrial complex with a high demand for electrical energy and which can be extended northwards into Southern Africa, as now demonstrated by the Cabora Bassa example, that will form the cornerstone of energy and water interdependence in Southern Africa – to the mutual benefit of all concerned.

Such a grid makes it possible to absorb and integrate the output of electrical generation from all available sources: thermal, nuclear and hydro-electric. Furthermore, the peak demands of energy, whether on a daily or annual load curve, can be met from the most economic source according to availability and efficiency. The daily load curve generally varies hour by hour and reaches peaks which may be of short duration but are much higher than the average for the day. Such peak surges arise due to various causes: rush hour traffic, sudden cold snaps, or the habits of consumers in a given locality. The annual load curve also has seasonal variations with peak periods. In modern practice very large turbo-generators are used in thermal stations, whether coal or nuclear fired, and these operate most economically when run at nearly constant output i.e. producing base load energy. Spare capacity must therefore be available to meet these sudden short duration surges in demand. The most efficient means of supplying such peak demands is by hydro-electric or gas-driven turbo-generators. These water turbines can be brought into service or on stream in a matter of minutes. A hydro-plant does not "consume" water. When the energy of the falling water has been transformed by the turbo-generators into electrical energy the de-energised water returns peacefully to the river. There is also no pollution.

Moreover, by virtue of the regulation of the river through the storage dam, the downstream flow is generally improved, creating more reliable flows for riparian users, i.e. new water. The massive storage at Kariba has made it possible to regulate the Zambezi at this point so that periodic minimum flows of 200 cubic metres per second would ultimately be ironed out to the average flow of the river of the order of 1 000 cubic metres a second. The additional reliable flow could be called "new water".

Apart from downstream re-use aspects, this regulation has an important bearing on reduction of floods and on navigation.

In developed industrial communities it is essential to have peak capacity available, because a failure can mean a regional blackout, an interruption of supply to industries with serious economic consequences. For this reason special tariffs are usually adopted for the availability of such generating capacity and the emphasis here is usually not on price per unit of energy sold (kilowatt hours), but on price per unit of power "available" (kilowatts).

In Southern Africa the water position is generally such that, in simplified terms, we think of the "wet" north and the "dry" south.

To the north of the Republic there are the mighty major rivers such as the Zaire (Congo), Rufiji, Rovuma, Shire, Cunene and the Zambezi.

At Cabora Bassa the average flow in the Zambezi is 40% greater than the total annual run-off in the Republic. However, though relatively dry there is hydro-potential in the Orange River and much more so in the Tugela and Transkei basins where the annual run-off is equivalent to about 25% of that of the whole Republic. There is also potential for pumped storage schemes in some of the Cape rivers.

Escom, in collaboration with the Department of Water Affairs, have launched very big water and hydro-electric projects in the Tugela basin, such as the Tugela-Vaal (Drakensberg) scheme where a pumped storage scheme with an installed generating capacity of 1 000 Megawatts is to be coupled with the transfer of $11 m^3/s$ of Tugela water to the thirsty Vaal River basin. Further downstream on the Tugela River conventional hydro and pumped storage schemes with a total installed capacity of 5 300 MW are being investigated.

On March 5, 1975, Paramount Chief Kaiser Matanzima made the following announcement in the Legislative Assembly of the Transkei relating to special investigations authorised to assess the hydro-electric potential of the Umzimvubu River and its tributaries:

'This investigation is being undertaken by the Republican Department of Water Affairs in conjunction with an investigation of the potential also of the Republic. Preliminary indications are that the Umzimvubu River and its tributaries – the Mvenyane, Kinira, Tina and Tsitsa – are hydrologically and topographically suitable for the development of hydro-electric power and that a major project which could be constructed with advantage in several stages is worth investigating.

'The eventual output of such a scheme will be far in excess of the requirements of the Transkei and a viable scheme will only be possible if the power generated is exported to the Republic, that is, fed back into the grid controlled by Escom to which I referred earlier on – possibly along the same route by which it is intended to import electricity. If this scheme is finally found to be feasible it will be of great benefit to the Transkei. During the construction period thousands of people will be employed in various tasks, many of whom will be trained in skilled occupations. Roads and power lines to be built for the project will contribute materially to the provision of infrastructure, and ultimately when capital costs have been recovered the Transkei should have a steady income running into many millions of rand per annum from electric power sold to the Republic.

'The planning and construction of so ambitious a scheme will not be possible unless technical and financial assistance is offered to the Transkei. In this regard I am most happy to announce that not only is the Republican Government financing the preliminary investigations, but it has already intimated that should the scheme in fact prove feasible it would enter into negotiations with this Government for the financing of the project. Naturally negotiations will have to be entered into with Escom also, and it is envisaged that the operation of the project would be under the direction of Escom for the foreseeable future.

'Mr. Chairman and Honourable members, on behalf of the Transkeian Government I wish to express the greatest appreciation to the Republican Government for its initial and proffered future assistance in this matter. It is tangible evidence of faith in the Transkei and is seen as a further opportunity for co-operation and the fostering of goodwill between our respective governments. May I leave this particular subject by remarking that it may well be that the oil crisis may prove a blessing in disguise for the Transkei, for it has certainly precipitated the whole issue.'

A hydro-electric project is capital intensive because of the heavy civil engineering structures involved such as dams, tunnels and other conduits. However, it must be remembered that these structures have a life of up to 100 years or more. Furthermore, once loans and credit have been extinguished the operating and maintenance costs are almost negligible. In many cases one or a group of such stations can be operated by automatic remote control.

These advantages add up to the fact that there is very little escalation or inflation effect on hydro-electric energy because of low running and maintenance costs.

Against this, the inflation effects on thermal stations are significant and not foreseeable. What will labour and replacement plants cost in 15 to 20 years ? What will uranium, coal, or oil (if available) cost in 15–20 years ? One cannot say, but the major conclusion would appear to be that wherever sources of hydro-electric energy are available these should be exploited.

The advantages of having a high proportion of hydro-power available in the system has been amply demonstrated in Zambia and Rhodesia where, by virtue of the Kariba project which serves both countries, the average national cost of electrical energy has been reduced in the 10 year period 1963–1973 from 0,605 Rhodesian cents to 0,390 Rhodesian cents – a truly impressive performance. The same relative drop in unit costs was experienced in Zambia over the period.

The real importance to and potential beneficial impact on regional economy and standards of life by injecting hydro-electric capacity into the national electric generating system, were not realised until recently. One might say that, with isolated exceptions, there was disinterest. And so in the main the mighty rivers of the wet north and the relatively smaller rivers of the dry south continued to flow to waste to the sea. What has come about to change this philosophy ?

There are a number of reasons which can be quoted as causing an upsurge in interest since 1960 among which are the following:

1 Until 1972 oil and coal were available and cheap. The price of oil has quadrupled over a few months and to some is not available, and factors concerning coal are emerging which indicate there will be a steady increase in costs in the foreseeable future.

2 From 1969 onwards the world was hit by rampant inflation which in places such as England has reached more than 20% per annum. There seems to be little indication or guarantee that the inflation rates will abate materially for some time.

   If forced to plan ahead for 20–25 years should one base the calculations on 6% or 8% escalation/inflation rates or what percentage ?

   The ratio of mechanical and electrical engineering plant to civil engineering structures is relatively small in hydro-electric engineering whereas it is the other way round for thermal stations. There is thus the question of "importing" inflation by purchasing abroad in countries where the average inflation rate of 15% per annum is the rule and not the exception.

   Assume there is a present cost advantage of say R10 million in favour of a hydro compared with a thermal station. If the average inflation rate is taken at 6 per cent per annum, then over 40 years the cost advantage improves to R52 million – more than 5 times. This sort of consideration provides powerful food for thought.

3 Since 1960 the economic benefits that stemmed from the efficient Kariba, situated right in the heart of Africa, provided persuasive stimulus to rethinking on the entire subject of water engineering in relation to hydro-electric generation.

4 The distance factor proved to be a deterrent in the past. Hydro-electric sites are usually remote from the markets for sale of energy and this caused fears about transmission losses. The examples of the Kariba and Cabora transmission systems and the emergence of the South African national transmission grid have had a powerful effect on forward thinking and inter-regional planning. The efficient services rendered by the Hendrik Verwoerd project in supplying valuable peak power has also affected the outlook.

5 The energy crisis sparked off in the main by the oil situation is in a large measure responsible for turning planning minds towards re-consideration of hydro-electric resources and regional co-operation in this field. Here the fact that there is no pollution from hydro-electric generation is a pertinent consideration particularly when viewed in the light of the recent scares in the United States and elsewhere regarding pollution risks by radiation leaks from nuclear power stations.

A conservative estimate of the total potential hydro-electric capacity of the Southern African rivers, excluding the Nile is 60 000 Megawatts of which approximately 8 000 Megawatts can be installed in the Republic, the Homelands, and in neighbouring territories. The total hydro potential of the Zambezi and its tributaries is of the order of 11 000 Megawatts, or about the same as the total 1975 installed capacity of the Republic. The mighty Zaire (Congo) River is rated at 40 000 Megawatts or thirty-three times the installed capacity at Kariba.

Taking the long term revenue to be as low as 0,5 South African cents per unit of energy, the potential energy earnings are of the order of R1 000 million a year and the potential savings of oil and coal come to around 150 million tons a year, which is five times the amount of coal burnt in the Republic in 1973 for thermal generation. Think what this means to the budgets and security of some of the developing countries in the region. But the long term revenue, on present forecasts, is more likely to be 2,0 cents per unit – which will quadruple the above financial benefits.

If the high voltage transmission grids that already exist in the Republic, Zambia, Rhodesia and Moçambique are extended to form a pan-African grid including Zaire, Malawi and Tanzania, the development of all this hydro-potential to the mutual benefit of all becomes possible. It can be achieved in a decade!

South Africa will not always do the importing of electricity. By virtue of her present economic dominance of the continent she is for the moment the only market. But with a pan-African power transmission grid, it will be possible to export and import electricity on the best use or most economic basis. There will be further by-products, such as the infrastructure requirements to build the necessary dams, tunnels and power stations. This would help materially to open up the presently remote areas.

This is not all. The pan-African grid will make it possible to integrate the hydro and thermal potential of all the territories in the region, which means that the coal reserves now lying fallow in countries like Botswana, Swaziland, Rhodesia, Zambia, Moçambique, and others could be put to economic use and further bolster the budgets of the region and make best use of the manpower resources of the whole region.

Moreover, the utilisation of say 40 000 megawatts of installed hydro-power in Southern Africa will save, in avoided evaporation of cooling water for thermal stations, about 500 000 megalitres a year or 300 million gallons a day – the equivalent of a sizeable river!

Taking the Zambezi as an example, the massive storage capacities and regulation of river flows from the combined reservoirs of Kariba, Kafue and Cabora Bassa will reduce the annual devastating floods in the lower reaches of the Zambezi to tolerable proportions.

Again, taking the Zambezi as an example, the steady discharges of water through the turbines of Kariba, Kafue and Cabora Bassa will regulate the flow downstream of Cabora Bassa to an almost steady average flow of 2 300 cubic metres per second, which will have a marked effect on inland transport, as it will make navigation of the river possible for great distances. One must try to visualise what this will or could mean to trade for countries like Malawi, Zambia, Rhodesia, Moçambique and perhaps even Botswana. The exciting prospect of creating inland harbours for Zambia, Rhodesia and Malawi using the Zambezi River and dam systems is dealt with in a separate section.

The creation of numerous fresh water lakes in Southern Africa will stimulate tourist potential, sporting and recreation facilities and with the introduction of fish farming in these reservoirs, create an assured protein diet for the indigenous populations. Under present circumstances it would be difficult to estimate the foreign exchange that accrued to Zambia and Rhodesia under these headings during the past decade, but most definitely it cannot be insignificant.

Finally, the biggest by-product of such visionary multi-purpose and inter-regional planning will accrue from the utilisation of the stored and regulated waters in the guise of "new", that is firmly reliable friendly waters, to irrigate the fertile soils in Southern Africa and thus create food banks and cash crops which will, in the first instance, make the region self-sufficient as regards food, and secondly help us to provide for the increasing needs, in the nutritional field, of less fortunate regions. Southern Africa can become the granary of the continent. Finally, the cash crops envisaged would make a material contribution to foreign exchange earnings.

Such a combination of human and electrical energies can make the Southern African region one of the strongest regions in the world with the power to develop and defend all its abundant resources.

In the first place, the Southern African countries will be able to export surpluses in food and cash crops to the needy and drought-stricken northern areas or to other needy Third World countries. Later on, as the Northern African countries begin to develop, on perhaps an equally co-ordinated basis, their plans for electrical grids and the uplift of nutritional standards assistance could be proferred from the south with technology, experience and other resources.

In general, the soils of equatorial Africa are not so productive as those in the more southern regions such as in Rhodesia, Zambia, Malawi, Moçambique and the Republic. But modern technology and sound management have made it possible to improve efficiencies and hence yield per unit of water.

Recent soil surveys carried out in Moçambique under the direction of the Portuguese authorities have revealed that in the lower Zambezi valley there are some of the most fertile soils in the world for the production of sugar, cotton, rice and beef. Free from annual devastating floods, because of the regulating capacity of Cabora Bassa, some 2½ million hectares can be brought under controlled irrigation agricultural development. These areas are located near the coast, and export of cash and subsistence crops or livestock may well be by rail or by fast barge or by hovercraft to Beira for transhipment.

There are about 8 million mouths to feed in the territory. The development, in phases, of some 2½ million hectares in a sparsely populated region will go far towards settlement of the population, improving nutritional and educational standards and earning foreign exchange from exported cash crops. On the backbone of such solid agricultural development it should be possible in due course to turn to utilisation of the mineral resources that appear to be present in reasonable quantities and the creation of industrial enterprises such as steel making.

The water, the land, the forests, minerals and other resources of a region or nation are all closely interrelated and will usually dictate the essential elements in the best plan of development for the commercial, agricultural and industrial economy that they will support. Thus basin by basin water resources blend thoroughly and intimately into, and reasonably control other types of development. To this extent their utilisation will promote, define or limit the economic life and livelihood of a nation.

The engineering, scientific, financial and managerial skills are available. Suitably marshalled and monitored these skills of the Western World could do much to help Africa to help itself.

2 Introduction

SOUTHERN AFRICA – A POSSIBLE GRANARY FOR AFRICA AND THE THIRD WORLD.

Food has been given more attention recently than ever before. This has been mainly brought about by the unexpected shortages of grain, the principal food in world trade, which have occurred since 1972. The alarming rate at which the world population is growing in relation to the global capacity to provide adequate quantities to maintain, let alone improve, nutritional standards has brought this question to the forefront as of great topical importance.

Food shortages that were caused by world-wide droughts and unfavourable weather conditions in 1972 and 1973 and the resultant lower crop yields in Russia, South-East Asia, North Africa and in parts of North America and Australia resulted in the colossal purchases of grain which became necessary by Russia and China and the drop in the world's food stores to the lowest level in 20 years.

The developing or poorer countries comprise more than two-thirds of the world's population, but they produce less than 44 per cent of the world's food. As regards the production of farm animal products the balance is even more unfavourable because these countries produce only about 22 per cent of the total quantity of meat, milk and eggs. Moreover, the imbalance position will get rapidly worse. Whereas the developed areas have maintained and improved their output of food per head of population, the developing territories have shown a plunging decline in output per head of population that has already dropped to below 1954 figures.

His Imperial Majesty, the Shahanshah of Iran is on record as saying this question is perhaps of greater importance, with greater potential ultimate impact on international relations, than the oil and energy crisis.

The quadrupling of oil prices over a few years span has of course forced up the prices of mechanisation, fertilizers and agricultural chemicals, which has affected adversely the agricultural development in the poor countries and is yet another factor in increasing the imbalance.

Hunger knows no boundaries. The whole question deserves urgent attention by all countries that are able to contribute towards a stable food supply to the needy Third World.

The countries which still have reserve capacity for agricultural development are South America, where only 16 per cent of potential arable land has so far been used, Australia and New Zealand with more than a third in reserve and Russia with exactly a third. Even the older countries in Europe can extend their arable land by 20%. North America has put 86 per cent of its potential culturable land under the plough. Here, however, there is some pessimism as to whether by 1985 the anticipated grain surplus of 30 million tons in the U.S.A. will be able to cope with more than one year's requirements of the poorer countries at the present population growth rate.

The Director of Agricultural Economics in the United States Department of

Agriculture in a recent review of the world food situation expressed the following pessimistic opinion:

> 'In the long run, looking into the twenty-first century, unless the population growth is checked, there is no solution to the world food problem. Projected at recent rates the population chart runs off the page: the numbers become not only unmanageable but inconceivable'.

In the hopes that a brake can by some means be placed on the rate of increase in population we have say a quarter of a century in which to buy time. This will require dedicated efforts by all the "have" countries during that period.

This brings the spotlight on to Southern Africa.

In Southern Africa the food position must also be viewed in terms of population numbers and the anticipated increase therein.

South Africa

The total population of 25 million is expected to be doubled by the end of this century and doubled again by 2020.

Despite limited resources and a generally unfavourable climate, South Africa has a record of rising performance in agriculture. The high level of agricultural science and of farming application can be compared with the best in the world.

As regards the amount of available tillable soil per capita, the Republic is in a weak position compared with other African and developed countries of the world. Experts accept that at least 0,4 hectares of cultivated land are needed to provide the food requirements of each person and already by 1973 it was estimated that there are available only 0,57 hectares of arable soil per person.

Another serious limiting factor is water, which will limit the areas which could be put under irrigation.

The Republic has at present just under a million hectares of land under irrigation which, the authorities estimate, can eventually be expanded by a further 340 000 ha.

Much research is being concentrated on finding underground water sources that will in the first place be used for livestock and hence animal food production.

In the Homelands altogether 76 per cent of the land has an annual rainfall higher than 500 mm, and 33 per cent of the land is classified as having high potential.

According to the Department of Bantu Administration and Development the Homelands produce only about one-seventh to one-sixth of the yields produced on comparable land in the rest of South Africa. Most Homelands do not even produce enough food to supply the basic requirements of their own people.

It is estimated that the newly independent Republic of Transkei has the potential to produce more than 3 million tons of maize, but its production in the 1972/3 season was only 0,1 million tons.

Here there is a great opportunity to apply the South African skills in agricultural technique to the development of water-land resources to achieve its proper potential yield to the mutual benefit of both countries.

South West Africa

In South West Africa agricultural and other development is seriously hampered by the scarcity of water. Only a third of South West Africa has a rainfall of more than 400 mm a year, and because of the erratic nature of the rainfall the cultivation of dryland crops is carried out on only 1,1 per cent of the territory. The Namib Desert covers about a fifth of the country, and in the interior there are no perennial rivers, except on the extreme northern and southern borders. The carrying capacity of the veld in general is very low. A high level of farm management is nevertheless maintained and large numbers of cattle and cattle carcases are exported to the Republic, but S.W. Africa in turn imports a lot of agricultural products from S. Africa.

Botswana

The greater part of Botswana is covered by the Kalahari Desert with its extremely low rainfall that increases towards the northern and eastern parts of the country. Owing to the poor rainfall only about 400 000 hectares are used for crops, less than 10 per cent of the total arable land in Botswana. Much of the arable land, however, is marginal because of erratic rainfall. Cattle farming is the main source of revenue from agriculture, but the turnover is low, namely 8 per cent. Owing to the lengthy droughts in the 1960's, Botswana could not provide the food requirements of its population of 630 000 in 1971 and food had to be imported. Conditions today are more favourable.

The limiting factors to dryland farming are:

(*a*) Uncertainty of amount and distribution of rainfall;

(*b*) inherent infertility and further loss in fertility in the soils available both in the eastern sector and elsewhere in the Protectorate;

(*c*) in the tribal areas, the ravages of pests and disease;

(*d*) in the tribal areas, the continuing low standards of husbandry: these are not to be criticised but to be realised and, wherever possible, altered.

Taken together, these constitute a definite challenge to the extension service. Time and patience, with the aid of practical demonstration of African lands, alone are likely to change the position.

Irrigation is not practised on a significant scale. Water is available from the Okavango, the Chobe and Limpopo Rivers. A limited amount of irrigation is practised along the Limpopo (Tuli Block) with fair to a high degree of success.

Lesotho

Although this country has a high rainfall of nearly 2 000 mm a year the country is poorly endowed with agricultural resources. The high rainfall combined with poor farming practices in this overpopulated territory with about 3 hectares a head for its million people has led to very serious soil erosion over virtually the whole country. About 85 per cent of the economically active population is engaged in agriculture, but yields are very low though there is potential for higher yields.

Swaziland

The possibility for agricultural expansion in this small country are very good indeed both as regards extra land available for irrigation and improved farming practices.

It has large perennial rivers and good soils for agriculture and forestry and a relatively high rainfall over the entire country. The yield per hectare is the same as that of South Africa.

One of the best irrigation projects in Southern Africa is the Swaziland Irrigation Scheme (SIS) run by the Commonwealth Development Corporation (CDC). The present area under irrigation is 12 750 ha of which 10 400 ha is planted to sugar and the remainder to citrus, rice and other crops.

It is estimated that at least 75 000 ha can be put under economic irrigation at today's prices. Storage sites are known to be available on the Komati, Lomati, and Usutu River tributaries.

The Komati River is part of an international river basin, the Incomati, involving water contributions from and to the Republic, Swaziland and Moçambique.

There is no doubt that by co-operation between these countries, on the best use philosophy, all will benefit and Swaziland will not only be able to remain self-sufficient in foodstuffs but will be able to make food contributions to a "granary bank" for Southern Africa.

Rhodesia

Of the Southern African countries, Rhodesia has the highest average yield of produce in kg per hectare: 2 962 compared with 1 704 in South Africa. Recent reports indicate however, that overpopulation in the tribal trust areas, which cover about half the area of the country, overgrazing and poor farming techniques have already converted much of the soil into "wasteland" virtually unfit for agriculture.

Rivers such as the Zambezi and Sabi and their tributaries provide more than adequate water resources for agricultural and industrial development. Great efforts will be required to raise agricultural production in the tribal trust lands to above subsistence levels.

During 1975 more than 20 storage works were completed in the European areas with a total storage capacity of some $12\,000 \times 10^3$ m^3 and in the same year 12 dams and weirs were completed in the tribal trust areas with a total storage capacity of $14{,}13 \times 10^6$ m^3.

Rhodesian exports of tobacco have made a major contribution to the country's economy for some years but it is not so well known that many other agricultural products are exported to Africa and elsewhere. Rhodesian quality beef production has reached very high standards and no doubt this sector will continue to expand and in selected specialised brands. Rhodesia will be able to contribute to a granary bank.

Zambia

A large variety of farm crops can be grown in its various regions which range from tropical to subtropical and temperate climates. Production is limited by poor soils over great parts of the country and a long dry season.

Although Zambia has a number of strong flowing rivers little irrigation has as yet taken place.

A group of white farmers produced in 1970 about half the marketed agricutural production. In recent years the government has by various methods tried to promote efficient agricultural production, but progress has been slow. There is no reason to doubt that with a dedicated approach Zambia can be made more than self-sufficient agriculturally.

Malawi

Malawi is a relatively small country, but more than half of its land area is suitable for cultivation. Its soils are among the most fertile in Southern Central Africa. Nevertheless, only about one-third of the arable land has so far been utilized, and the country's considerable potential for irrigation farming has hardly been developed. The Malawian agricultural industry must, however, be regarded as one of the most efficient in Southern Africa, and agricultural products worth millions of rands are exported annually after the local requirements have been satisfied.

Angola

The agricultural potential of this huge well-watered land is considerable. However, at present only about 2 per cent of the country's arable land is utilized actively.

Moçambique

Like Angola, Moçambique has a number of rivers and agricultural potential is high: agriculture is generally better developed than in Angola but even here only about 5 per cent of the arable land is cultivated.

With the completion of the Cabora Bassa project a new era has dawned for Moçambique as regards meaningful and hopefully rapid agricultural development.

Cabora Bassa is generally regarded as a power project for the export of energy, but perhaps the real significance of this project will be in its contribution to food and fibre production for the 8 million people who exist on a subsistence economy.

Soil surveys have revealed immense agricultural potential in the lower Zambezi. The ultimate potential for cultivation within the Zambezi valley is assessed at some 2,5 million hectares of which 1,5 million hectares will be irrigated and the remainder dry-farmed. It is understood that plans under consideration allow for cattle, citrus and food crops in the highlands and sugar, cotton and jute in the lowlands.

The massive storage capacities of Lakes Kariba and Cabora Bassa will ensure that the annually recurrent destructive floods are eliminated or considerably mitigated. Every additional dam built on the Zambezi will strengthen this insurance. Thus it is possible to settle the population in this lower region and to develop the agricultural potential to its fullest extent. The revenues earned from energy exports to the Republic will make this economically feasible in due course when the loans and export credits have been extinguished.

The agricultural potential of the Incomati River, which is fed by tributaries such as the Komati, Sabi and Krokodil Rivers, is also considerable. Here collaboration between Moçambique, Swaziland and the Republic will be necessary in order to regulate this international river system.

The potential in this country for contributions from surplus food and fibre products is immense.

The table below by Mr Nsibandze of Swaziland gives an interesting comparison of the relative agricultural performances of the countries in Southern Africa described herein.

AGRICULTURAL STATISTICS

(a) Productivity of the land

Country	*Area (ha)*	*Population in 1970 (x 10^3)*	*Harvested area (10^3ha)*	*Food production (10^3 tonnes)*	*Yield (kg/ha)*
Angola	124 670	5 943	520	400	769
Botswana	60 037	652	22	12	545
Lesotho	3 035	1 084	110	59	536
Malawi	11 848*	4 672	1 100	1 150	1 045
Moçambique	78 303	8 056	420	430	1 024
Rhodesia	39 058	5 414	520	1 540	2 962
South Africa	122 104	21 129	5 650	9 630	1 704
South West Africa	82 429	659	34	14	412
Swaziland	1 736	447	69	120	1 739
Zambia	75 261	4 564	1 285	612	2 147

(b) Index of food production (relative to the average for 1961–65 = 100) in recent years

	Angola	*Botswana*	*Lesotho*	*Malawi*	*Moçambique*	*Rhodesia*	*South Africa*	*South West Africa*	*Swaziland*	*Zambia*
1970	106	109	101	129	118	107	129	108	174	121
1971	102	125	103	143	122	126	148	111	189	123
1972	112	128	79	152	130	141	154	112	198	133

*Includes Lake Malawi (2 440 ha).

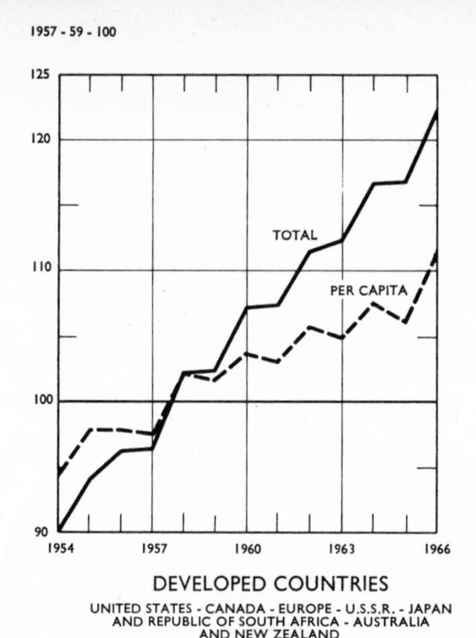

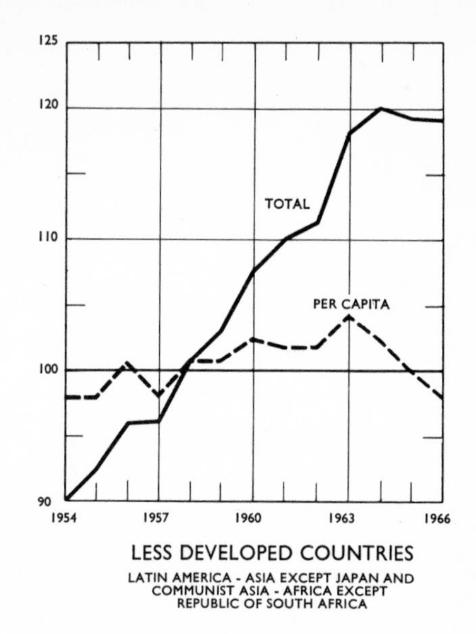

WORLD FOOD PRODUCTION - 1954 - 66

SOURCE: UNITED STATES DEPARTMENT OF AGRICULTURE

World food production 1954–1966

It is clear that by 1985 the Republic, like the U.S.A., will find it difficult to produce surpluses in foodstuffs for dissemination to other needy countries.

But Southern Africa in co-operation can save itself and produce enough for its own needs and to spare for the Third World.

The proven agricultural skills of the South Africans and Rhodesians, the well-watered northern areas, education of the African farmers, crash action as regards soil surveys are all factors to be harnessed to this ideal.

The Southern African power transmission grid that is in the making, will undoubtedly be a meaningful catalyst in this context. Cheap energy will make it possible to irrigate areas now considered unsuitable for topographical reasons. High-lying land can be irrigated with supplies delivered by pumps using cheap electricity.

If we visualise specially constructed silos near coastal areas where the surplus contributions from the different countries can be stored for distribution to the needy areas we have the same concept as that of the World Bank. In the case of the Bank, money is mixed in an international fund and no borrower enquires from which particular nation his money originates.

Once the food products with the necessary calorie or protein values are "mixed" in international silos, the hungry will not prevaricate about the origin of life-preserving material.

The most effective and economic locations for such large storage complexes would be on the coast with good harbour facilities to ensure rapid despatch to Third World countries on a long-term basis or in crises when famines or shortages threaten.

In general the points to consider would include the following:

(*a*) The sources and types of grains
(*b*) Sea routes
(*c*) Draught and berthing facilities
(*d*) Needs of the nations likely to be served and the facilities for local storage in their countries
(*e*) Cross-country distances
(*f*) Land transport facilities
(*g*) Foundation problems for massive silos
(*h*) Social and economic considerations.

There will of course have to be close co-operation between the materials handling engineer and the structural and civil engineers in the design of such storage complexes. From the structural and civil engineering point of view the silo complexes are better when their plan arrangement is wide and short, rather than narrow and long and this also reduces river frontage works and is more efficient in the use of expensive land.

Extensive research work is being conducted in Britain to determine the best arrangement for large complexes of the type envisaged. Many parameters have to be investigated. The length of storage period has a great influence on design approach. Another important factor is what segregation of grain types is considered necessary, i.e. use of interbin spaces as separate bins or for natural spill in and out.

The greatest problem will be pest control. Grain is a live commodity and subject to biological deterioration either from insects or from mould growth. If left standing in bulk, insects and other pests breed rapidly in the heart of the mass and the warm moist air arising from the hot-spots so caused create condensation near to the cooler surfaces and this produces a layer of mould on top of which shoots start sprouting. Barley, for example, at 23% moisture and 25 °C (77 °F) will spoil within a week, whereas at 11% moisture and 7 °C (45 °F) it is known to store satisfactorily for over twenty years.

There are two approaches to this problem. One is to provide reasonable aeration from the bottom of the store and if stability is not achieved fumigation with methyl bromide circulated in the general aeration system or with phosgene gas introduced at the top and allowed to diffuse downwards with the aeration system switched off.

The alternative approach to aeration is to seal the contents by making the silos airtight. Apparently with only 2% oxygen insects die. The sealing method is difficult to achieve.

It is understood that exercises are being carried out to determine the best solution in terms of capital and running costs.

The question can and *must* be resolved, and the pattern can be set in Southern Africa.

There are a number of suitable harbour sites on the east and west coasts of Southern Africa: Richards Bay and Saldanha Bay in the Republic of South Africa, Beira, Maputo and Ncala in Moçambique, and Lobito and Luanda in Angola.

If a harbour such as Beira in Moçambique is selected and the Zambezi is made navigable to the Indian Ocean from the headwaters of Lake Cabora Bassa, grains and other perishables from Zambia, Rhodesia, Malawi, possibly Botswana, could be transported at relatively high speed and low cost to the international silos at the coast. This would apply equally to the excellent natural harbour facilities at Ncala to the north of the Zambezi mouth.

Naturally such a concept will involve also great strides as regards transport infrastructure and new and novel ideas must be considered in order to make the movement of high value, perishable goods rapid and inexpensive.

The dams of Southern Africa have begun to show the way.

3 Introduction

NAVIGABILITY OF THE ZAMBEZI RIVER AND ITS TRIBUTARIES

The large lakes in Africa such as Victoria, Tanganyika, Myoga, Albert and Malawi are all navigable. Long stretches of the Rivers Nile, Zaire (Congo) and the Shire are navigable for craft of varying size. The Zaire River is some 4 800 km long. It is navigable from its mouth to Matadi, a distance of about 150 km, the chief port. From there upstream are a series of rapids which, as regards transport, have been avoided by a railway about 360 km in length. Thereafter it is again navigable in the main stream for about 1 600 km with services also on some of the major tributaries. Eventually by part rail, part river Lake Tanganyika is reached, from where the cargoes get to Dar-es-Salaam.

The Shire River is referred to later in the context of the Zambezi River.

At present the Zambezi is navigable only for about 160 kilometres as far as Vila Fontes from its mouth.

The regulated flow through the turbines and sluice gates of the dams at Kariba, Itezhitezhi, Kafue and Cabora Bassa with massive combined reservoir storage capacity will regulate downstream flow to the river mouth at the Indian Ocean in such a manner that improved draft for rivercraft will be available which could both lengthen the distance for navigation and permit increase in vessel size which could be used successfully.

The creation of these vast inland lakes and the regulation of the rivers lift the conceiving eye to the long range potential for navigation with special reference to the immense potential economic benefits to Moçambique and the landlocked countries Malawi, Zambia, Rhodesia and Botswana resulting not only from cheaper but quicker transport of high value, perishable products – both for export and for import.

The water surface of Lake Cabora Bassa stretches to the Zambian and Rhodesian borders and with co-ordinated operations between Kariba and Cabora Bassa could be kept at a reasonably constant level. Thus a stretch of lake 270 km long to the dam from say two such ports located on each of the shores of these countries is available for rapid air-cushioned vehicles or barges.

A foreword to *Jane's Surface Skimmers 1975–76* records:

'For the hovercraft industry the road to recognition has been long and hard. But in the past twenty months it has been acquiring an air of stability which has the unmistakable look of permanency about it.

One of the biggest factors in the industry's change of fortune is that inflation, devaluation and rising costs have so eroded the economics of new displacement ferry boats that hovercraft which are smaller, faster and often less expensive are rapidly emerging as the better buy.

One big attraction is that the craft could be smaller. A hoverferry needs only one-third of the payload capacity of the vessel replaced, but being three to four times faster it would be able to convey at least the same volume daily over the same route or even a greater load if required.

Crewing requirements are much lower: fewer maintenance staff are needed; fuel bills are reduced. There can be no longer any doubt about the popularity of large hovercraft ferries with the public. During the first nine months of 1975 the average joint load factors of British Rail Seaspeed's two SR.N4's and Hoverlloyd's three SR.N4 MK II's on the English Channel route were 60,1 per cent for passengers and 73,05 for cars.'

Jane's also points to the "new generation" hovercraft now being planned, such as the BH. 88 which is claimed will be five knots faster than the existing craft (SR.N4) but will show a 40 per cent reduction in power requirements and a 60 per cent saving in fuel. This latter figure is one of the most significant of all, since it will mean that the fuel used by the BH. 88 will be about 2,0 lb/payload ton-mile, (0,55 kg/tonne-km), some 15 per cent less than orthodox ships per unit of payload.

So much for the future plans which are maturing actively and apparently rapidly and which are a guidance to forward-looking engineers for conceiving projects in infrastructure development which will bring the greatest economic benefits to the greatest number.

Making the Zambezi and its tributaries navigable is such a concept. It has, however, to be approached in a practical manner having regard to the terrain and the longitudinal water profile of the river in relation to flow of water.

At present the world's biggest hovercraft is the Mountbatten (SR.N4) Class Mark I. It is a passenger/car ferry of 190 tons with a speed of 40–50 knots in waves up to 10 ft. (3,04 m) in height.

But in the naval field Vosper Thornycraft have designed a 500 ton ocean going escort hovercraft capable of a maximum continuous speed in excess of 50 knots.

Although there has been dynamic recent progress as regards more sophisticated and practical designs and performance relating to reliability, greater payloads, higher cruising speeds and lower costs per tonne kilometre, there is still much trial and prototype testing to be done before it will be possible to switch to navigability of the Zambezi by this type of craft.

However, most encouraging developments of a more immediate practical significance relate to the recent development of the Hoverbarge which, while not as fast as the hovercraft, operates at some 20 to 30 knots.

The Air Lubricated Barge concept seeks to provide a solution to the marine transport problem that lies between existing slow displacement craft carrying bulk

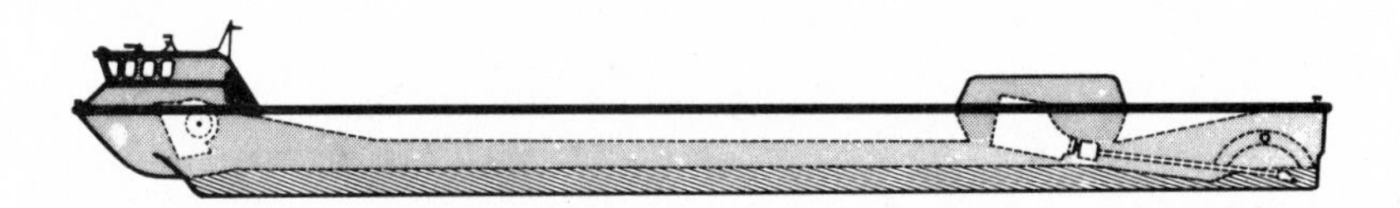

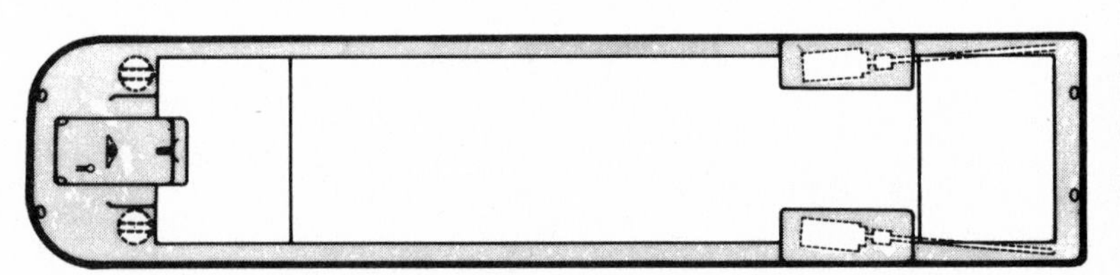

PAYLOAD	100 Ton	400 Ton
CRUISE SPEED	25 Knots	25 Knots
MAX. SPEED	28 Knots	28 Knots
LENGTH	55,5 Metres	76,2 Metres
BEAM	11,3 Metres	15,2 Metres
ALL UP WEIGHT	200 Ton	600 Ton
INSTALLED POWER	4250 HP	9000 HP
FLOATING DRAUGHT	1,8 – 2,4 Metres	1,8 – 2,4 Metres

GENERAL ARRANGEMENT
OF AIR LUBRICATED BARGE (ALB)-TYPE 525

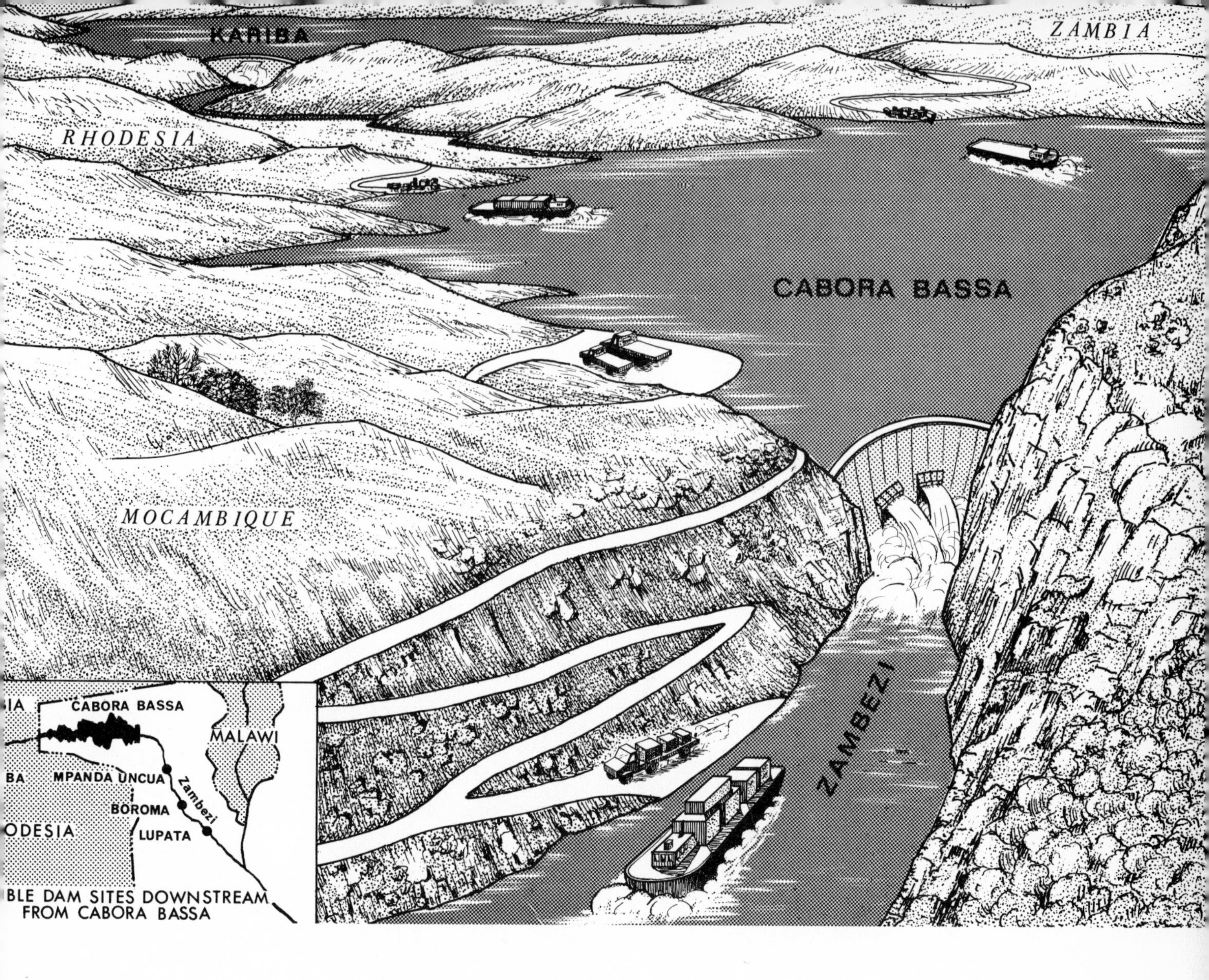

Artist's impression of hover-barge transport on the Zambezi River from inland and coastal ports

cargo and the high performance marine vehicle (hovercraft or hydrofoil) suitable for carrying passengers and cars on premium routes.

For its size and speed it is unique in providing relatively high performance at extremely low power, i.e. fuel consumption. The simple construction involving weldable light alloy and reliable robust diesel engines provides an economic first cost and low running costs.

A recent design that holds out high promise is the ALB Type 525 (Air Lubricated Barge), a diesel powered hoverfreighter designed by Hovermarine to carry general or containerised cargoes of up to 400 tons on rivers and inland waterways at speeds of 25–28 knots.

By adding cabin modules to the basic hull, the craft can be adapted to various passenger carrying roles.

One must look also at the almost universal trend towards containerisation of cargo. This makes the "roll on roll off" technique a practical proposition applied to hoverbarge traffic that will make it possible to bypass the series of dams on the Zambezi and to transfer loads from one lake level to the next. The construction of locks at the dams which are generally located in mountain gorges is out of the question.

Another approach would be to construct large hoisting gear to lift or lower the craft complete with cargo from one level to another, but this would tend to limit

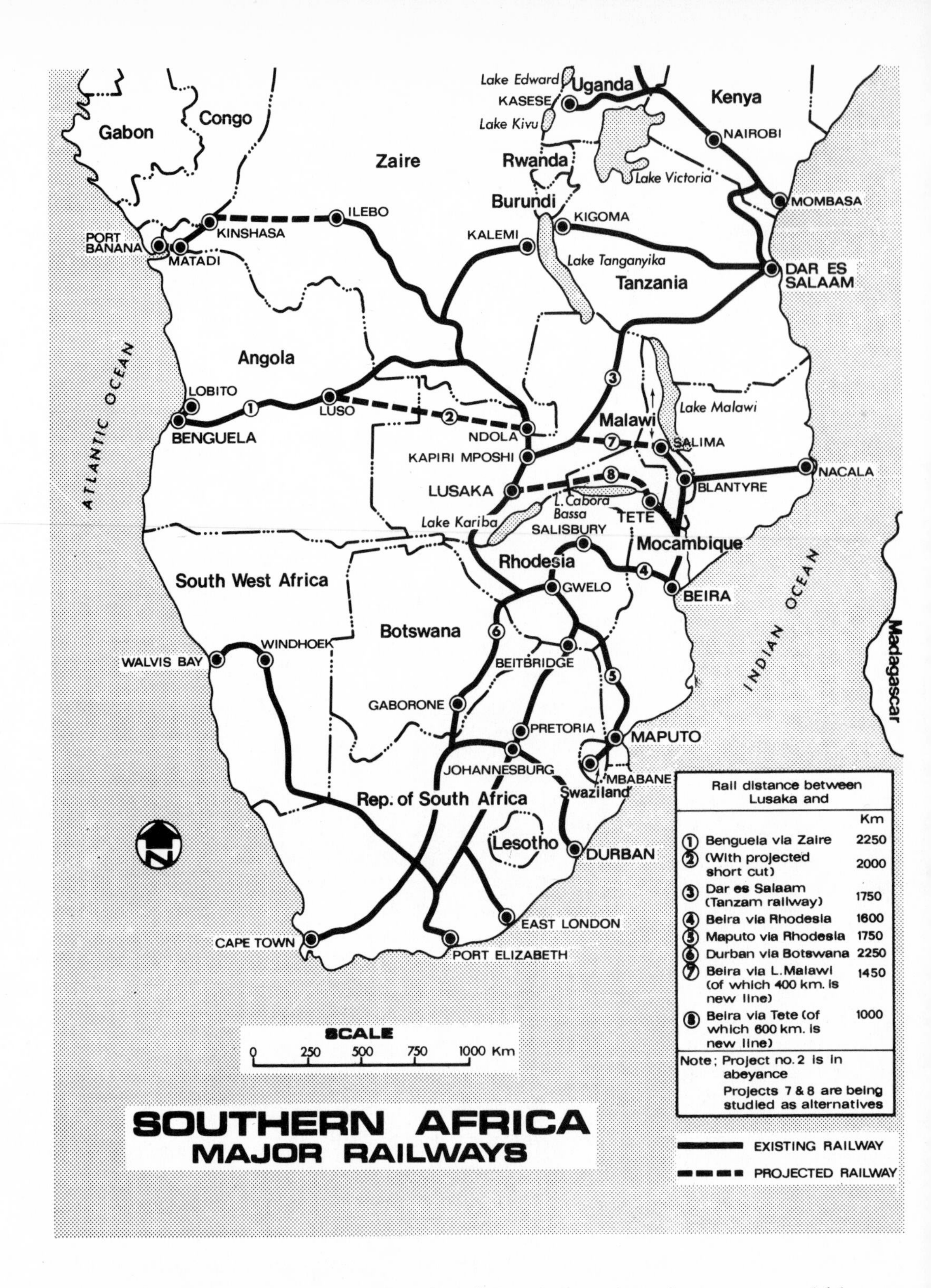

Rail distance between Lusaka and	Km
① Benguela via Zaire	2250
② (With projected short cut)	2000
③ Dar es Salaam (Tanzam railway)	1750
④ Beira via Rhodesia	1600
⑤ Maputo via Rhodesia	1750
⑥ Durban via Botswana	2250
⑦ Beira via L.Malawi (of which 400 km. is new line)	1450
⑧ Beira via Tete (of which 600 km. is new line)	1000

Note; Project no. 2 is in abeyance
Projects 7 & 8 are being studied as alternatives

the size of craft that could be utilized and the civil engineering costs would be very high in such mountainous terrain.

The answer would appear to be to confine the hoverbarges to their own lakes and at the dam terminus to "roll off" the cargo and passengers on to road vehicles which would descend (or ascend) to the next lake by roads with suitable gradients and "roll on" to the next hovercraft, and so on.

At some sites it will be possible to hoist or winch the craft up or down ramps at relatively less expense and so eliminate the use of extra craft and crews.

In the case of the Zambezi, traffic will generally be cargo orientated, at least in the early stages of operation, and passenger accommodation need only be minimal. However, with such beautiful lakes and scenery along the route the position is likely to change, probably rapidly, and tourist traffic should increase.

In the light of possible waterway transport along the Zambezi it is interesting to compare rail infrastructure alternatives. For purposes of comparison the Zambian terminus of Lusaka is used.

Waterway

1	Waterway distance from Zambian hoverport to Beira is 950 km plus rail distance from Lusaka to hoverport – say 275 km . . .	1 225 km

Rail Routes and distances

2	Lusaka – Lobito via Zaire	2 250 km
3	Lusaka – Lobito with short cut under investigation from Ndola to join the Benguela line near the Angolan town of Luso	2 000 km
4	Lusaka – Dar-es-Salaam (Tanzam railway)	1 750 km
5	Lusaka – Beira via Rhodesia	1 600 km
6	Lusaka – Maputo via Rhodesia	1 750 km
7	Lusaka – Durban via Botswana	2 250 km
8	Under investigation: Lusaka – Beira via Lake Nyasa (of which 400 km would be a new line)	1 450 km
9	Lusaka – Beira via Tete (Moatize) (of which 600 km would be new line) under investigation	1 000 km

Note: Projects Nos. 8 and 9 are being studied as alternatives to Project No. 3.

It is noted that the only comparable distance to the waterway should be the unsurveyed route No. 9 via Tete (Moatize).

Hoverbarge traffic on the Shire River could commence without waiting for additional dams or weirs to be built on the Zambezi, bearing in mind the improved flow conditions downstream of the junction of the Zambezi and Shire Rivers due to the regulation of flows by the reservoirs on the Zambezi and Kafue Rivers and on Lake Malawi.

The Shire joins the Zambezi River downstream of Mutarara bridge where the river level is approximately 50 metres above sea level and the slope of the river to the sea, a distance of some 200 km, would be about 1:4 000. The junction is below the last weir or dam to be built in a downstream direction on the Zambezi River.

Hoverbarges could thus operate on Lake Malawi up to Luvondo (Kamazu) barrage. The cargo would be "rolled off" and transported past the rapids and gorges by road or rail to the tailwater of the Shire at Hamilton Falls (Chikwawa). There it would be "rolled on" to other hoverbarges for an uninterrupted journey

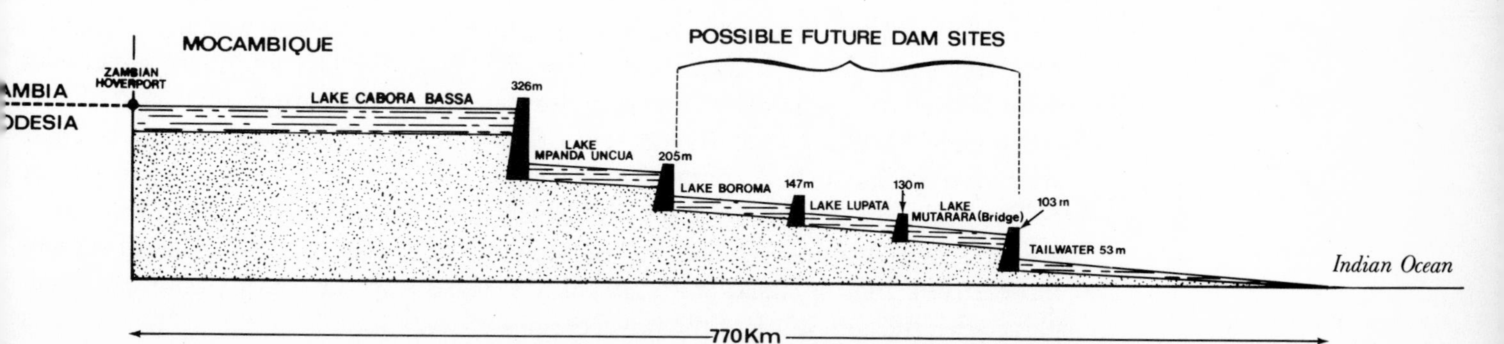

ZAMBEZI RIVER
EXISTING AND POTENTIAL DAM SITES AND FERTILE AREAS SUITABLE FOR IRRIGATION

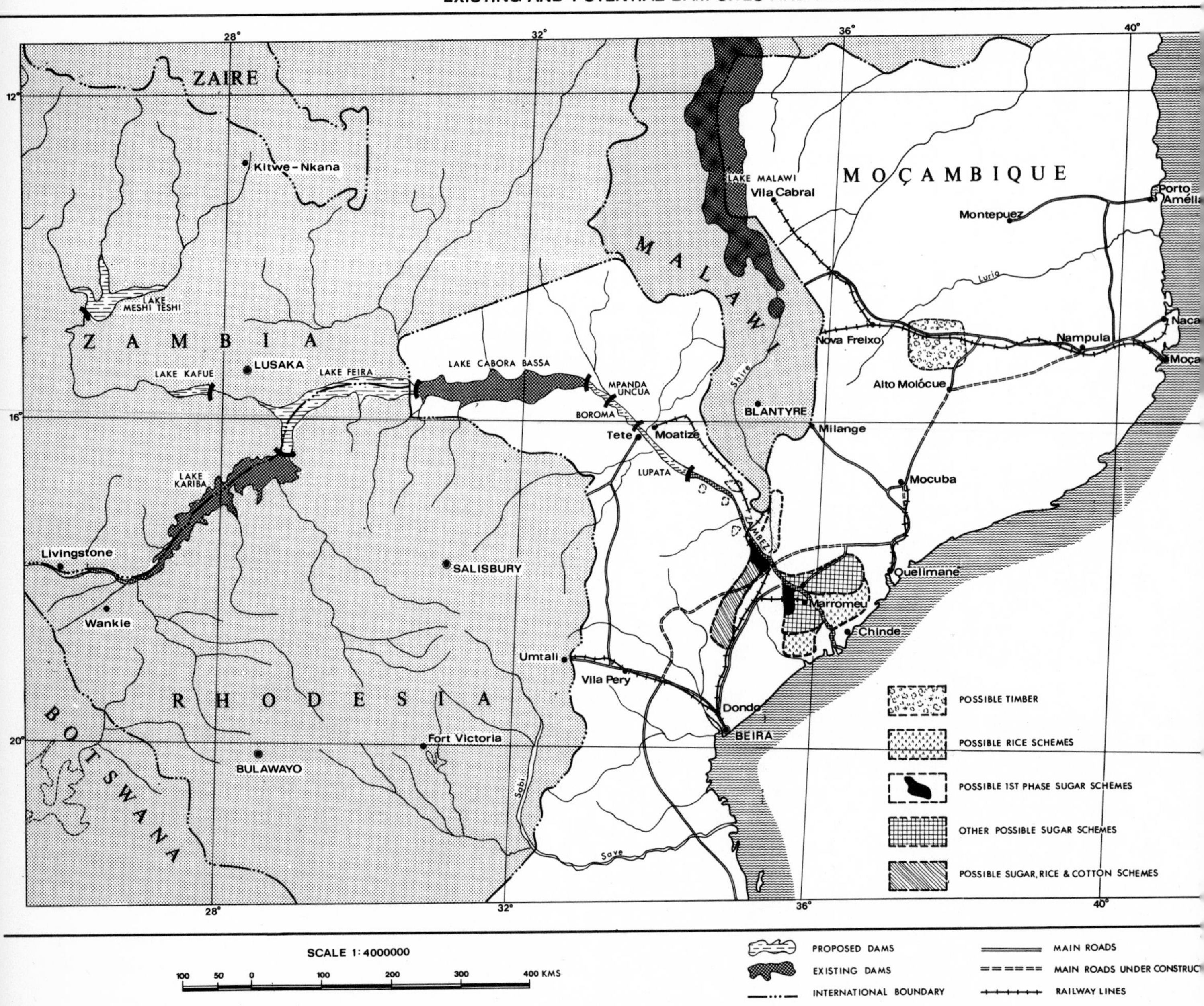

to the sea and to ports in the Indian Ocean either to the north or the south of the river mouth.

It is understood that some time before Moçambique became independent the Portuguese authorities had commissioned consulting engineers to study the possibilities of barge traffic from Vila Fontes at least as far upstream as Tete which is near Moatize where the present railway from Beira ends and where the valuable coal fields are located. This would have involved the construction of only the relatively small dams at Mutarara and at Lupata, where it might be possible to incorporate locks in the structures which are not located in gorges such as at Cabora Bassa and Mpanda Uncua. Hydro-electric power could be generated also at these sites to supplement the Cabora Bassa output.

The costs per tonne/kilometre for railways must escalate faster than for hover-barges when account is taken of rail maintenance, the operation of numerous locomotives, rolling stock, stations, sidings, signalling fuel, and labour etc. compared with the simplicity of the hoverbarge system.

The average 800 tonne train from Lusaka to Beira via Rhodesia, a distance of 1 600 km, takes 15 days for the return journey compared with 4 days for the hoverbarges, allowing for rail or road connections from Lusaka to Lake Cabora Bassa and the by-pass road traffic between the lakes in the gorge areas as far as Mutarara.

It is of course not contemplated that hovercraft traffic will *replace* the railway systems. The concept envisages that the waterway traffic will be supplementary to the rail systems as regards conveyance at speed of high value, perishable products. Existing rail systems will have to be modernised or remodelled and new links are being surveyed.

Assuming for the moment that the costs per tonne-kilometre for prime cargo are identical for rail and hovercraft, the costs per tonne from Lusaka to Beira would be R38 by waterway and R50 by rail via Rhodesia (route 5 – existing line).

Referring to the plan of the Zambezi valley it is seen that the large areas of fertile land suitable for agricultural development, both by irrigation and by rainfall cultivation, lie in the lower stretches of the Zambezi. Perishable and high value freight can be conveyed by hoverbarge to Beira or other ports at speed and this traffic can be inaugurated before the upsteam dams are built in order to bring rivercraft up to Tete and eventually up to the headwaters of Lake Cabora Bassa.

Feasibility studies along these lines may show that the same techniques may be applied to make the entire Zaire (Congo) River navigable by using roll-on roll-off techniques to by-pass the Inga rapids.

CHAPTER 1 Uganda

OWEN FALLS HYDRO-ELECTRIC SCHEME

HISTORY

Although the Owen Falls dam at Jinja, Uganda, is located barely one degree north of the equator it is included in this volume because of the potential for creating links to the north as well as to the south by virtue of the vast capacity of the reservoir controlled by the dam.

Electrically it will be possible to link with transmission grids both to the north and to the south, and because the Southern African power grid is rapidly becoming a fact and is expanding steadily to the north, the indications are that the link with the south will probably mature first.

Waterwise it will be possible to store enough water in Lake Victoria to regulate the northward flow of the White Nile to provide for the needs of downstream Sudan and Egypt for more than a hundred years.

The Owen Falls are named after Major E. R. (Roddy) Owen, D.S.O., the Lancashire Fusiliers, who came to Uganda with Portal's mission in 1893; he died of cholera in the Sudan three years later. There is no evidence that Owen ever visited the falls; his name was given to them quite arbitrarily by a fellow officer who was engaged in the survey of the country. Owen was an accomplished rider and won the Grand National on Father O'Flynn in 1892.

In 1907, the late Sir Winston Churchill visited Uganda as Under-Secretary of State for the Colonies and in his book *My African Journey* wrote:

'We must have spent three hours watching the waters and revolving plans to harness

Artist's impression of Owen Falls Hydro Electric Scheme

H.M. Queen Elizabeth II investing the author with "The Most Distinguished Order of St. Michael and St. George" on April 27, 1954, at Entebbe The Governor of Uganda, Sir Andrew Cohen is in attendance

and bridle them. So much power running to waste, such a coign of vantage unoccupied, such a lever to control the natural forces of Africa ungripped, cannot but vex and stimulate imagination. And what fun to make the immemorial Nile begin its journey by diving through a turbine!'

In 1935 the East African Governments commissioned two eminent firms of consulting engineers to report on the utilisation of the water-power resources of the Victoria Nile. However, at that time it did not appear to the consulting engineers that there was adequate demand for the power to justify what was then thought to be a costly scheme.

In 1947 the Government of Uganda called for a report on the potential demand that might be expected within a reasonable distance of the Owen Falls.

In the same year firms of civil, mechanical and electrical engineers were appointed to study the scheme afresh.

In 1948 the Uganda Electricity Board was set up. Early in the year the report of the consulting engineers was presented to the government and instructions to proceed with the work were given in June 1948. As a result of additional surveys, borings, and model tests the generating capacity was increased to 150 000 kilowatts: ten sets of 15 000 kilowatts.

Tenders were invited early in 1949 and the civil engineering contract was awarded in September 1949.

The project was inaugurated formally by Her Majesty, Queen Elizabeth II on 29th April 1954. In her speech she said:

'This power will serve industries which are already in being and others which will be established in the future. Without power there can be no economic development and without power no country can go forward in the modern world. But let us not forget that economic development and the building-up of industries are not ends in themselves. Their object is the raising of the people's standard of living. We welcome this great work because, by increasing the wealth of this country, it enables the people – and above all the African people – to advance. It will help you, the citizens of Uganda, to reach the higher levels of health and prosperity towards which you so rightly aspire. I confidently believe that your children and grandchildren will look upon this scheme as one of the greatest landmarks in the forward march of their land. Not only will this dam produce power for Uganda, it is also designed to store water for Egypt. This part of the scheme, which is an example of practical co-operation between nations, is of broad scope and of the greatest importance to the future welfare of all the dwellers of the Nile valley'.

DIMENSIONS

The Owen Falls power station and dam are sited on the Nile, about three kilometres below the outlet from Lake Victoria at Ripon Falls.

The area of Lake Victoria is 67 000 square kilometres, approximately equal to that of Scotland. Its catchment area of 267 000 square kilometres is bigger than the whole of Uganda.

The storage capacity of the lake is 207 000 million cubic metres, which makes it the biggest reservoir in the world.

The average flow of the Nile at Ripon Falls is equivalent to 630 tonnes of water a second; the maximum recorded flow – records have been kept since 1899 – is nearly twice this.

At Owen Falls 1 m^3 of masonry stores 1 million m^3 of water. Between Lakes Victoria and Albert, a distance of 375 km, the Nile falls a total of 516 metres. This represents total available power amounting to 2 million kilowatts at an average flow of 632 cumecs.

The total length of the Owen Falls dam is 831 metres and the maximum height 26 metres. The length of the road over the dam is about half a kilometre. The main dam, curved in plan, is a mass concrete gravity dam, and crosses the Nile at a point where the river is reduced in width by a low promontory projecting from the west bank. Six ground sluices are incorporated in the centre of the dam, with a combined capacity of some 1 275 cubic metres per second, or slightly in excess of the maximum recorded discharge from the lake at Ripon Falls.

Ten turbines have been installed with a combined capacity of 150 000 kilowatts. The volume of water passing through each turbine is equivalent to 90 tonnes a second. The size of turbines and generators was determined by transport limitations. Each one is provided with single bulkhead gates for normal operation, and with emergency gates in two halves to operate in front of the main gates.

The turbines have propeller blades which can be adjusted while the sets are running in accordance with the prevailing load and head of water.

There are six sluices which are designed to throw the water 30 metres away from the dam to prevent erosion of the river bed near the main structure.

The volume of water passing through the sluices is continuously recorded and regulated in accordance with the agreement with the Egyptian Government.

The average rainfall on the catchment area of 268 000 km^2 is 1 075 mm, but only 7 per cent of the run-off flows down the White Nile. Despite this, the volume of water available at Owen Falls is considerable, ranging from the minimum of 300 cumecs to the maximum of 1 229 cumecs.

Electricity is generated at 11 000 volts and stepped up to 33 000 volts and 132 000 volts according to the distance over which the supply is to be carried. In the first instance a 132 000 volts line was constructed to Tororo to serve local industries. This was extended later to Nairobi to export power to Kenya.

CONSTRUCTION

The power station is located on the downstream side of the promontory on the west

Cross-section of main dam (left)

Cross-section of ground sluice (right)

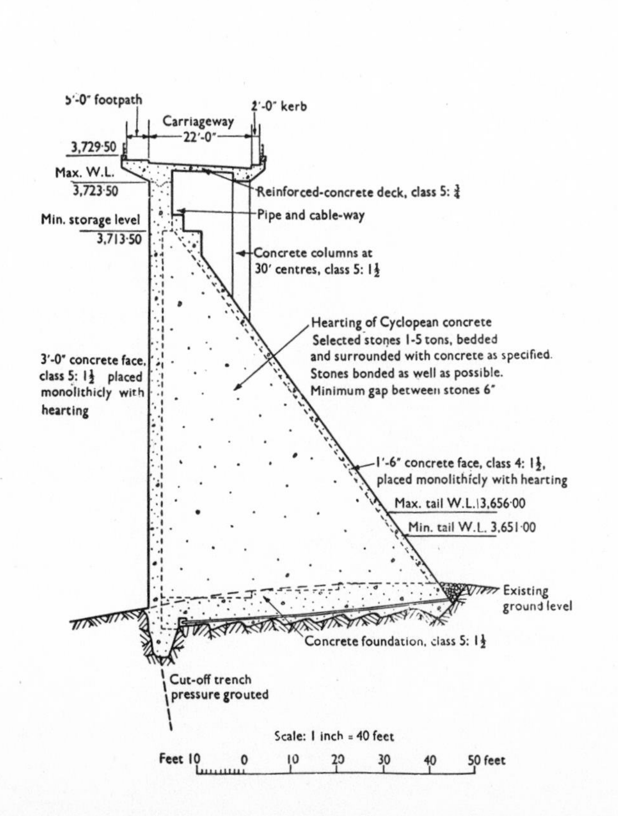

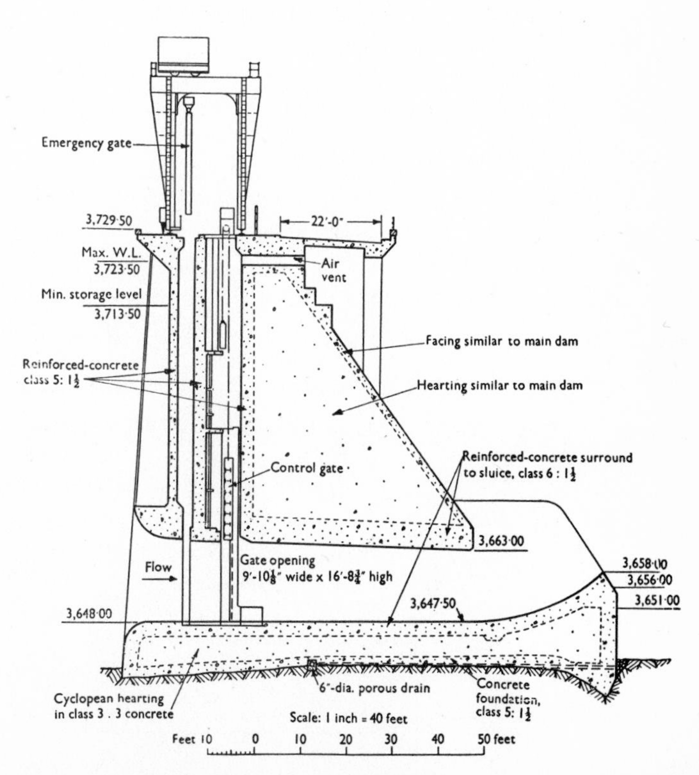

bank in such a position that full use could be made of the existing wide embayment in the Nile at that point to facilitate the discharge through the draught tubes of the turbines. The mass concrete intake dam is integral with the power house, and the turbines are fed with water from a pool 162 metres wide excavated in solid rock and confined between two headrace dams.

The new maximum water level is above the roadway that was suspended from the railway bridge and it was therefore necessary to construct a new reinforced concrete roadway consisting of a 6,7 metre wide carriageway and a 1,5 metre wide footpath across the headrace and the main dam. This consisted of precast sections.

The dam forms a complete barrier to the passage of migratory fish. The only fish which move up or down the Victoria Nile are of little or no economic value, and it was not considered necessary to incur the heavy expenditure involved in constructing a fish pass at the dam.

Construction difficulties may be classified under three broad headings: water control and cofferdam construction, having regard to the unpredictable nature of foundation conditions in the river bed; labour problems; and the phasing of deliveries of construction materials in the difficult post-war years, having regard to the long hauls involved. The total tonnage of temporary and permanent materials brought to site by East African Railways and Harbours, from Mombasa, a distance of more than 1 200 km, amounted to about 80 000 tons, of which 36 000 tons represented cement from Europe. The control of the Nile during the construction of the dam was exercised by control of the outflow from Lake Victoria at Ripon Falls and in three phases at the dam site which can best be illustrated by a series of drawings.

At Ripon Falls two islands divided the river into three channels. About half of the river flowed through the west channel. The west channel was blocked by constructing a vertical tower made of timber to a height of 22 metres and 3 metres square, filling the tower with large rubble contained by torpedo netting and then blasting two timber legs on the water side causing the tower to collapse across the channel and reduce the flow sufficiently to complete a rockfill dam behind the fallen tower.

A series of sluices was constructed on the east bank. Thus complete control could be exercised over the flow downstream. This made it possible to comply with the agree-

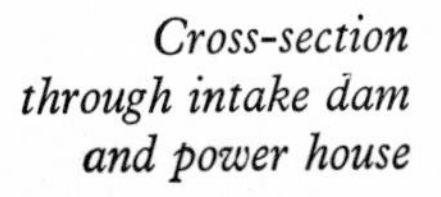
Cross-section through intake dam and power house

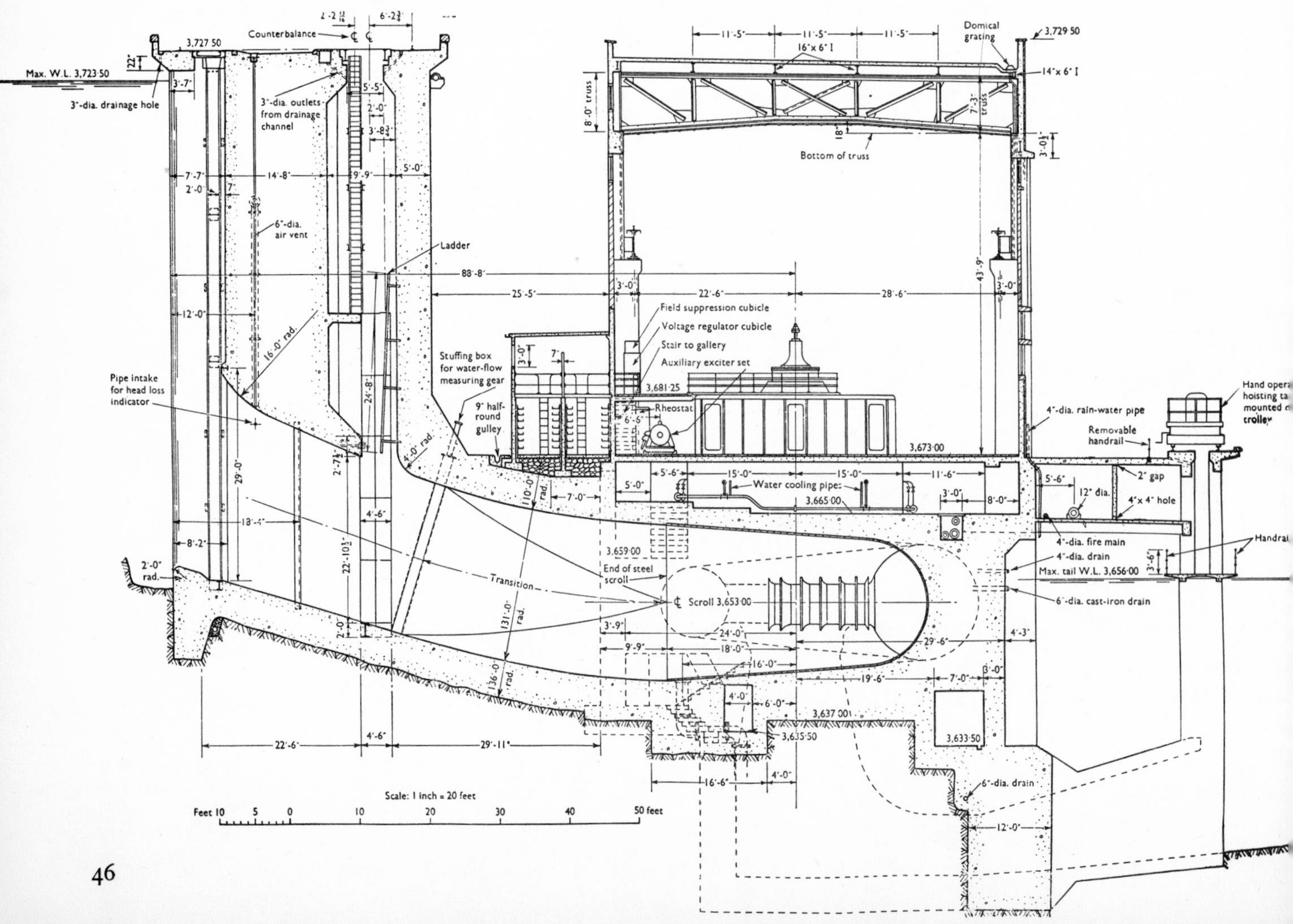

METHOD OF CONSTRUCTION OF MAIN DAM

RIVER NILE

SLUICE DAM

OUTLINE OF DAM

A

A

ROAD BRIDGE

SITE PLAN

SECTION A-A

(DAM AFTER COMPLETION)

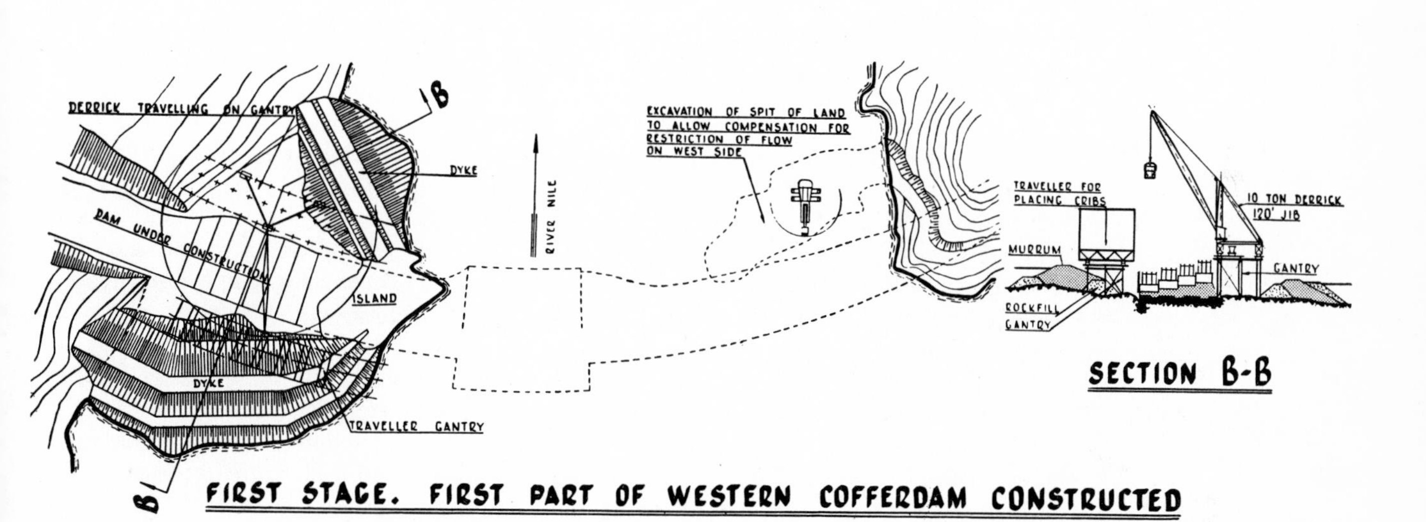

FIRST STAGE. FIRST PART OF WESTERN COFFERDAM CONSTRUCTED

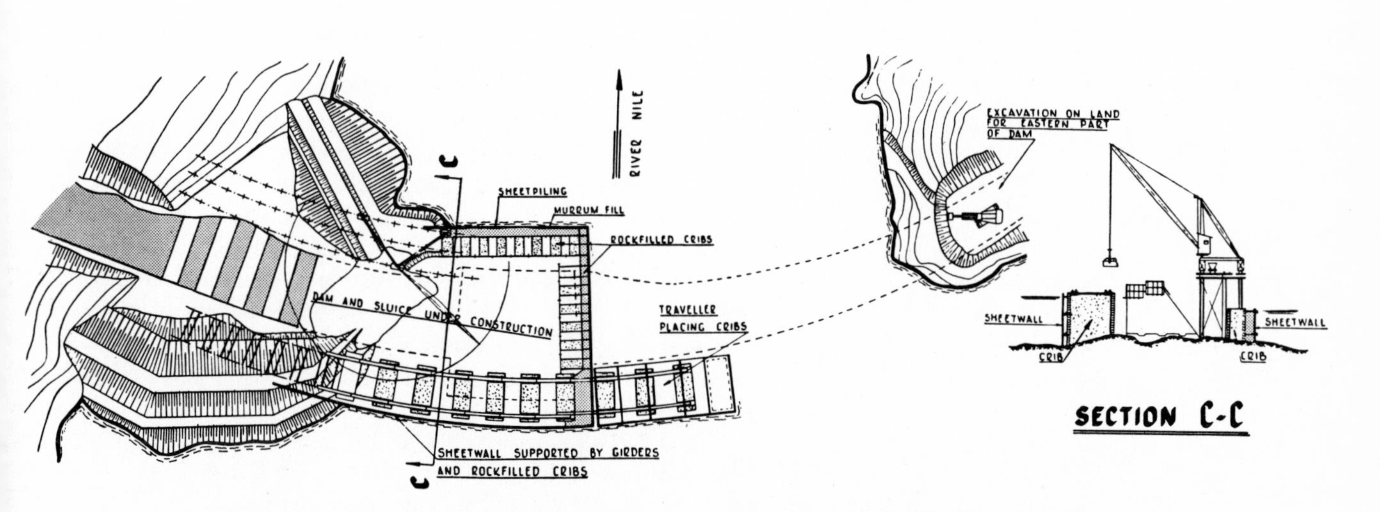

SECOND STAGE. SECOND PART OF WESTERN COFFERDAM CONSTRUCTED

ment reached with the Egyptian Government to the effect that, for a 24 month period, the flow could be reduced to, but not below, 600 cumecs. This made it possible to work out the most economic diversion arrangements as the cofferdams needed to be designed only to a level corresponding to this flow.

It was noted that by December 1953, shortly before control of the Nile ceased, the level of Lake Victoria had risen by 112 mm., representing an additional storage of 8 700 million m^3 available to Egypt when run-of-river conditions were re-introduced.

Meantime, in the first phase at site two dykes of rubble and sealed with "murram" (Laterite) were constructed to join at a convenient existing island in the river. Inside the dewatered cofferdam the foundations were prepared and the first sections of the dam concreted, leaving gaps in the dam wall through which the water could be diverted later. (Stage 1.)

It would have been difficult to carry out the Owen Falls project if the impervious indigenous reddish soil called "murram" had not been so readily available. It was used to seal all the cofferdams. While construction proceeded inside cofferdam 1, a spit of land on the east bank was blasted away to facilitate passage of water without raising the river levels.

In phase two, the cofferdam was extended by placing huge cribs made of eucalyptus poles about 13 metres in length, prefabricated in sections, weighing up to 45 tons with the bottom section tailormade to fit the exact contours of the riverbed. These poles had to be brought in from Kenya as suitable timber was not available in Uganda. They were transported and placed by means of a large structural steel "traveller" made up on site and which we christened the "organ". (Stages 2 and 3.)

The crib section, lifted at the rear end of the traveller by four 15-ton chain blocks travelling on two joists running the length of the structure and supported by its framework, was moved forward until vertically over its destined position. It was then lowered and wedged into a frame which travelled vertically on four members attached to the traveller framework, and which were capable of being lowered or raised to suit the depth of the river (though they were never allowed to rest on the bottom). The two downstream members were supported by a diagonal 28 mm wire rope, which transmitted to the main structure the force of the current on the crib section whilst being lowered, the traveller having been anchored to the two previous cribs so as to sustain this load. It was calculated that the maximum horizontal force would be of the order of 20 tons.

When the bottom section of a crib had been wedged to the positioning frame, and was all but completely immersed, buoyancy having released the traveller of most of the vertical load, the next section would be placed on top and strapped to it. In this way, the joining of sections below water level was avoided, and a crib, weighing say 120 tons, could be placed quite quickly in position.

Once placed, the cribs were filled with rock, the wedges pulled, and the positioning frame released and raised to the surface. Two reinforced-concrete bearing pads were constructed on each filled crib to carry the beams spanning from crib to crib. The railway tracks were then extended so that the traveller could be moved forward 10 metres in readiness to repeat the series of operations. Once experience was gained the complete cycle of operations regularly took only 8 to 10 days.

Owing to the heavy current it was found necessary to construct a second sheet-pile wall, 3 metres upstream of the main line of piles, to protect the divers while engaged on sealing operations.

Originally it had been intended that divers should excavate any loose material overlying the rock and form a seal by means of a triangle of clay. Unfortunately, it was found that for a length of about 30 metres no rock could be proved, and a test boring at the south-east corner showed a depth of 8 metres of broken material. Moreover, the divers discovered a number of seams running in a north-south direction. To seal this area it was therefore decided to form a piled box, 7 metres wide, which was filled with about 2 metres of murram. A 20 metre long grout curtain was formed to cut off any water which might penetrate below the murram blanket.

The downstream arm and the cross-wall were formed by smaller cribs (6 metres by 5 metres in plan) placed by two cranes manufactured on site, having two fixed jibs each and capable of lifting 25 tons. The crib to be placed was assembled on the one which had just been placed. After it had been lifted the crane, with the crib suspended, was moved forward until it was able to lower the crib into its correct position. They were held in position whilst being lowered by wire ropes attached to the upstream arm of the cofferdam, one leading downstream and two holding the crib tight to vertical guides attached to the previously placed crib. These small cribs were timber-sheathed on the outside, and provided

METHOD OF CONSTRUCTION OF MAIN DAM.

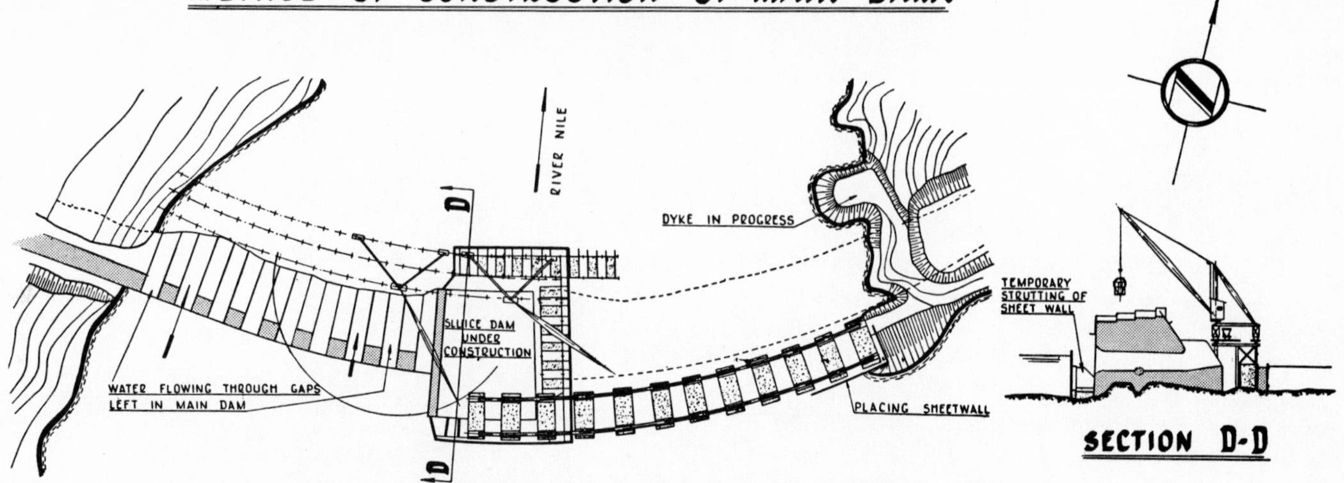

THIRD STAGE. CONSTRUCTING EASTERN COFFERDAM.

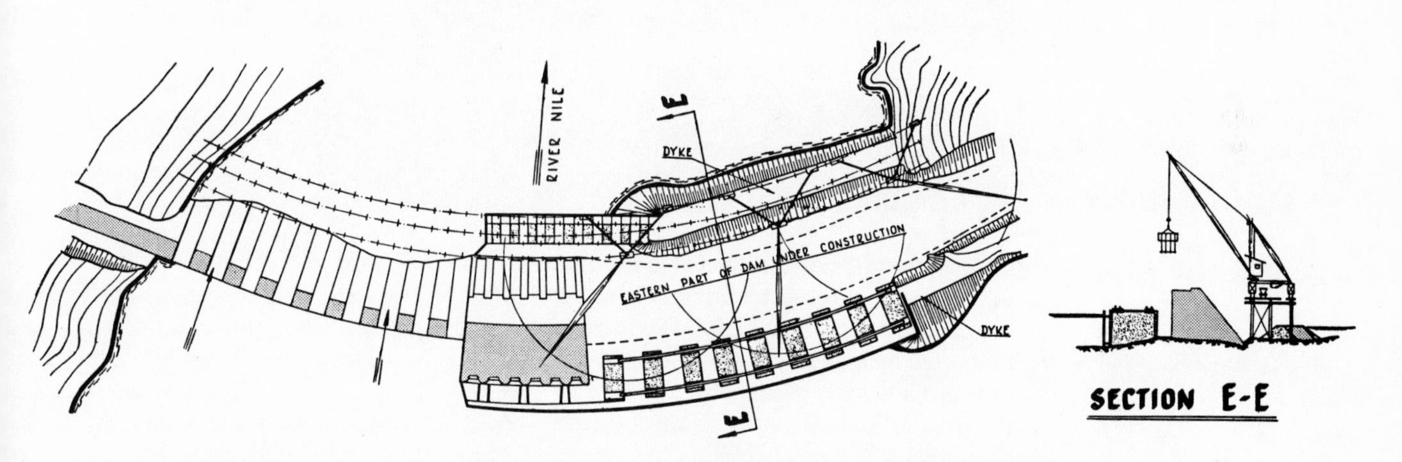

FOURTH STAGE. EASTERN COFFERDAM CONSTRUCTED.

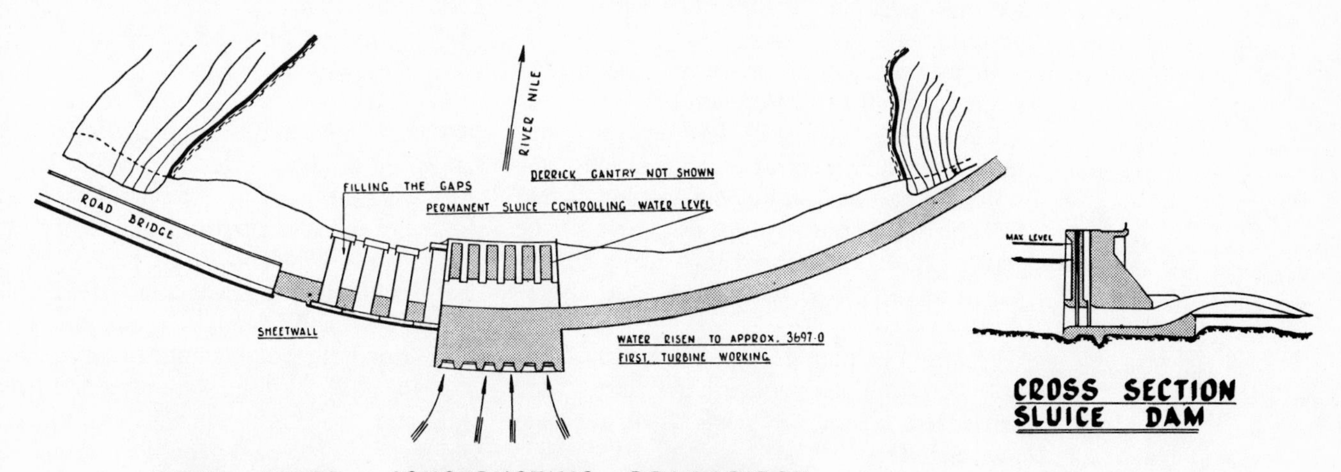

FIFTH STAGE. CONSTRUCTING ROADBRIDGE AND CLOSING GAPS.

The Nile at Ripon Falls at the spot to be blocked by the falling tower

(opposite page – top) *The Rock Tower constructed on the west bank of the Nile at Ripon Falls (a)*

(bottom) *Rock Tower, after blasting of two legs, falling to block the west channel at Ripon Falls (b)*

with cantilevered pile guides so that, after placing, sheet-piles could be driven 2½ metres outside the timber sheathing, providing a corridor which was sealed with murram after the river bed had been cleaned by divers.

Owing to the force of water and the irregular river bed, two of the downstream cribs moved about 1 metre in spite of support from cables taken to the upstream cribs. This, however, did not seriously affect the programme, and the sheet piles were realigned to suit the new position of the cribs.

Inside cofferdam 2, the sluice dam was constructed with the sluice openings in.

During the third period cofferdam 1 was removed and crib construction proceeded towards the east bank. The Nile was now forced through the temporary openings left in the west bank section of the dam. The rest of the dam was completed within the third cofferdam.

During this period some interesting bore-hole grouting and gravimetric tests were carried out on the east bank where the dam was to abut.

As a result of this work it was concluded that the east abutment could be regarded as a natural earth and rockfill dam. Hence the dam was not taken into the east bank as far as originally planned and a short thin concrete cut-off section was deemed to be sufficient. This saved £150 000 compared with the cost of the investigations: only £5 000.

The cribs for the upstream wall of cofferdam No. 3 were placed whilst the dam was under construction in cofferdams Nos. 1 and 2. As soon as these were in position a travelling 6 ton derrick was erected on them, which was used for placing the walings and sheetpiles. The timing was such that all preparatory work for placing the piles was finished at the same time as the river was diverted through the gaps left on the western side of the main dam. The complete diversion was accomplished gradually by the pile placing, which had to be done under a head of water ranging from 1½ to 2½ metres. In order to guide each pile over the bottom waling, while being tapped home by a McKiernan-Terry automatic hammer, a wire rope was taken from the toe of the pile upstream to a return block and back to a winch on the cribs.

It was found necessary to extend the 7 metre box of cofferdam No. 2, and after divers had cleaned the area along the toe of the sheeting, clay and clay bags were placed to form a seal, and the grout curtain was extended westward. At the eastern end, a murram bank was tipped out from the shore.

The downstream arm, where the water was shallow, was formed by a murram bank with a rubble filter, a few piles being driven into the bank to make good to the downstream piling of cofferdam No. 2.

It took about 5 weeks to dewater this cofferdam, owing to the extensive leak through jointed rock at the south-east corner. This was staunched by a combination of grouting and tipping of a further murram blanket upstream of the cofferdam.

The main dam is founded on an amphibolite ridge which runs across the river and falls away very sharply on the upstream side. It was therefore a great advantage to keep the upstream piles as close as possible to the dam. The problem was overcome by removing the cribs in front of the sluices after alternative supports for the piles had been provided by constructing between the cribs, concrete buttresses reinforced with rails and designed to transmit the load from the walings to the dam.

Later, when the whole of the front face of the sluice dam was sufficiently high, the walings were re-strutted to the concrete using round poles, after which the concrete buttresses were in their turn removed, and the space between dam and piles was cleared down to a level 3 metres below sluice-invert level.

In order to permit first-stage water-raising before sluices Nos. 5 and 6 were ready, a temporary sheet-pile cofferdam, attached to the foundation concrete, was erected round these intakes and strutted from the downstream portion of the dam, which, since it did not contain any gate guides, was much simpler to construct and permitted the guides and grouting to be done after water-raising. The removal of the piling, walings and struts upstream of the sluice dam was done by flooding between dam and piling, with gates in sluices Nos. 1–4 closed.

Since the first three openings could be blocked without raising the water above desirable operating levels it was decided to use a limpet dam to close these openings one after the other. This limpet was constructed using sheet-piles with timber bearers and was lowered into position, sliding on previously prepared fixed ways. In order to overcome the friction during the last few feet of lowering, the limpet was ballasted and hammered down with a pile hammer. After placing this upstream limpet a similar downstream limpet was placed and ballasted with about 30 tons, following which the opening was dewatered and concreted.

A different system had to be adopted for openings Nos. 4 to 9 in order to save time and because of the greater head involved. Sets of walings were placed across the openings and sheet-piles driven in front of them. In order to delay water-raising as long as possible, only the piles in front of the concrete piers and one

Night shift at Owen Falls Dam 1954

double pile in the centre of each opening were driven down to rock in the first instance. Even so, the effect of the pressure of water was such that when it came to closure, a 1,2 m taper was necessary, this in spite of stiffening plates having been welded into the bosoms of the piles.

Before completing the entire sheet-pile curtain-wall it was necessary to restrict flow of water at Ripon Falls to the maximum possible extent, and to bring sluices Nos. 1–4 into commission.

Great difficulty was experienced in closing opening No. 9 owing to the fact that a 30 cm by 30 cm R.S.J. 17 metres long, which had previously been used as a strut alongside the sluice dam, had dropped to the river bed during dismantling, and had ended up across the bottom waling of this opening where it interfered with the completion of pile driving. At this stage it was impossible to extract the piles in question and it was therefore necessary to remove the joist by a diver, working next to sluice No. 1 (closed during the operation) attaching slings from two derricks to its upstream end. In the act of wrenching it free some pile toes were damaged and the waling tilted so that it became practically impossible to complete a reliable curtain. To overcome these problems of sealing the opening a waling grillage was lowered into two recesses left in the concrete walls as keyways, about 5 metres downstream of the dam face. Against this framework, 10 metre-long piles were placed and the space between this diaphragm and the curtain wall was filled with rubble and 7,5 cm stone so as to transfer the hydrostatic pressure from the curtain to the diaphragm. Pipes with welded bracings were placed within the stone to permit this portion of the dam to be "colcreted" after placing the concrete downstream of the diaphragm. After water-raising the considerable leakage under the piles was sealed by placing gabions (8 cm stone bound up in wire netting) and rubble, followed by stone graded gradually to smaller sizes, and finally murram.

During this short period the flow in the Nile was stopped completely by closing the sluices at Ripon Falls.

A downstream dike of rubble and murram enabled the temporary openings to be dewatered and concreted.

The control room, the heart of operations, is able to establish radio communication with mobile maintenance units.

In the event of insulation failure in the alternator windings, carbon dioxide is automatically injected to extinguish any fire which might result. At the same time the machine is automatically stopped.

The rock on which the dam and power house are founded is amphibolite: it is so hard it will cut glass.

The amount of rock and earth excavated for the scheme was 252 000 cubic metres; the amount of concrete used was 176 000 cubic metres.

The Uganda Cement Industry's plant at Tororo, which came into production early in 1953, provided 16 000 tonnes of cement for the project; the remaining 36 000 tonnes were imported from Europe.

Material and equipment brought from the United Kingdom and elsewhere amounted to 69 000 tonnes. All of it had to be carried 1 300 kilometres by rail from Mombasa.

The maximum labour force engaged was close on 3 000, of whom 350 were Europeans – British, Danish, Dutch, and Italian – and 127 Asians.

The African workers were accommodated in a specially built camp with a church, hospital, cinema, shops and playing fields.

CONSULTING ENGINEERS AND CONTRACTORS

The Civil Consulting Engineers were Sir Alexander Gibb and Partners and the Mechanical and Electrical Consulting Engineers were Kennedy & Donkin, both of London.

Main contractor:- The Owen Falls Construction Company formed by the following firms:-

- Christiani & Nielsen Limited
- Dorman Long & Co. Ltd.
- Edmund Nuttall Sons & Co. (London) Ltd.
- Hollandsche Beton Maatschappij N.V.
- Internationale Gewapendbeton Bouw N.V.
- J. L. Kier & Co. (London) Ltd.
- Nederlandsche Aanneming Maatschappij N.V.
- Nederlandsche Beton Maatschappij Nato N.V.

Principal sub-contractors:-

- Electrical installations: Electrical Installations Ltd.
- Heating and ventilation, and hot- and cold-water services: Norris Warming Co. Ltd.
- Lifts: Marryat & Scott Ltd.
- Furnishings: Waring & Gillow, Ltd.

CHAPTER 2 Zaire

INGA HYDRO-ELECTRIC PROJECT

HISTORY

The River Zaire ranks second in the world after the Amazon, both as regards its catchment area (3 800 000 km^2) and its annual discharge. The geographical position, astride the equator, provides it with an exceptionally stable hydrologic regime: based on hydrologic records covering an uninterrupted period of 72 years, the average annual flow is 40 000 m^3/sec. and for 98% of the time flow exceeds 26 400 m^3/sec. These figures may be compared with the average annual flow of 2 800 m^3/sec at Cabora Bassa dam site.

The longitudinal profile is also exceptional: after flowing through the central basin, nearly horizontal and navigable over a length of 1 730 km, the river flows to the sea through a series of rapids, through which the drop is 225

Panoramic view showing the layout of Inga I and Inga II hydro-electric projects

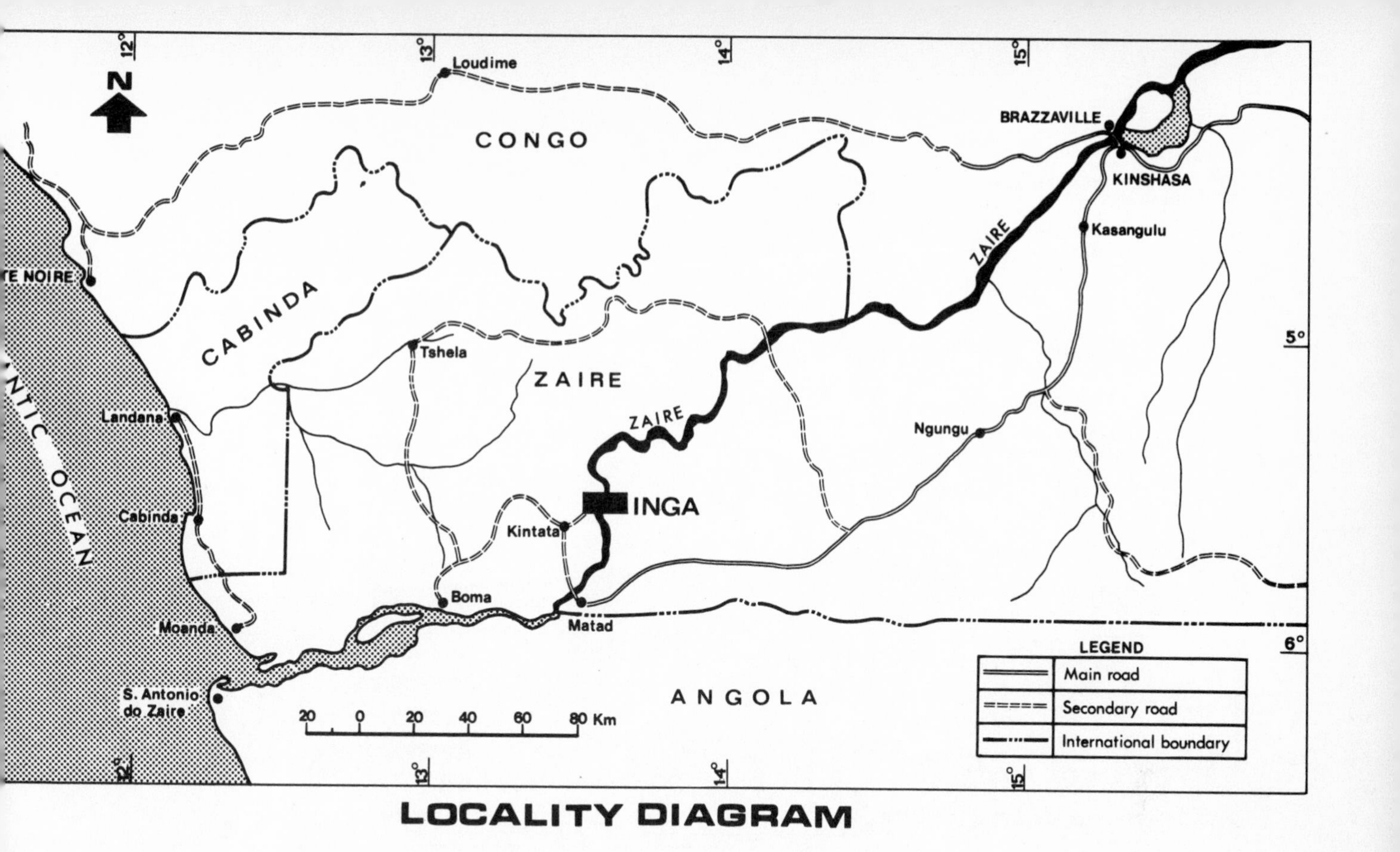

LOCALITY DIAGRAM

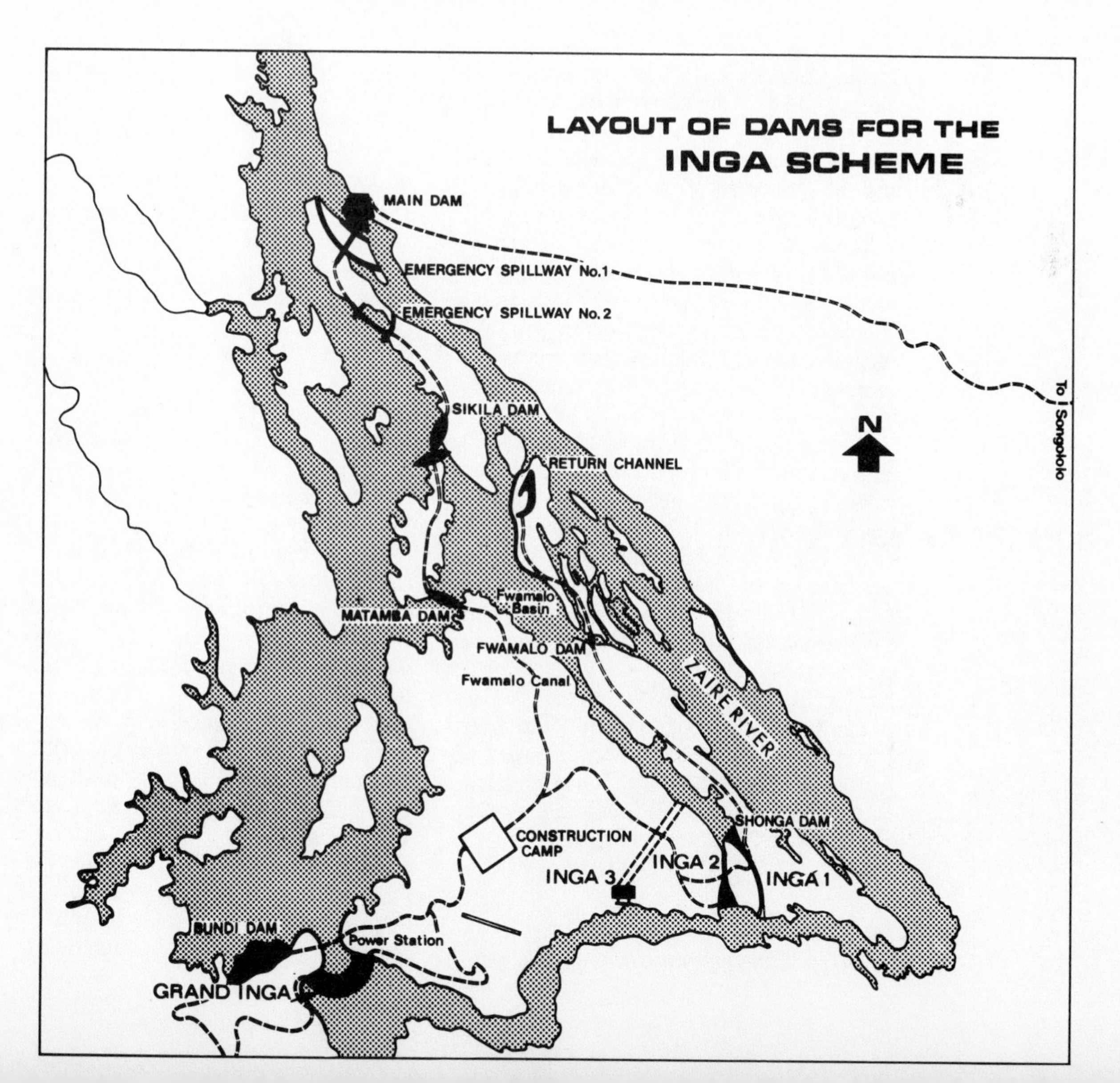

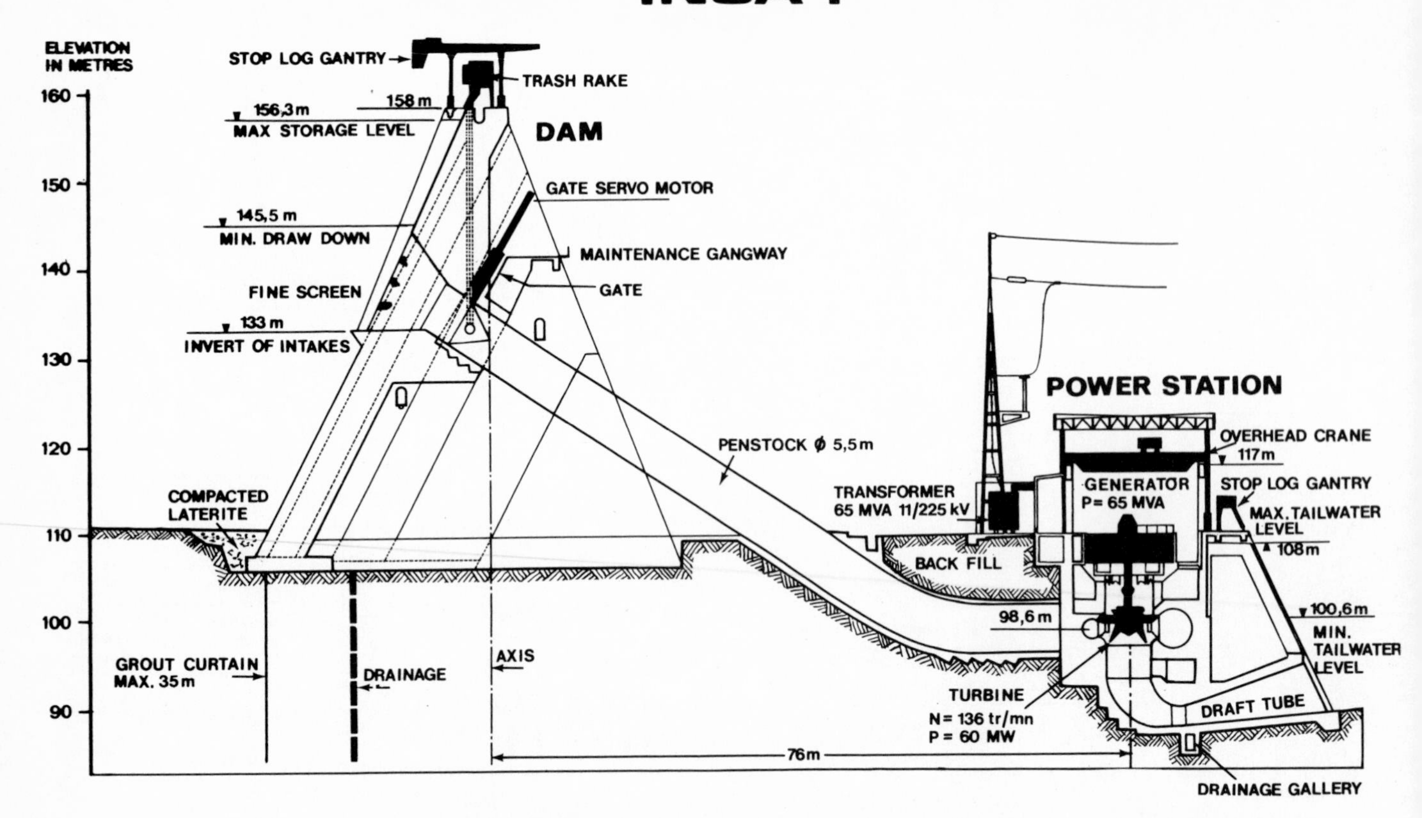

metres between Kinshasa and Matadi. The most impressive rapids are those at the Inga site at latitude 5°30′ south, where there is a fall of 98 metres between the Island of Sikila and the confluence of the Bundi River, which are only 15 km apart in a straight line.

These natural characteristics of the site make of Inga one of the greatest single natural sources of hydro-electric power: the potential annual production is of the order of 305 TWh* equivalent to the total electric power consumption of France and Italy combined in 1973.

However, the very magnitude of such an undertaking could only have delayed its construction had local topography not enabled a staged development to be undertaken.

The N'Kokolo Scheme does not call for the construction of a dam across the main river. It uses a dry valley the bottom end of which has been closed off by the Shongo Dam: this enables the exploitation of a fraction of the flow to be drawn from the river 8 km upstream under a head of 60 metres. The Inga I power station, already in operation and the Inga II power station, under construction, are based on this scheme and will be capable of providing 12 TWh as from 1980.

* Twh = Tetrawatthours = milliard kWh.

INGA STAGE ONE DEVELOPMENT

DIMENSIONS

Shongo Dam	
Type	Buttress
Maximum height	52,5 m
Length	600 m
Volume of concrete	245 000 m³
Penstocks	
Diameter	5,5 m
Length	6 × 86 m
Power Station	
Type	Surface
Length	135 m
Width between rails	15 m
Tailrace	
Discharge capacity	780 m³/s
Length	1 230 m
Volume of excavation	800 000 m³

MECHANICAL AND ELECTRICAL EQUIPMENT

Power Station	
Six units (of which one in reserve)	350 MW
Guaranteed capacity	300 MW
Generation	2,4 × 10⁹kWh
Turbines (Francis)	
Nominal head	50 m
Discharge capacity	140 m³/s
Power	60 MW
Speed	136 r.p.m.

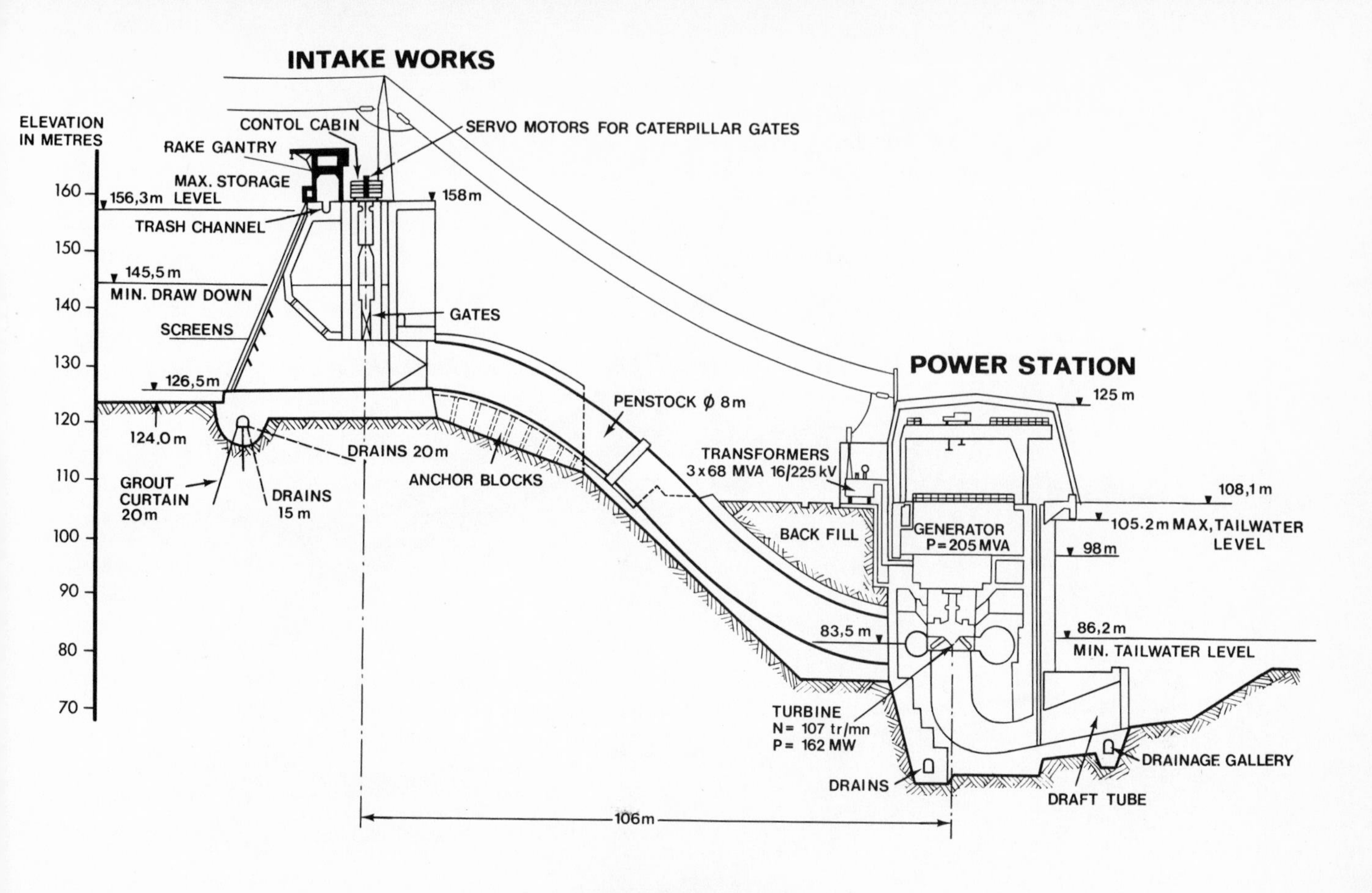
INTAKE WORKS
ELEVATION IN METRES
CONTOL CABIN
SERVO MOTORS FOR CATERPILLAR GATES
RAKE GANTRY
MAX. STORAGE LEVEL
156,3m
158m
TRASH CHANNEL
145,5m
MIN. DRAW DOWN
GATES
SCREENS
126,5m
124.0m
POWER STATION
125 m
PENSTOCK ϕ 8m
DRAINS 20m
ANCHOR BLOCKS
TRANSFORMERS
3x68 MVA 16/225 kV
GROUT CURTAIN 20m
DRAINS 15 m
108,1 m
105.2m MAX, TAILWATER LEVEL
BACK FILL
GENERATOR P=205 MVA
98m
83,5 m
86,2m
MIN. TAILWATER LEVEL
TURBINE
N= 107 tr/mn
P= 162 MW
DRAINAGE GALLERY
DRAINS
DRAFT TUBE
106m
160
150
140
130
120
110
100
90
80
70

INGA 2

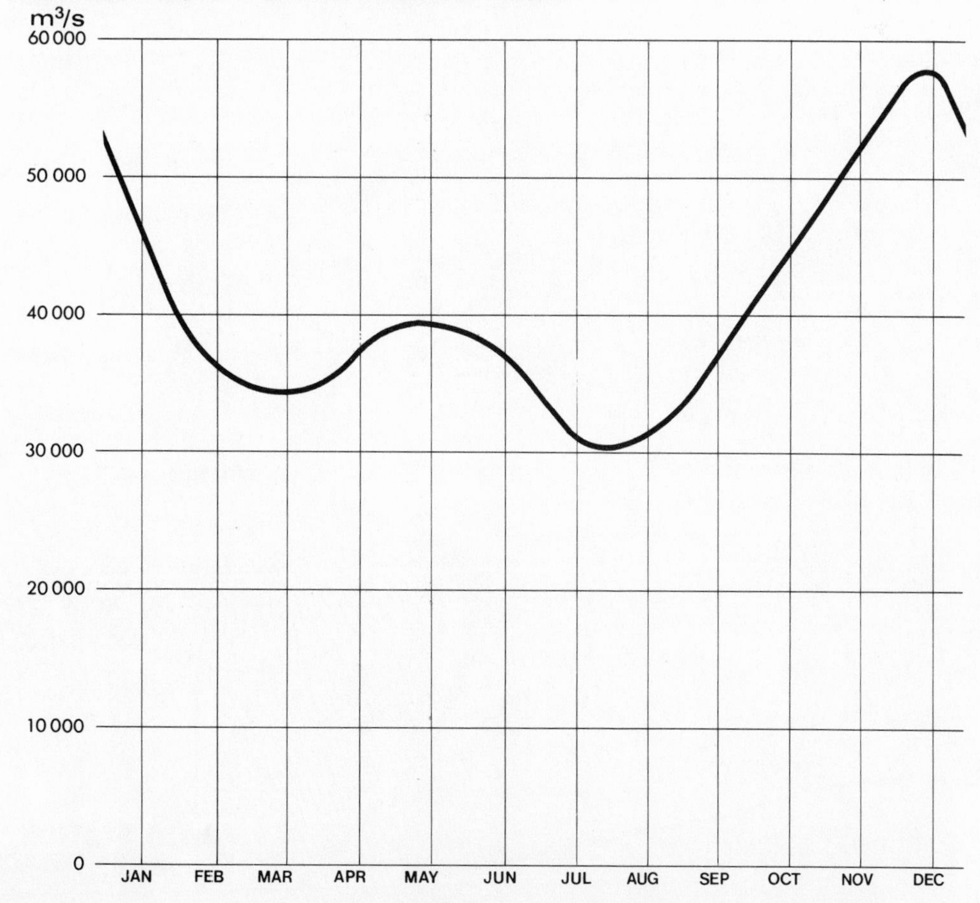
AVERAGE FLOW of ZAIRE RIVER at INGA
m³/s
60 000
50 000
40 000
30 000
20 000
10 000
0
JAN
FEB
MAR
APR
MAY
JUN
JUL
AUG
SEP
OCT
NOV
DEC

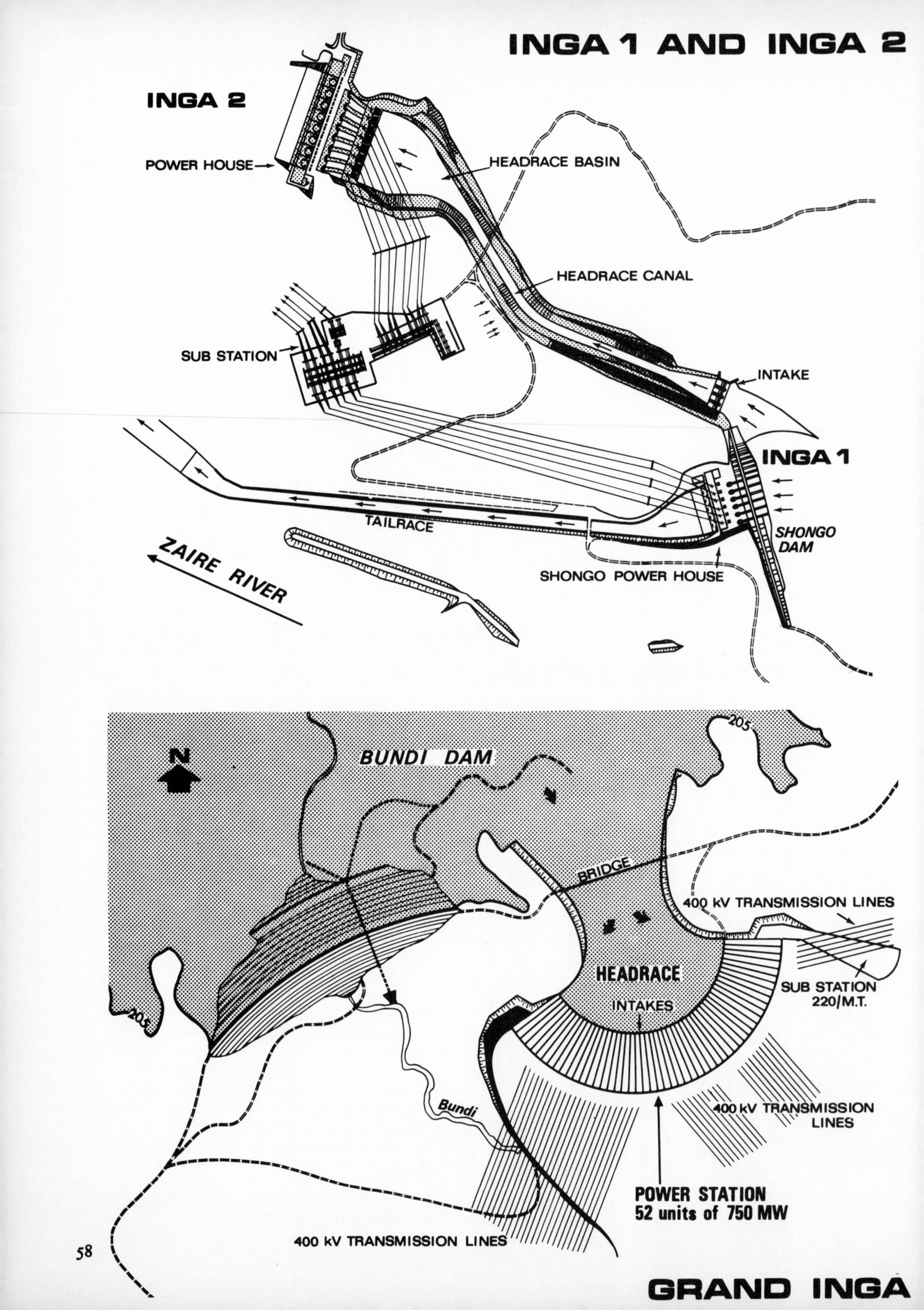
INGA 2
POWER HOUSE
HEADRACE BASIN
HEADRACE CANAL
SUB STATION
INTAKE
INGA 1
TAILRACE
SHONGO DAM
SHONGO POWER HOUSE
ZAIRE RIVER
N
BUNDI DAM
205
BRIDGE
400 kV TRANSMISSION LINES
HEADRACE
INTAKES
SUB STATION 220/M.T.
Bundi
400 kV TRANSMISSION LINES
POWER STATION
52 units of 750 MW
400 kV TRANSMISSION LINES

Forbay Canal and Inga I barrage in relation to the River Zaire

Generators	
Power	60 MVA
Power Factor	0,90
Voltage	11 kV
Transformers	
Capacity	65 MVA
Voltage	11/225 kV
Transmission	
2 lines to Kinshasa	
Voltage	225 kV
Length	280 km
1 line to Matadi and Boma	
Voltage	132 kV
Length	140 km
Supplied by two transformers:	
Voltage	225/132 kV
Capacity	50 MVA

INGA STAGE TWO DEVELOPMENT

River Intakes	4 × 14,5 m diameter
Headrace canal and basin:	
Discharge capacity	2 200 m³/s
Length	1 000 m
Volume excavated	2 200 000 m³
Concrete (3 gravity dams and one retaining wall)	129 000 m³
Powerhouse Intakes	
Intake dam length	176 m
Concrete volume	157 000 m³
Penstocks	
Eight conduits diameter	8 m
Length	105 m
Power Station	
Type	Surface covered
Length	297 m
Width between rails	22,50 m
Total volume of concrete	190 000 m³

Power Station Equipment

Installed capacity . . .	1 400 MW
Guaranteed availability .	1 100 MW
Generation of energy . .	$9{,}6 \times 10^9$ kWh
Turbines	
Francis Nominal Head .	56,20 m
Discharge	315 m^3/s
Nominal power per turbine	162 MW
Maximum power per turbine	178 MW
Speed	107 r.p.m.
Generators	
Rating	205 MVA
Power Factor	0,85
Voltage	16 kV
Transformers	
Capacity	3 × 68 MVA
Voltage	16/225 kV
Power Station Cranes	
2 × 310/30/5 tonnes	
Span	22,5 m
Switchyard	
Voltage	225 kV
2 lines to Banana	
2 lines to converter station for the Inga Shaba line (D.C.)	

It is possible to carry on this development by constructing Inga III which would consist of one or possibly two power stations capable of doubling or even tripling this output.

The Grand Inga Scheme involves damming the main river so as to divert its waters into a valley close to that of the Bundi River and thus exploit the whole of the river flow under a head of 150 metres. This scheme will bypass the projects referred to above as the water will be diverted upstream of its intake and will return the water downstream of its tailraces; it will therefore be necessary to abandon the river intake structures and to construct new intakes on the by-pass so as to supply water to the original facilities. The Grand Inga project can be sub-divided into thirteen identical stages, to be undertaken only when demand warrants it, each of which could generate 23 TWh a year.

The proposed installed capacity of 52 units with a capacity of 750 MW each, i.e. a total of 39 000 MW will make it the biggest single hydro-electric enterprise in Africa and among the biggest in the world to be rivalled only by proposed installations in Brazil, Paraguay and Argentina.

The project should produce relatively inexpensive electricity because the terrain is most favourable as regards the low ratio of civil engineering to plant work. Moreover the Grand Inga project can be phased over many years to meet the growing energy needs of Southern Africa. This should make it the key project in a series already constructed or planned for Southern Africa and would complete the closure of the growing Southern African electrical transmission grid which is already more than 60 per cent completed. Such a pan-Southern African link will make it possible to send Inga power to any part of Southern Africa at prices which will go far to free the economic development plans of Southern African territories from the scourge of oil and energy crises, and, indirectly, will help these countries to feed themselves and others.

CHAPTER 3 **Tanzania**

KIDATU HYDRO-ELECTRIC SCHEME

HISTORY

The Kidatu Hydro-electric Scheme on the Great Ruaha River some 280 km south-west of Dar-es-Salaam was constructed between 1970 and 1975.

The main components of the Kidatu power-plant are:

(*a*) Kidatu earth-rockfill dam.

(*b*) Kidatu concrete spillway dam on the right river bank, with a height of 26 m and three radial gates with a discharge capacity of 6 400 m³/s. The spillway is combined with a spillway chute, a stilling pool and a spillway channel.

(*c*) Headrace, comprising a tunnel intake, a 9,6 km long unlined tunnel having a cross section of 70 m², and a surge gallery.
A diversion tunnel which is combined with the upstream part of the headrace tunnel.

(*d*) An underground power station comprising penstocks, machinery hall, control room and transformer hall.

(*e*) A tailrace comprising a surge gallery, a 1 km long unlined tunnel having a cross section of 70 km², and a short tailrace canal.

(*f*) An outdoor switchyard and a 220 kV transmission line to Dar-es-Salaam.

(*g*) Roads, camps, offices, health centre, etc.

The first stage of the hydro-electric development involved the installation of 2 × 50 MW turbo-generators.

At a later stage, when the river is planned to be regulated by a large dam at Mtera, 170 km upstream of Kidatu, a further 100 MW will be installed, bringing the total capacity up to 200 MW.

KIDATU DAM

PLAN

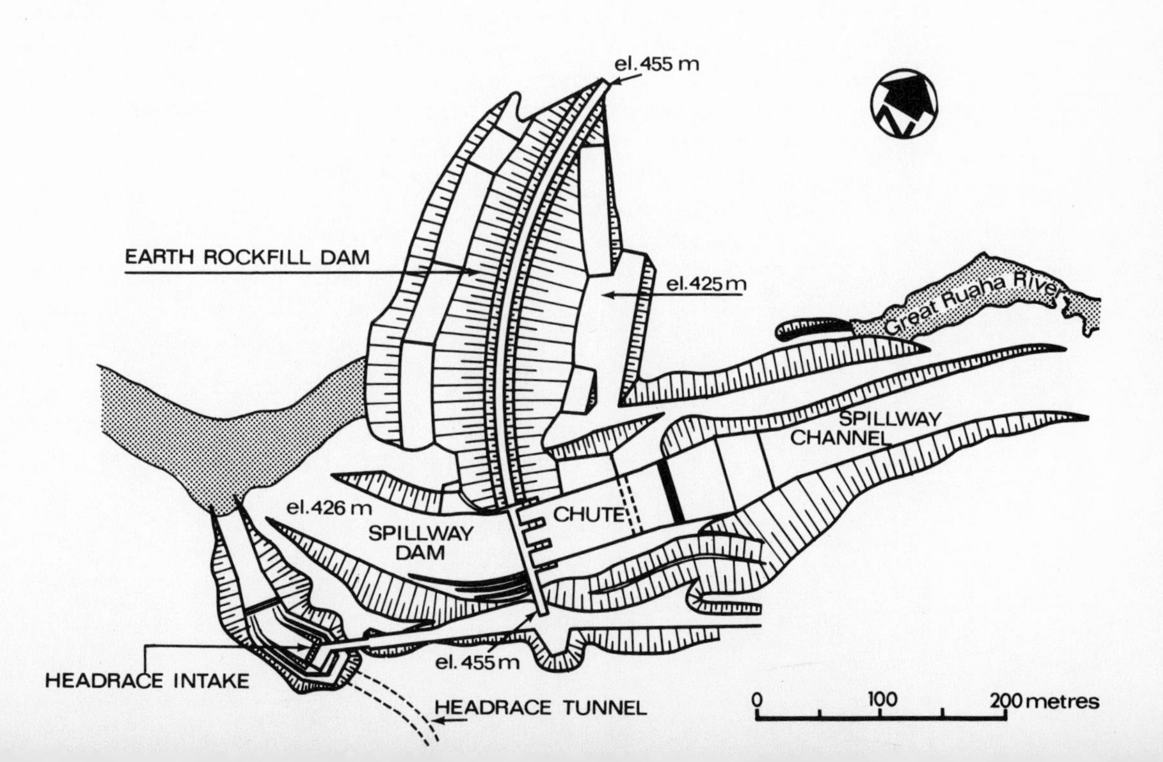

KIDATU DAM

PLAN AND LONGITUDINAL SECTION OF SPILLWAY DAM

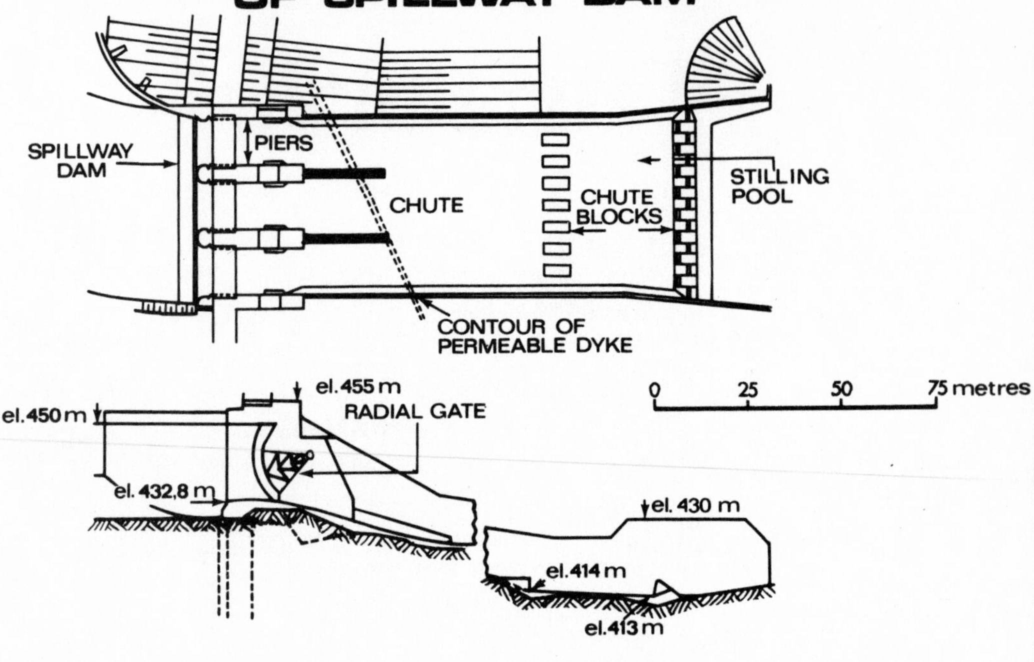

DIMENSIONS

Principal data for the project:-

Catchment area	80 000 km²
Mean flow (1945–74) . .	153 m³/s
Maximum recorded flow (1968)	2 450 m³/s
Surface area of reservoir at Full Supply Level 450 m above sea level is . . .	10 km²
Volume between F.S.L. and 433 m which is minimum storage level is . . .	125×10^6 m³
The earth-rockfill dam:	
Height above river bottom .	40 m
Crest length	350 m
Volume of fill	800 000 m³
Spillway Quantities:	
Excavation in soil . . .	700 000 m³
Excavation in rock . . .	300 000 m³
Concrete	40 000 m³

CONSULTING ENGINEERS AND CONTRACTORS

Owner: Tanzania Electric Supply Company Limited (TANESCO), Dar-es-Salaam.

Engineer: SWECO, Swedish Consulting Group, Stockholm, Consulting Engineers, Architects and Economists.

Site Supervision: SWECO

Contractors: KICON, Kidatu Consortium Joint Venture, formed by Skanska Cementgjuteriet, Sweden.

Main Civil Works: SENTAB, Sweden
Stirling Astaldi, Italy.
Royal Netherland Harbour Works Co., The Netherlands.
Mwananchi Engineering and Contracting Co., Tanzania.

Gates: INGRA-METALNA, Yugoslavia.

Turbines: INGRA-LITOSTROJ, Yugoslavia.

Generators: INGRA-RADE KONCAR, Yugoslavia

CHAPTER 4

Malawi

SHIRE HYDRO-ELECTRIC PROJECTS

HISTORY

There are as yet no major dams in Malawi. However, the hydro-electric potential of the Shire River fed from the third largest lake in Africa, Lake Malawi, is considerable and should be considered both in the context of being able to meet all foreseeable demands within Malawi up to the year 2000 from agricultural and industrial developments and for export to neighbouring territories once the pan-African electrical transmission grid has been established.

The upper Shire River falls only 7 metres in 132 km between the lake and Matope. Similarly the Lower Shire falls very little beyond Chikwawa. It is in the middle Shire, where in 84 km the river plunges 384 metres in a series of rapids known as the Murchison Cataracts, that sufficient head is available for hydro power.

Water quantity as well as head affects the amount of hydro-power available.

CONSTRUCTION

The Kamuzu barrage constructed at Liwonde was completed in 1965. It was constructed to permit control of the level of Lake Malawi in order to ensure that an adequate discharge will always be available in the Shire for hydro-electric generation. This marked the first step in harnessing the Shire River.

It is economically impossible to harness the full theoretical power of the Shire River, but the Malawi Electricity Supply Commission and their Consulting Engineers have concluded that most of the potential could be

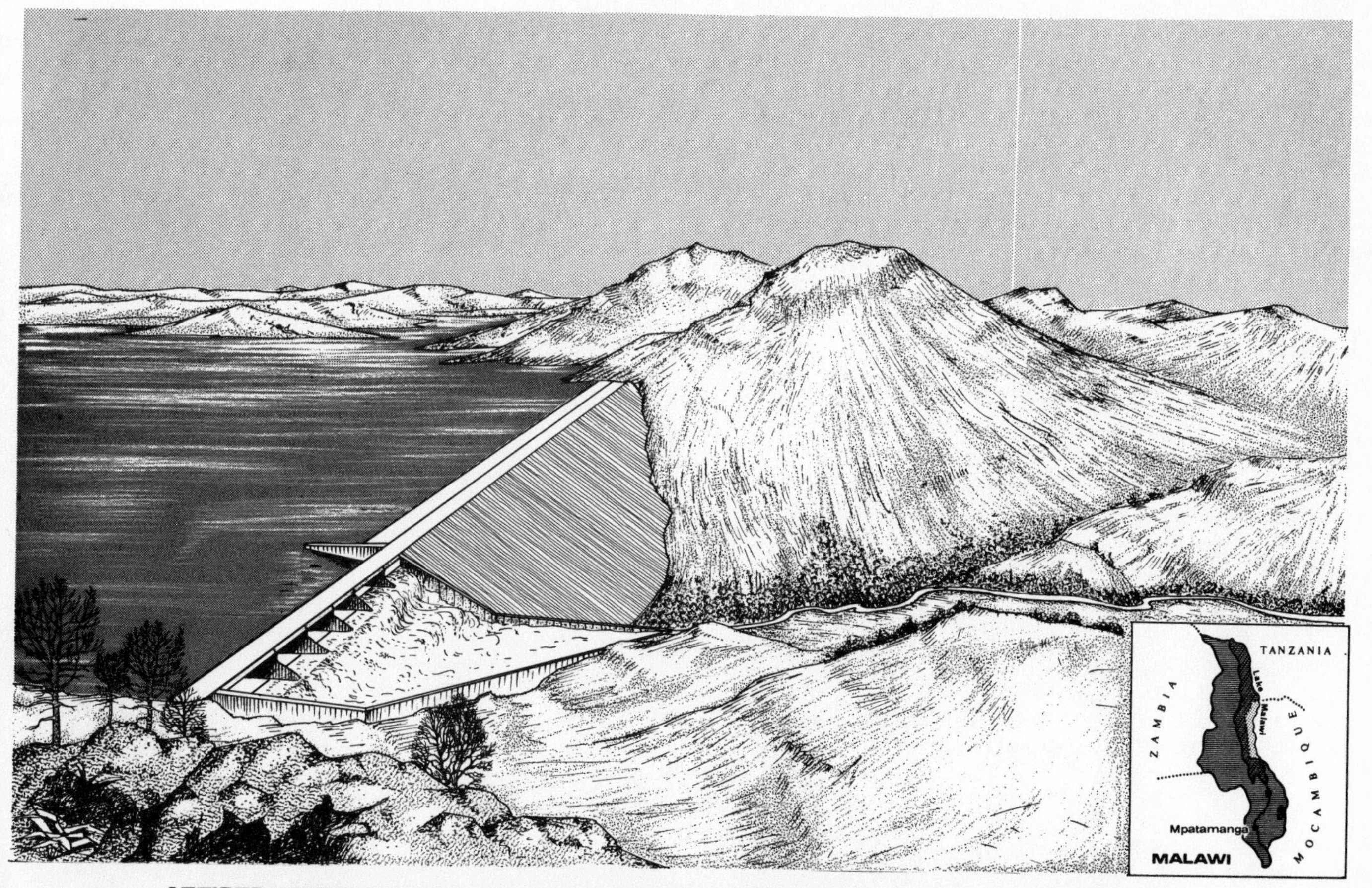

ARTISTS IMPRESSION OF A MAJOR HYDRO-ELECTRIC DEVELOPMENT AT MPATAMANGA

The Kumuzu barrage controls, Lake Malawi

realised by constructing hydro-electric power stations at six sites in the middle Shire.

The minimum reliable discharge of the river is 170 cumecs.

Work on the second stage at Tedzani is complete and it is planned to commence work on the first phase of the second stage development of the Nkula Falls Hydro-electric scheme. This will be another rockfill barrage having an overall length of 650 metres and a height of 15 metres and equipped with five 30 × 15 metre radial gates.

In the central and northern regions several of the rivers which discharge into Lake Malawi are also known to possess valuable hydro-electrical potential.

The rate of electrical load growth forecast by Malawi ESCOM and their Consulting Engineers is estimated at from negligible proportions in 1970 to 2 500 million kWh firm energy by the year 2000 which will still leave surplus energy of about 1 000 million kWh from the river's resources.

The Consulting Engineers to ESCOM are Messrs. Kennedy & Donkin (Africa).

THE SHIRE RIVER

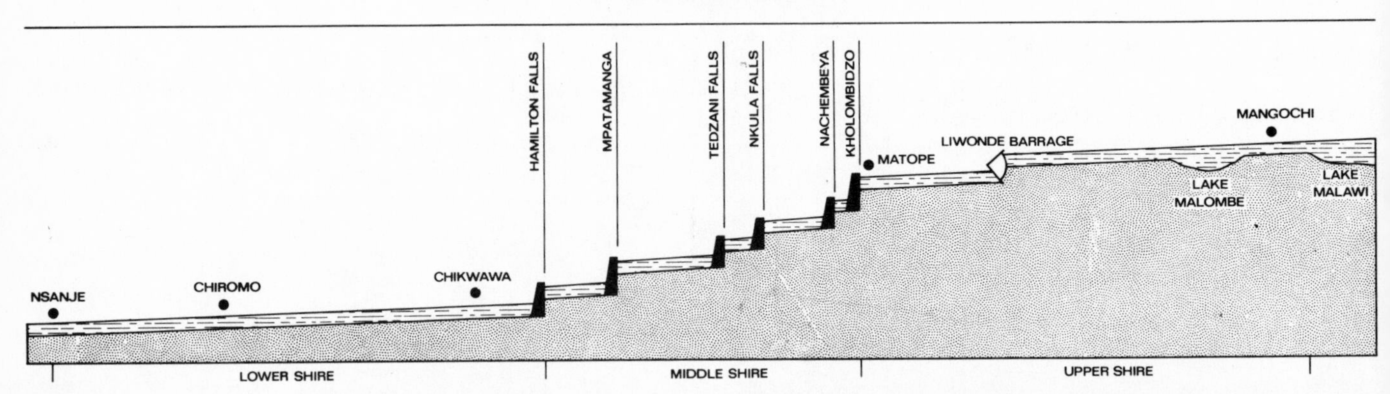

The Upper Shire falls only 7 metres in the 132 km.between the Lake and Matope. Similarly the Lower Shire falls very little beyond Chikwawa.

It is in the Middle Shire, where in 84 kms. the river plunges 383 metres in a series of rapids known as the Murchison Cataracts, that sufficient head is available for hydro-power.

Water quantity, as well as head, affects the amount of power available. Liwonde Barrage, completed in 1965, was constructed to permit control of the level of Lake Malawi in order to ensure that an adequate discharge will always be available in the Shire for hydro-electric generation.

It is economically impossible to harness the full theoretical power of the Shire River but detailed studies by ESCOM (Malawi) and their Consulting Engineers have led to a number of practical proposals whereby construction of hydro-electric power stations with dams or barrages, at the six sites shown ensures that most of the available potential can be developed.

Completion of Liwonde Barrage and Stage 1 of the Nkula Falls Hydro-Electric Scheme marked the first step in harnessing the Shire River. The next development was at Tedzani Falls.

CHAPTER 5 Zambia

KAFUE GORGE HYDRO-ELECTRIC PROJECT

HISTORY

This project is operated in conjunction with the Itezhitezhi reservoir, which provides the regulation of Kafue flow for economic utilisation of the Kafue hydro-electric plant with some water reserved for upstream agricultural and industrial use.

In April 1967 the Government of the Republic of Zambia decided to develop the hydro-electric potential of the Kafue River and to start the first stage of construction comprising a 600 MW power plant at the Kafue Gorge.

During the dry season of 1967 exploratory drilling was undertaken at the sites of the dam and the underground works. Preliminary works followed soon after, including construction of access roads, two diversion tunnels to divert the river round the dam site during construction, an access tunnel to the power house, and the first stage of camps to house eventually about 5 000 workers.

The award of the main contract was made in April 1968. Excavation of the underground machinery hall began in September 1968; in December the river was diverted from its natural course and construction of the dam wall started.

In October 1971 the first generating unit was commissioned and was handed over to ZESCO for commercial operation. By April 1972 all four units were in commission.

DIMENSIONS

The first stage project generates 600 MW through four 150 MW units but provision is made for adding two more units at a later stage.

The main components of the project are:-

Gorge Dam	An earth-rockfill dam with a spillway consisting of a concrete structure with four radial gates.
Headrace	comprising tunnel intake, tunnel with surge gallery, penstock intakes and six vertical penstocks.
Underground power station	comprising machinery and transformer halls and shaft for cables, lift and ventilation.
Tailrace	comprising surge gallery and tailrace tunnel.

330 kV Outdoor Switchyard.
Administration Building, Permanent Workshop, Stores and Roads.
Housing for the operating staff.

Principal technical data:

Earth/rockfill Dam	
Crest length . . .	1 230 ft. (375 m)
Crest width . . .	20 ft. (6 m)
Maximum height above bedrock . .	165 ft. (50 m)
Volume of rockfill .	990 000 cu. yds. (757 000 m^3)
Volume of impervious fill	260 000 cu. yds. (199 000 m^3)
Volume of filter material	325 000 cu. yds. (248 000 m^3)
Spillway	
Number of gates . .	4
Dimensions of gates .	46 ft. wide × 38 ft. high (14 m × 11,5 m)
Discharge capacity .	150 000 cusecs (4 250 m^3/s)
Reservoir	
Maximum water level at Kafue road bridge	3 204 ft. (976 m)
Surface area . . .	200 000 acres (80 940 ha)
Live storage capacity	600 000 acre ft. (740 m^3 × 10^6)
Hydrology	
Catchment area . .	59 000 sq. miles (153 000 sq. km.)
Rainfall	32-50 inches per annum (813 – 1 270 mm per annum).
Maximum recorded flood	70 000 cusecs (1 982 m^3/s)
Mean flow . . .	12 100 cusecs (343 m^3/s)
Minimum recorded flow	400 cusecs (11 m^3/s)

Headrace Tunnel

Length	32 200 ft. (9 815 m)
Cross sectional area .	140 sq. yds (117 m^2)
Volume of excavation	1,5 million cu. yds (1,15 $m^3 \times 10^6$)

Machine Hall

Length	440 ft. (134 m)
Width	50 ft. (15 m)
Volume of excavation	74 000 cu. yds. (56 580 m^3)
Volume of concrete .	12 000 cu. yds. (9 175 m^3)

Transformer Hall

Length	409 ft. (125 m)
Width	55 ft. (16,8 m)
Height	57 ft. (17,4 m)
Volume of excavation	47 500 cu. yds. (36 318 m^3)
Volume of concrete .	5 000 cu. yds. (3 823 m^3)

Tailrace Tunnel

Length	4 530 ft. (1 381 m)
Cross sectional area .	140 sq. yds. (117 m^2)
Volume of excavation	210 000 cu. yds. (160 556 m^3)

Tailrace and Draft Tubes

Length	5 500 ft. (1 676 m)
Cross sectional area .	30–140 sq. yds. (25 – 117 m^2)
Volume of excavation	230 000 cu. yds. (175 858 m^3)

Penstock Shafts

Height	1 410 ft. (430 m)
Cross sectional area .	15 – 37 sq. yds. (12,5 – 31 m^2)
Volume of excavation (6 shafts)	57 000 cu. yds. (43 582 m^3)

Machinery

4 Francis Turbines, vertical shaft	
4 Generators, capacity	150 MW each
Total Head . . .	1 270 ft. (387 m)
Discharge . . .	6 200 cusecs. (174 m^3/s)
Transformer voltage	330 kV
Weight of transformers (stripped) .	75 tons (76 tonnes).

CONSTRUCTION

The layout of the project includes a diversion dam across the gorge located some 26 km downstream of the Kafue road bridge, a six mile (10 km) headrace tunnel, an underground power station and a 1,4 km tailrace tunnel.

The gorge reservoir will serve for annual regulation of the outflow from Itezhitezhi and the Kafue flats and it will thus be drawn down regularly each dry season.

The south bank where the power house structures are located is entirely a granite basement formation whereas on the north bank the basement is covered by Katanga formations containing micachlorite schists, dolomites etc. less suitable for tunnelling.

The main dam has an inclined core of laterite soil reposing on a fill of rock excavated from the tunnels. Inverse filters surrounding the core prevent piping and serve to reduce pore pressure. An upstream rockfill to increase the stability in case of rapid drawdown and a rip rap layer to protect against wave erosion are also included.

The design of the dam and the operation of the power station aim to preserve the ecological balance of the Kafue flats as far as possible.

CONSULTING ENGINEERS AND CONTRACTORS

Employer:	The Government of the Republic of Zambia represented by: Director of Electrical Engineering, Ministry of Power, Transport and Works.
Engineer:	SWECO, Swedish Consulting Group, Stockholm Consulting Engineers, Architects and Economists through their member firms.
For Civil Eng. and Mech. Works:	VBB, Vattenbyggnadsbyran, Stockholm.
For Electrical Works:	BECO, Bergman & Co., Stockholm.
Main Contractor:	Energoprojekt, Belgrade with sub-contractors.
For Civil Eng. Works:	Konstruktor, Split Hidrotechnika, Belgrade Tunelogradnja, Belgrade ZECCO, Lusaka.
For Gates, Cranes, Penstock Steel Lining, Transformers, Generator Switching Station cables and control equipment:	G.I.E., Milan.
For Turbines and Valves:	Kvaerner Brug, Oslo.
For generators and switchyard:	Alsthom, Paris.
Contractors for Preparatory works:	
Camp, first stage	Delkins Ltd., Lusaka.
Roads, access tunnel and diversion tunnels	Burton Construction Ltd., Lusaka. Skanska Cemengjuteriet, Stockholm.
Site Supervision:	SWECO Kafue Gorge.

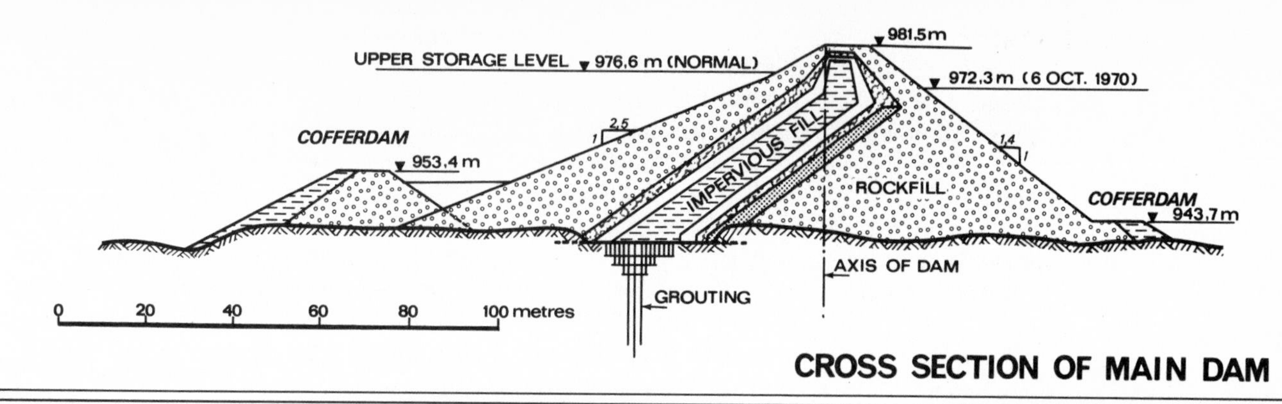

CROSS SECTION OF MAIN DAM

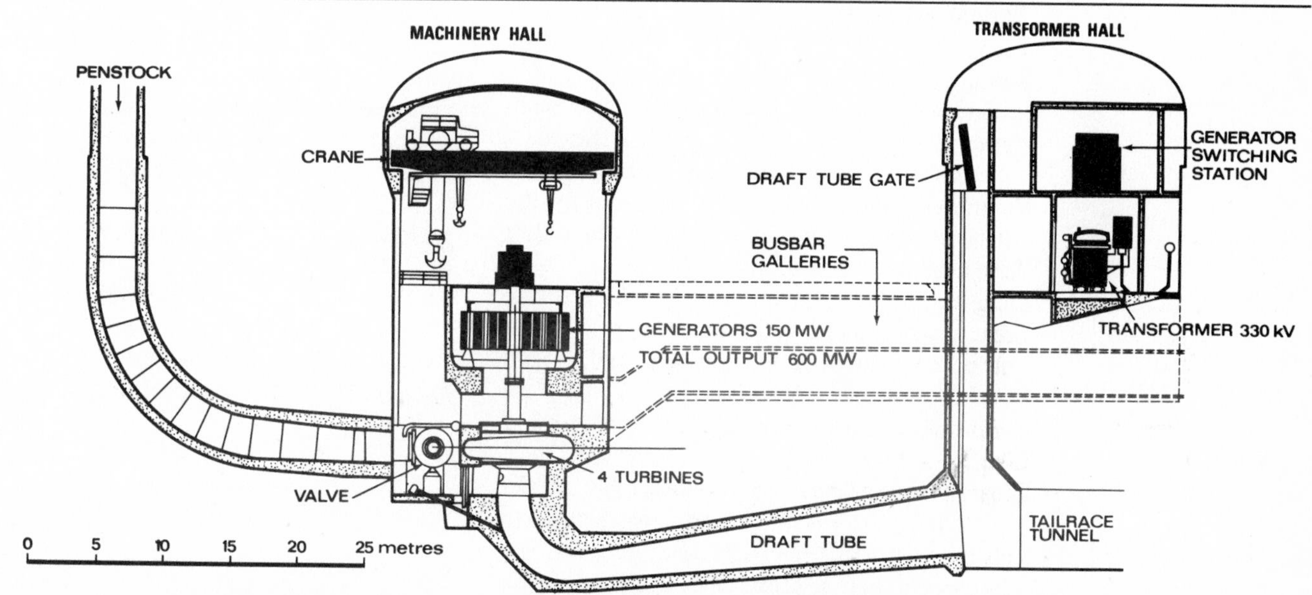

CROSS SECTION OF MACHINERY AND TRANSFORMER HALLS

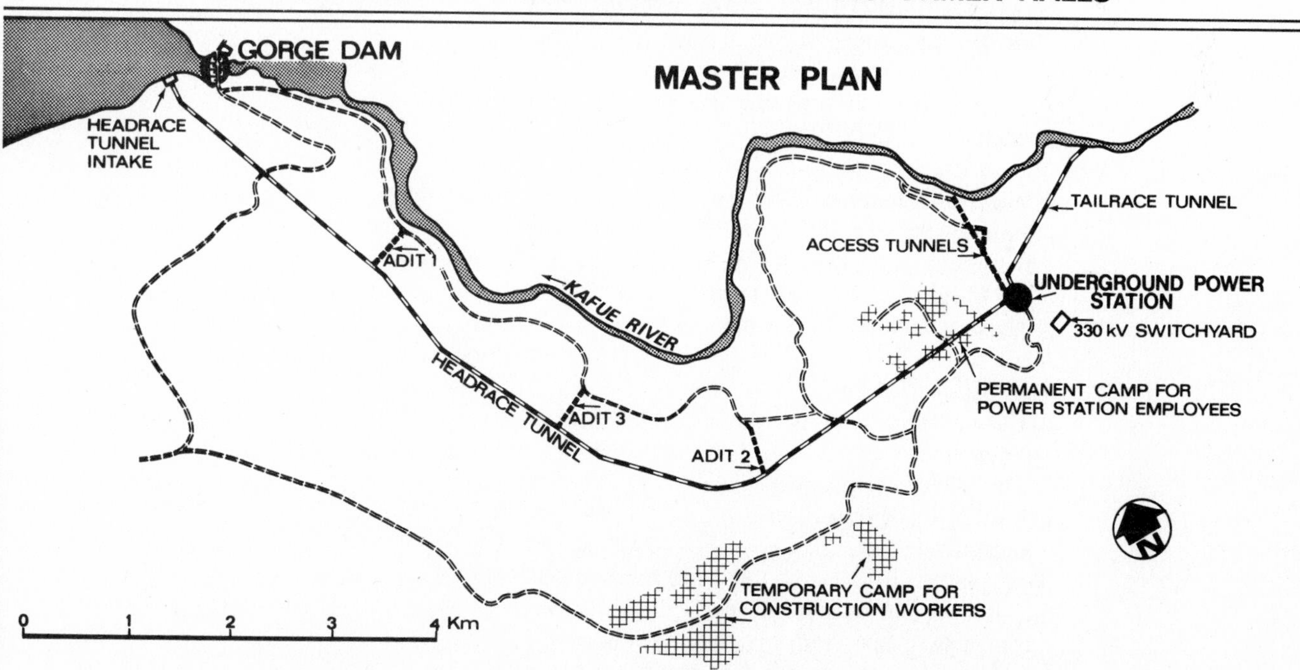

KAFUE GORGE HYDRO-ELECTRIC POWER PROJECT

ITEZHITEZHI PROJECT

HISTORY

The Itezhitezhi dam site is on the Kafue River some 240 kilometres west of Lusaka and is situated in a game park. Access to the site is by a surfaced road from Lusaka 320 kilometres long.

The project is being developed to provide a storage reservoir for regulating the flow of the Kafue River in order to enable a larger and more firm output of power from the Kafue Power Station downstream.

The water potential of the Kafue River that may be developed economically corresponds to an annual energy production of some 11 000 GWh. The development of the Kafue River potential was planned to be achieved in several stages. The first stage was the completion of the 600 MW underground power station at Kafue Gorge with provision for the installation of additional 300 MW to utilise the upper 396 m head at the Gorge.

The second stage was the regulation of the river by a storage dam at Itezhitezhi and thereafter the harnessing of the lower 198 m head at the Gorge.

The Kafue drains about one fifth of Zambia or an area of approximately 155 400 km^2. Large swamps extend upstream of Itezhitezhi at Lukanga and Busanga, and downstream the river meanders over the Kafue flats for 400 km. From there it plunges through the Kafue Gorge to the Zambezi plain.

In the northern part of the catchment area the average precipitation is some 1 500 mm, reducing to 750 mm in the southern part.

The estimated 1:10 000 year flood at Itezhitezhi is 5 114 m^3/s.

Of the average flow of 11 000 m^3 $\times$ 10^6 it has been estimated that two-thirds can be used for the generation of electric power, the remainder to be used for agricultural, domestic and industrial consumption and also in evaporation losses and spill.

DIMENSIONS

The crest level of the earth/rockfill dam is 1 035 m.

The upper supply level is . . 1 029,5 m
and the lower supply level . . 1 006,0 m.
Capacity U.S.L. . . . . . 5,6 m^3 $\times$ 10^9.
Capacity L.S.L. . . . . . 0,8 m^3 $\times$ 10^9.

At normal retention level the reservoir covers 36 500 ha. At drawdown the submerged area reduces to 10 000 ha, leaving a dead storage of 740 m^3 $\times$ 10^6.

CONSTRUCTION

Impounding reached 1005,29 m in October 1976. Two diversion tunnels were constructed on the south bank and the main dam was constructed between cofferdams. The spillway is located in a saddle on the northern ridge. The diversion tunnels had a capacity of 298 m^3/s during the construction period.

The earth/rockfill structure has an inclined impervious core of laterite soil reposing on a rockfill support.

The main body of the dam was founded on mudstone. The height for the first stage is 67 m, but there is provision for raising the dam at some future date.

The volume of fill is 6 m^3 $\times$ 10^6.

The spillway discharge capacity through four Tainter gates is 4 260 m^3/s.

The estimated cost of construction in 1969 was Kwacha 41 million, but since then escalation has increased the capital sum.

CONSULTING ENGINEERS AND CONTRACTORS

Client: The Zambian Electricity Supply Company (ZESCO).
The Fngineer: SWECO of Sweden.
The Contractor: Impregilo and Recchi.

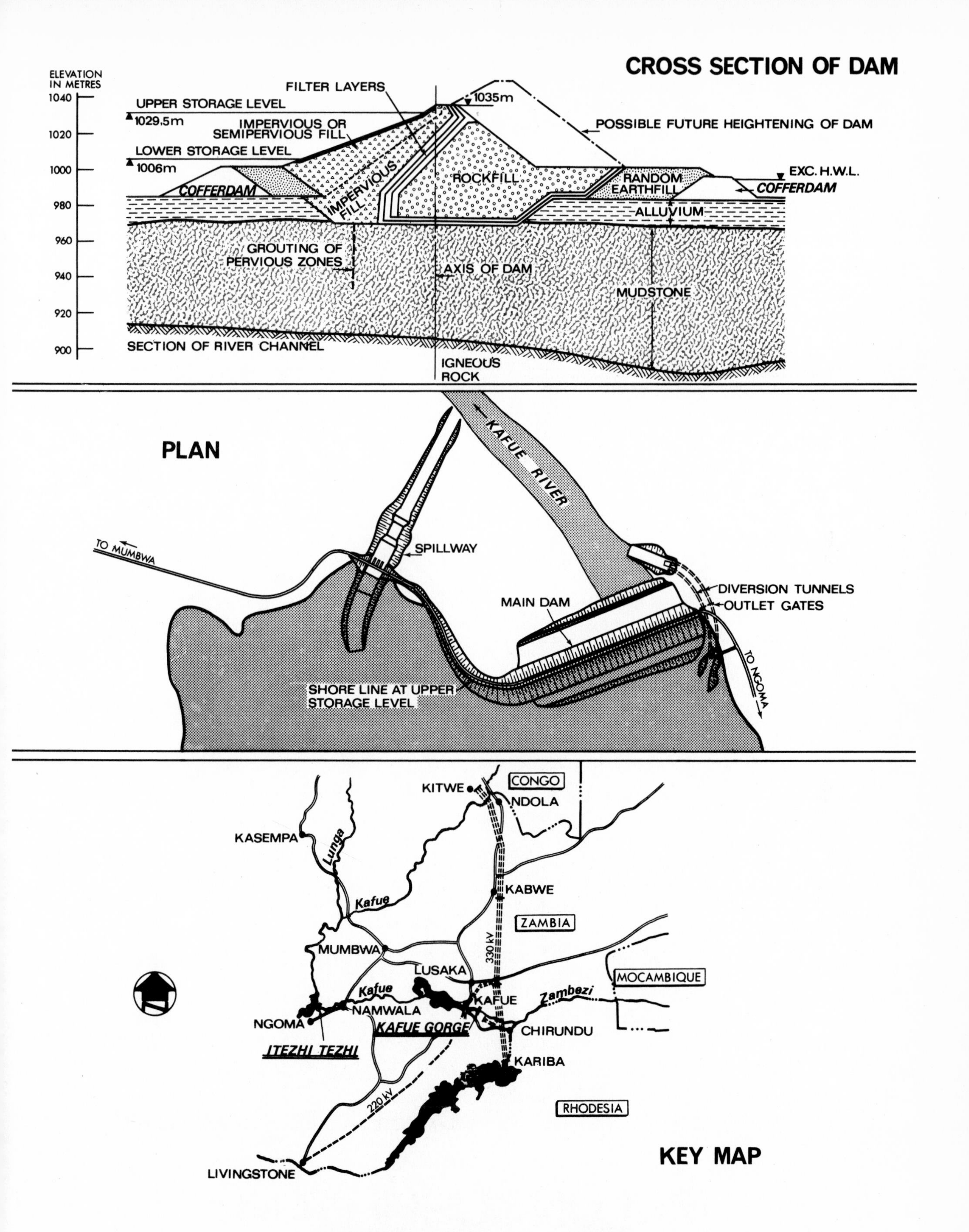
CROSS SECTION OF DAM
ELEVATION IN METRES
1040
1020
1000
980
960
940
920
900
FILTER LAYERS
UPPER STORAGE LEVEL
1029.5m
1035m
IMPERVIOUS OR SEMIPERVIOUS FILL
POSSIBLE FUTURE HEIGHTENING OF DAM
LOWER STORAGE LEVEL
1006m
COFFERDAM
IMPERVIOUS FILL
ROCKFILL
RANDOM EARTHFILL
EXC. H.W.L.
COFFERDAM
ALLUVIUM
GROUTING OF PERVIOUS ZONES
AXIS OF DAM
MUDSTONE
SECTION OF RIVER CHANNEL
IGNEOUS ROCK
PLAN
KAFUE RIVER
TO MUMBWA
SPILLWAY
DIVERSION TUNNELS
OUTLET GATES
MAIN DAM
TO NGOMA
SHORE LINE AT UPPER STORAGE LEVEL
KITWE
CONGO
NDOLA
KASEMPA
Lunga
KABWE
Kafue
ZAMBIA
MUMBWA
330 kV
LUSAKA
MOCAMBIQUE
Kafue
KAFUE
Zambezi
NGOMA
NAMWALA
KAFUE GORGE
CHIRUNDU
ITEZHI TEZHI
KARIBA
220 kV
RHODESIA
KEY MAP
LIVINGSTONE

ITEZHITEZHI DAM

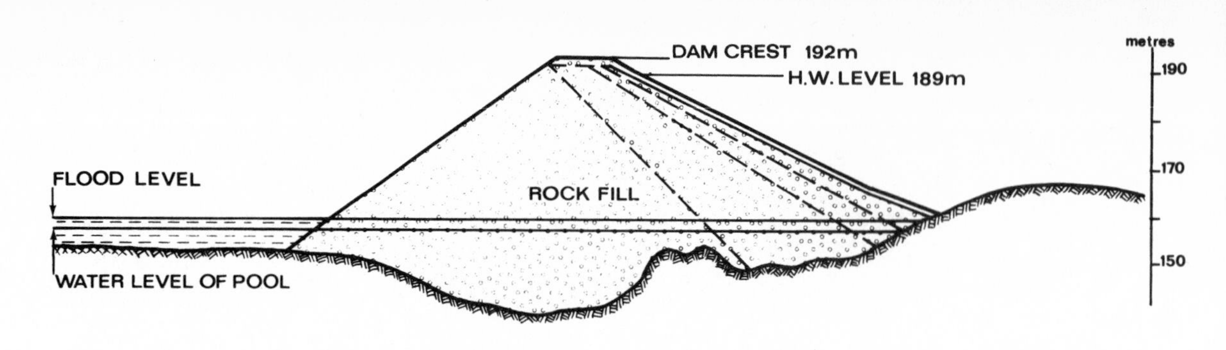

SECTION A-A

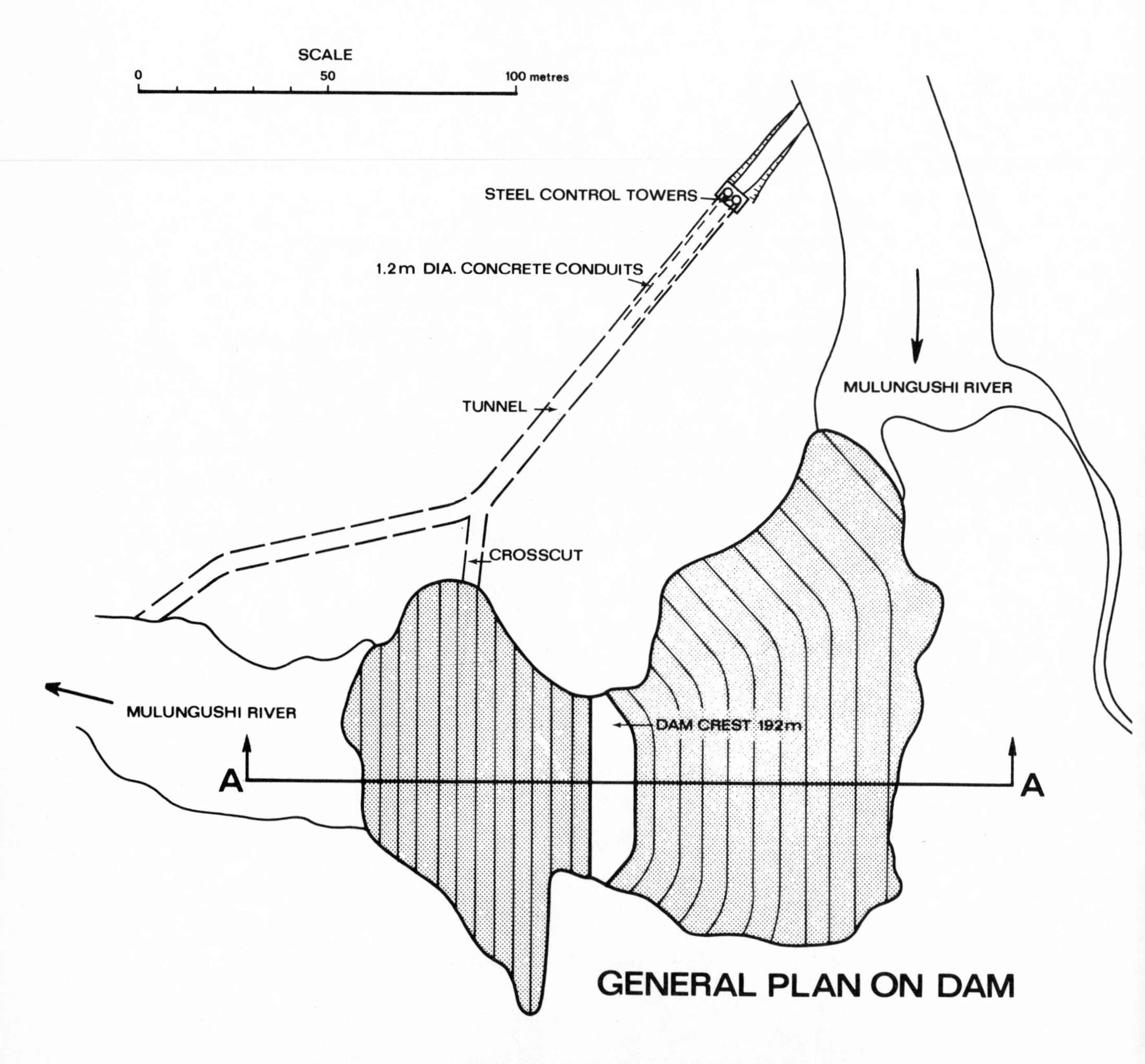

GENERAL PLAN ON DAM

MULUNGUSHI DAM

MULUNGUSHI DAM

HISTORY

The Mulungushi Hydro-electric Scheme was officially opened by the Prince of Wales during his visit to Southern Africa in 1925. This dam and the sister project were built at the head of two spectacular gorges where the Mulungushi and Lunsemfwa Rivers plunge more than 330 metres down to the Luangwa valley in a sparsely populated and heavily wooded country.

The Mulungushi Dam was constructed by the owners, the Rhodesian Broken Hill Development Co. Ltd. (now Nchanga Consolidated Copper Mines Limited – Broken Hill Division) with an installed capacity 20 MW with a gross head of 352 metres to provide a firm water supply to the Mulungushi Hydro-electric Scheme. It was completed in 1925.

DIMENSIONS

Catchment Area	3 000 sq. km.
Maximum Height	46 metres
Crest Length	35 metres
Storage Capacity	250 million cubic metres
Spillway Type	Free overflow
Spillway Capacity	60 cubic metres/second
Type of Embankment	Zoned rockfill with asphalt concrete cover to upstream face.
Volume of Embankment	120 000 cubic metres.

CONSTRUCTION

The embankment was constructed of rockfill obtained from the adjoining hills, which had to be placed in a narrow deep gorge in one dry season, under considerable difficulty as there was no economical means of diverting the summer flow of the river. Shortly after its completion the bituminous concrete waterface failed and it was repaired by dumping some 120 000 cubic metres of soil under water. It has performed satisfactorily ever since.

During the drought period 1933–35 the dam came perilously near to running dry, and in 1973 after an exceptionally severe drought it did, in fact, empty.

The average rainfall in the Mulungushi catchment is about 965 mm per annum, but the run-off is low, averaging about 10% of the rainfall. This is due to the nature of the countryside which is flat, well covered with trees and vegetation. A further feature is the frequent occurrence of small open patches known locally as dambos, that act as sponges, retaining some of the precipitation during the rains and discharging it slowly during the autumn.

CONSULTING ENGINEERS AND CONTRACTORS

Designed by	Rand Mines Limited
Constructed by	Rhodesian Broken Hill Development Co. Ltd.
Subsequent modifications and inspection	Watermeyer, Legge, Piesold and Uhlmann (formerly F. E. Kanthack & Partners.)

MITA HILLS DAM

HISTORY

The Mita Hills dam was first investigated by the late Dr. F. E. Kanthack in 1927 and a detailed scheme to construct it was put forward in 1931, but it was only in 1955 that the decision was made to construct it. Its purpose was to augment the hydro-electric generating capacity at Mulungushi.

During the Second World War the first stage of a hydro-electric scheme was constructed on the Lunsemfwa River, but it operated on run of river until the Mita Hills Dam was built in 1958.

DIMENSIONS

Maximum Height	47 metres
Crest Length	340 metres
Catchment	8 450 sq. km.
Storage Capacity	700 million cubic metres
Spillway type	Two radial gates (9,2 m × 9,2 m)
Spillway capacity	800 cubic metres/second
Volume of Materials	800 000 cubic metres
Installed capacity	18 MW

CONSTRUCTION

The embankment is constructed of rolled and compacted earthfill with a sloping area with rip rap on the water face and grass on the downstream face. Its spillway is in an adjacent neck in the Mita Hills range and comprises two large mechanically operated radial gates, each 30 feet wide and 30 feet high. (9,2 m × 9,2 m).

Completion of this dam made it possible to increase the capacity of the Lunsemfwa plant from its then existing 12 MW to its planned maximum of 18 MW.

The catchment and rainfall are generally very similar to those of the Mulungushi catchment which it adjoins, and the Mita Hills Dam also emptied in 1973 during the exceptional drought in Southern Africa.

CONSULTING ENGINEERS AND CONTRACTORS

Owned by	Nchanga Consolidated Copper Mines Limited – Broken Hill Division
Designed by	Watermeyer, Legge, Piesold & Uhlmann (formerly F. E. Kanthack & Partners)
Constructed by	Impresit (site preparation) Cementation Company (drilling and grouting) Burton Construction (Embankment & Spillway) Dorman Long & Co. (Spillway gates)

North-easterly view along Mita Hills Dam wall showing intake tower

North-westerly Mita Hills Dam view showing radial spillway gate

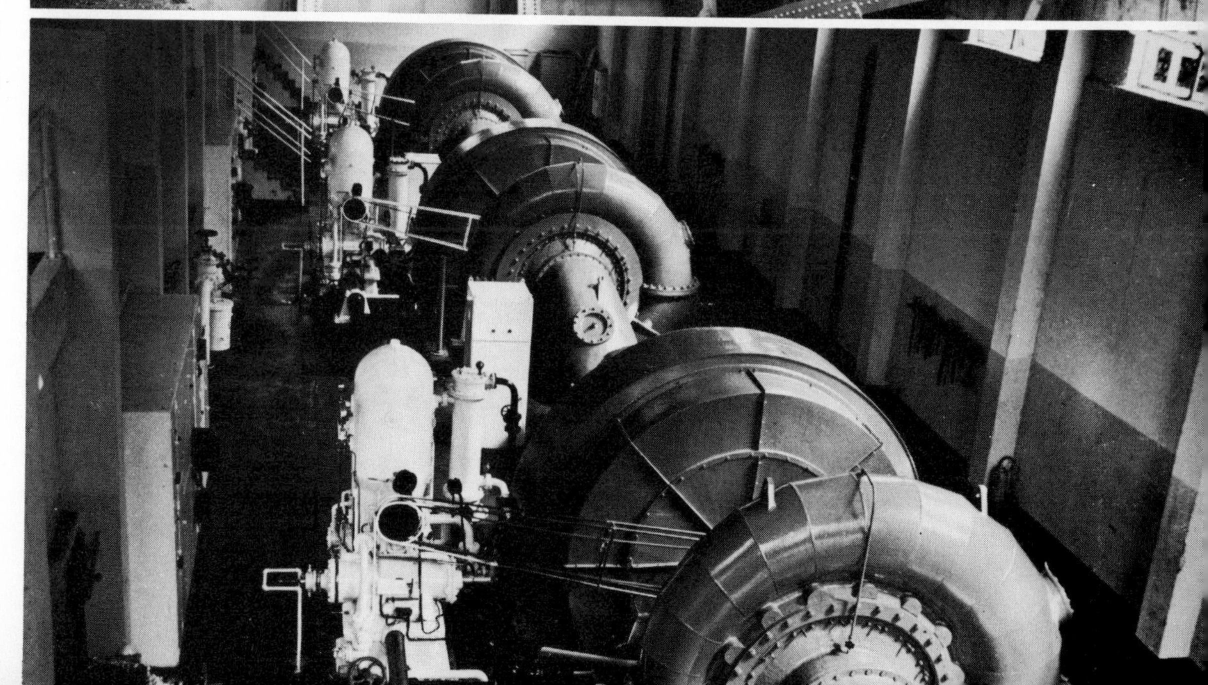

View of Lunsemfwa Power Station turbine hall

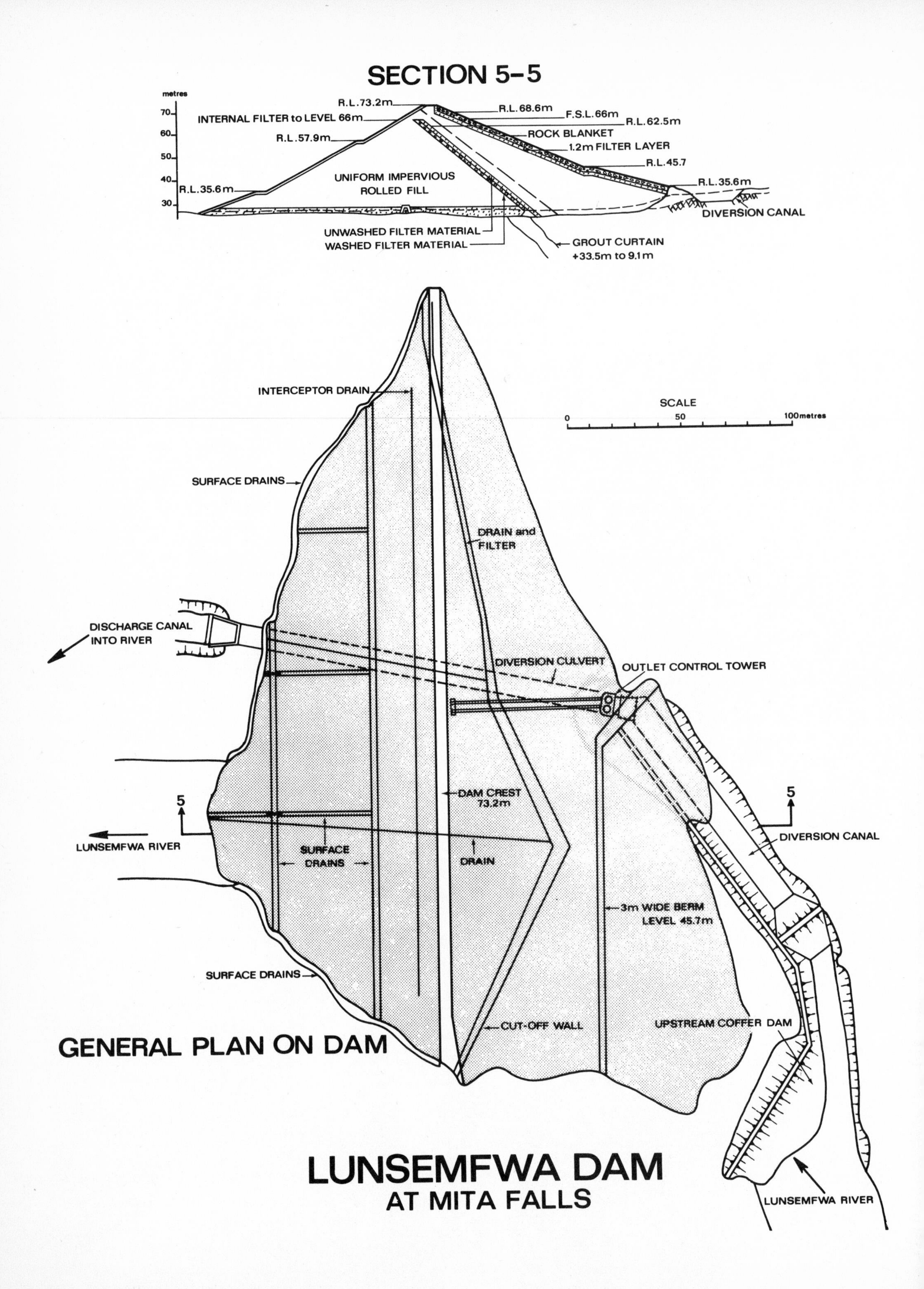
SECTION 5-5
metres
70
60
50
40
30
R.L.73.2m
INTERNAL FILTER to LEVEL 66m
R.L.57.9m
R.L.35.6m
R.L.68.6m
F.S.L.66m
R.L.62.5m
ROCK BLANKET
1.2m FILTER LAYER
R.L.45.7
R.L.35.6m
UNIFORM IMPERVIOUS
ROLLED FILL
DIVERSION CANAL
UNWASHED FILTER MATERIAL
WASHED FILTER MATERIAL
GROUT CURTAIN
+33.5m to 9.1m
INTERCEPTOR DRAIN
SCALE
0
50
100metres
SURFACE DRAINS
DRAIN and
FILTER
DISCHARGE CANAL
INTO RIVER
DIVERSION CULVERT
OUTLET CONTROL TOWER
DAM CREST
73.2m
5
5
LUNSEMFWA RIVER
SURFACE
DRAINS
DRAIN
DIVERSION CANAL
3m WIDE BERM
LEVEL 45.7m
SURFACE DRAINS
CUT-OFF WALL
UPSTREAM COFFER DAM
GENERAL PLAN ON DAM
LUNSEMFWA RIVER

LUNSEMFWA DAM
AT MITA FALLS

CHAPTER 6 # Moçambique

CABORA BASSA

HISTORY

In 1858 Dr. David Livingstone thrusting his way up the Zambezi in his steam launch *MaRobert* came upon a wholly unexpected obstacle – a ravine where the river ran so swiftly and in such a long succession of rapids and cascades that the *MaRobert*, struggle as she might, was forced back. A dream was born; a huge waterway into the very heart of the dark continent.

The site of the rapids he reached was called "Kebrabassa" which, it is understood, means "the end of the work". The title was apparently given to the gorge by the slaves who rowed the boats upstream from the coast, past Tete, to the entrance of the gorge from where they could proceed no further by boat. Their task was over. They turned round to drift and row downstream.

In 1957 the Portuguese Government es-

CABORA BASSA: LAYOUT PLAN

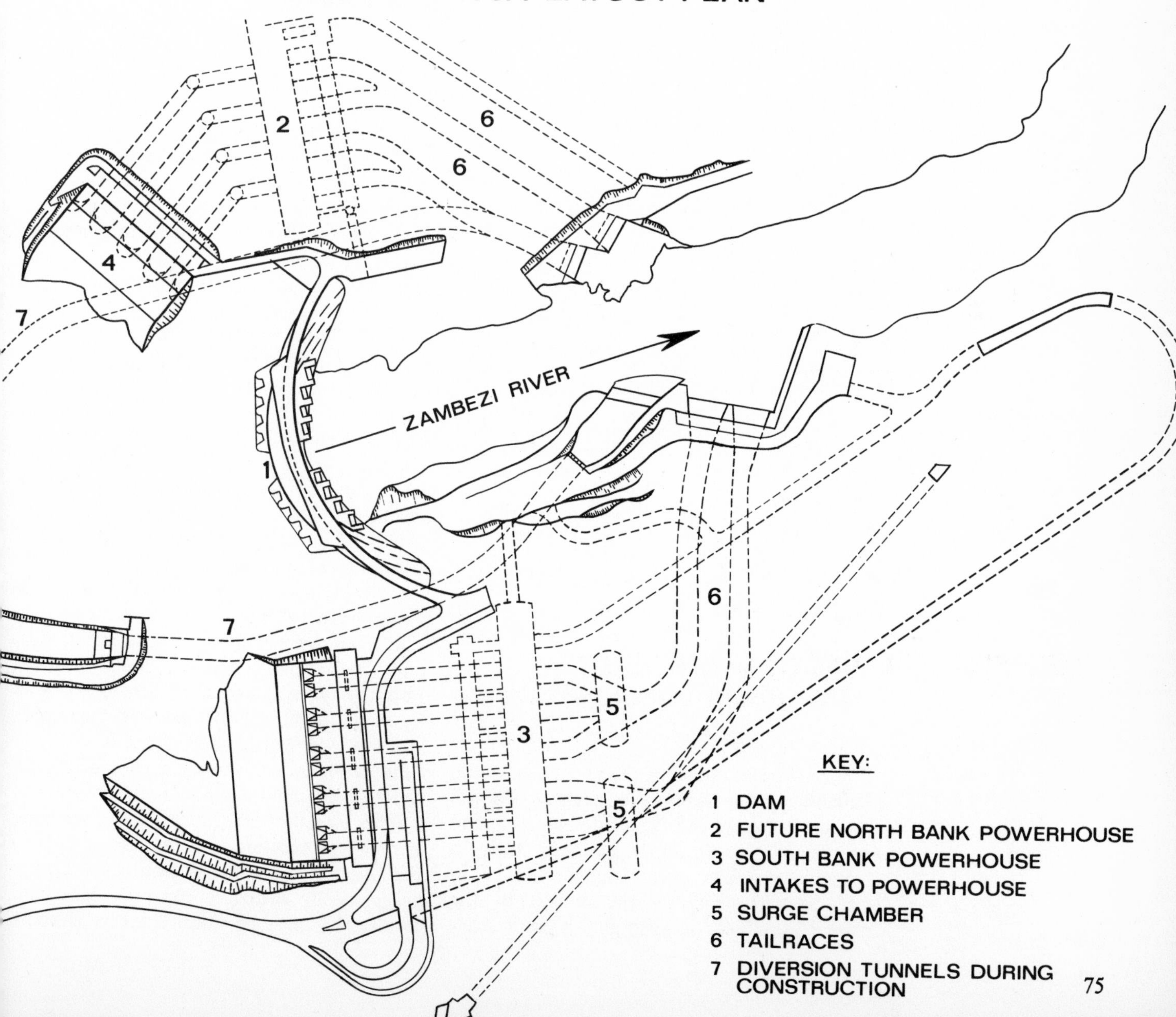

CABORA BASSA

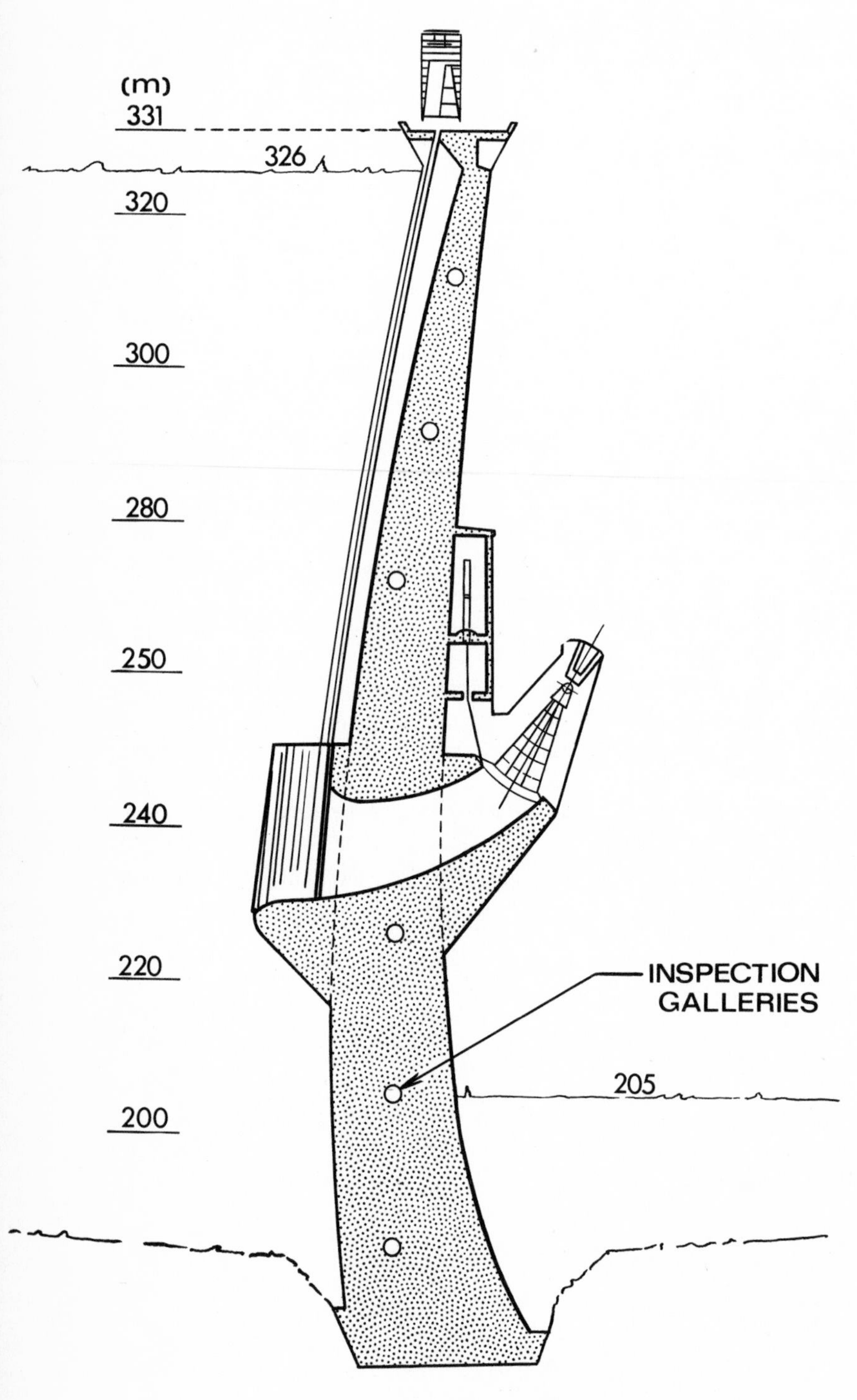

SECTION THROUGH DAM AND SLUICE GATES

tablished a commission called the Missao do Fomento e Povoamento do Zambeze (MFPZ) to study the possibilities of development and settlement in the upper Zambezi valley in Angola and the lower Zambezi valley in Moçambique. One of the reports by the MFPZ produced in 1959 gave a preliminary assessment of the advantages of a major power development at Cabora Bassa. The Zambezi studies came under the Cabinete do Plano do Zambezi (CPZ).

The Portuguese strategic outlook on the scheme was set out in a document published in Primeiras Jornadas de Engenheiro de Moçambique in 1965 entitled "O Aproveitzmento de Cabora-Bassa" by Hidrotecnica Portuguesa.

On January 15, 1966, the Ministry for Overseas Development (Ministro do Ultramar) created a Department styled "Grupo de Trabalho para o Zambezi" (GTZ) to finalise the technical and financial studies necessary to a phased programme to implement the Cabora Bassa complex.

It was with these organisations that negotiations leading to the final implementation of Cabora Bassa Stage I were to proceed through the years.

During 1965 the South African and Portuguese Governments began to give consideration to the possibility of developing the first stage of the Cabora Bassa site for the production of relatively cheap hydro-electric energy for export to the Republic. The first stage, as in the case of Kariba, would be to construct a concrete double curved arch dam and an underground power station on the south bank. A second stage, at a later date, would mean the doubling of the generating installed capacity by constructing another underground power station on the north bank. The dam is located at approximately 33°E 15°S.

The philosophy of the Portuguese Government was to export power and energy initially to South Africa at a realistic cost rate that would be acceptable to the Republic in terms of their own cost of production but would be adequate, as in the case of upstream Kariba, to give a surplus royalty to Moçambique, to be channelled into a Development Fund. This could be utilised to promote agricultural development in the fertile downstream regions for the indigenous populations who were eking out a subsistence economy in an area that was subject to devastating annual floods, and which situation would be greatly ameliorated by the attenuation of these flood flows by virtue of the vast storage capacities provided in lakes Kariba and Cabora Bassa.

It was intended to be a straight commercial deal to the mutual benefit of both parties.

The Zambezi catchment area receiving rainfall and yielding run-off or river flow to the Cabora Bassa site, situated in a narrow gorge ideally suited for an arch dam, and amidst the most beautiful scenery, is some

1,2 million square kilometres. Water donations come from Zambia, Zaire, Angola, Botswana, Rhodesia and Moçambique. The engineering philosophy is that as she receives water donations from so many African countries so should her benefits, direct and indirect, flow one day to many African countries.

At the site the average annual river flow is approximately twice that of the flow at upstream Kariba of some 2 273 cubic metres per second.

As regards control of the river during construction, despite this greater average flow, the major difficulties in handling floods at Kariba were not anticipated or experienced at Cabora Bassa. This is because the vast storage capacity at upstream Kariba Lake and at Kafue made it possible, by manipulation of the spillgates in conjunction with turbine discharges, to hold back the worst of the annual floods and to regulate the discharge at Cabora Bassa during the construction period to manageable and reasonably predictable proportions.

Negotiations, both as regards agreement with the South Africans and the manner of execution of the project, became protracted and it was not until 1969 that final agreement was reached with the South African Government. However, before that, in 1966, the Portuguese authorities made arrangements for three international consortiums of contractors to be established for purposes of competition as regards offers for the construction and financing of the project. The three international consortiums so created could be broadly described as American, British and South African led.

Competition waxed fast and furious and it was not until 17th September 1969 that contracts were finally signed in Lisbon between the State of Portugal and the Government of South Africa, the Electricity Supply Commission (ESCOM) and the winning consortium Zamco.

The Zamco Consortium consisted of 15 member companies from West Germany, France, Italy, Portugal and South Africa.

The Zamco Consortium was sub-divided into two major sub-consortiums: one for civil engineering (CEW) and the other for mechanical and electrical engineering equipment (EM).

The EM Consortium was further sub-divided into three sub-consortiums:

GAE Responsible for the generating equipment.

TSE Responsible for the terminal converter stations. Electricity is generated at Cabora Bassa as alternating current (AC) and is converted to Direct Current (DC) at the Cabora Bassa converter station be-

CABORA BASSA

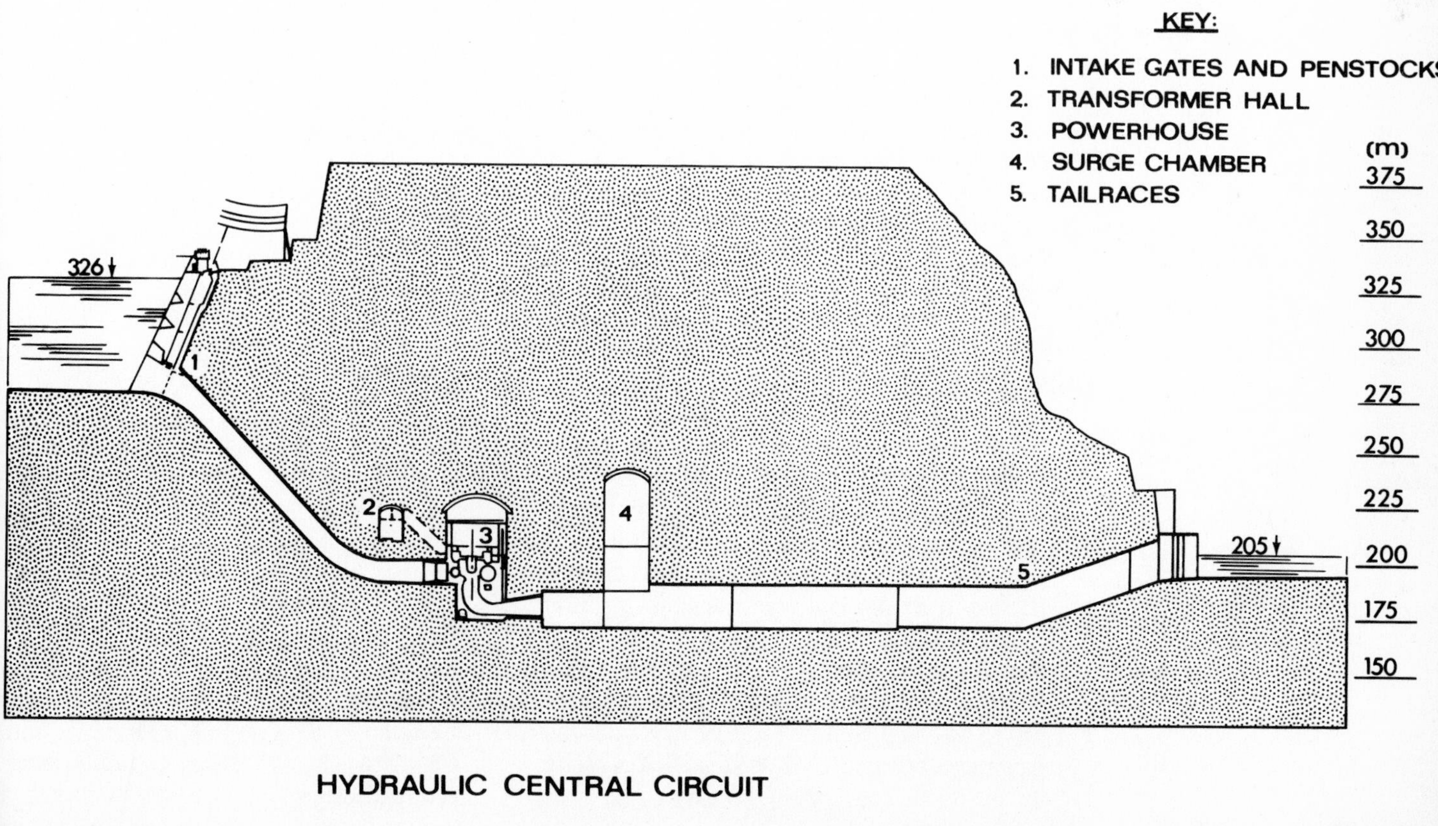

KEY:

1. INTAKE GATES AND PENSTOCKS
2. TRANSFORMER HALL
3. POWERHOUSE
4. SURGE CHAMBER
5. TAILRACES

(m) 375 350 325 300 275 250 225 200 175 150

326 205

HYDRAULIC CENTRAL CIRCUIT

Underground Power Station during construction

fore being transmitted to South Africa over a distance of some 1 400 kilometres terminating at the Electricity Supply Commission's (ESCOM) converter station called "Apollo" at Irene, near Pretoria.

TL Responsible for the double transmission line for the DC current between Cabora Bassa and Apollo.

South Africa had assumed financial responsibility for all work carried out inside the Republic and hence the Zamco contracts had to be divided into two parts: one relating to work carried out in Moçambique and the other to work carried out in the RSA.

The stage I project was to be completed in 3 phases, the first phase comprising the dam and south bank underground station and three turbo-generators each of 400 000 kilowatts capacity, the converter stations at Cabora Bassa and Apollo and twin transmission lines 1 400 kilometres long, was scheduled to be completed by April 1975. The second stage, with the addition of a fourth turbo-generator was to be completed by January 1977, and the final stage, bringing the rated installed capacity up to 2 000 megawatts, by January 1979.

At a Press Conference in Johannesburg on September 23, 1969 the costs and the national participation for financing the work in Moçambique and the Republic was set out as follows:

Cost	
Moçambique Contract . .	R305,0 million
South African (ESCOM) Contract	R 47,0 million
Total:	R352,0 million
Financing	
Moçambique Contract:	
Export Credits:	
France	R 63,0 million
W. Germany	R 63,0 million
Italy	R 39,2 million
South Africa	R 25,0 million
Total export credits:	R190,2 million
S. African Contract–ESCOM	R 47,0 million
Sub-total:	R237,2 million
Loans	
IDC and S.A. Government to State of Portugal . .	R 55,0 million
Portuguese private bank loans and cash . . .	R 59,8 million
Total:	R352,0 million

These figures demonstrate what efforts had gone into the marshalling of such a financing mixture of international credits and loans.

The Consulting Engineers for the Portuguese Government responsible for the engineering was Hidrotecnica Portuguesa.

DATA

(*a*)	DAM	
	Double curved concrete arch	
	Maximum height above foundations .	160 m
	Crown length . .	303 m
	Dam crest altitude .	331 m
	Storage level (Maximum flood) . . .	326 m
	Maximum drawdown level . . .	275 m
	Eight segmental gates each with a discharge capacity of 1 650 cumecs.	
	Total discharge capacity	13 200 cumecs
	Volume of concrete	450 000 m³
	Excavation for foundations	210 000 m³
(*b*)	RESERVOIR	
	Length	270 km
	Capacity . . .	52 000 million m³
	Average inflow . .	2 800 cumecs

Maximum flood inflow greater than .	30 000 cumecs
Area	2 660 km²
Catchment area . .	1 200 000 km²

of which 137 000 km² are in Moçambique and 140 000 km² in Angola

Comparison between Lakes Cabora Bassa and Kariba:

Description	Kariba	Cabora Bassa
Live (usable) storage .	44 000	48 000 million m³
Dead (unusable) storage .	116 000	4 000 million m³
Flood storage	25 000	14 000 million m³
Total storage	185 000	66 000 million m³
Surface area	5 200	2 660 km²

(*c*) INSTALLED GENERATING CAPACITY

South Bank

5 × 400 megawatt turbo-generators .	2 000 megawatts (MW)

North Bank

Probably a similar installation giving Cabora Bassa a total installed capacity of	4 000 MW
The permanent firm output from the river at this point is	2 100 MW

The potential installed capacity at Cabora Bassa compared with other hydro-electric schemes:

Kariba Rhodesia .	1 100 MW
Tumut 3 Australia	1 500 MW
Churchill Falls Canada . . .	4 500 MW
Krasnoyarsk USSR	5 000 MW
Grand Cooly USA	3 600 MW

Staged delivery of power to South Africa at the Apollo

Panoramic view of Lake Cabora Bassa filling on completion of the Dam

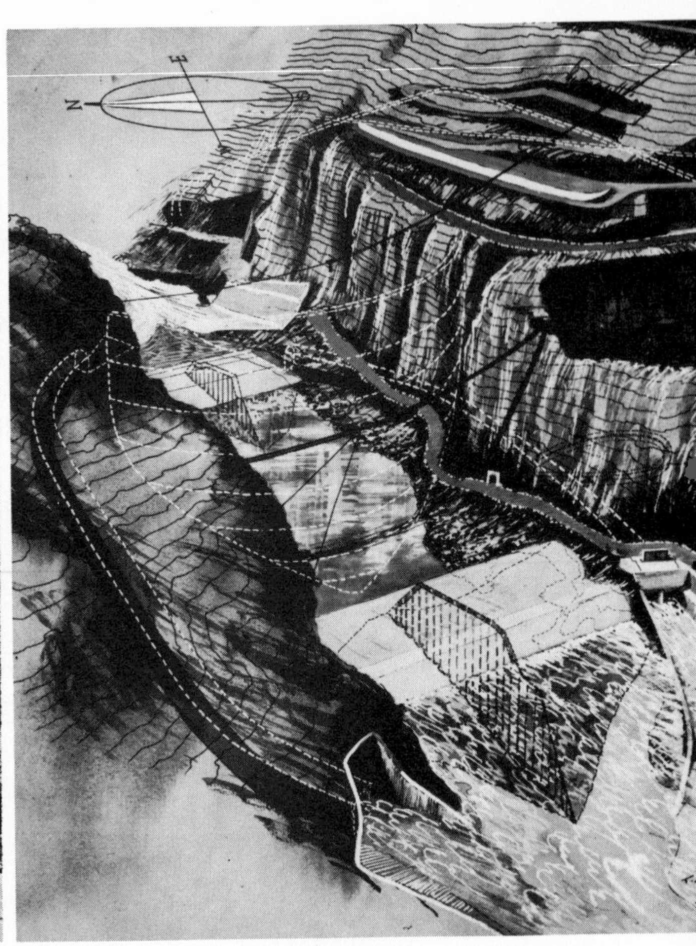

1

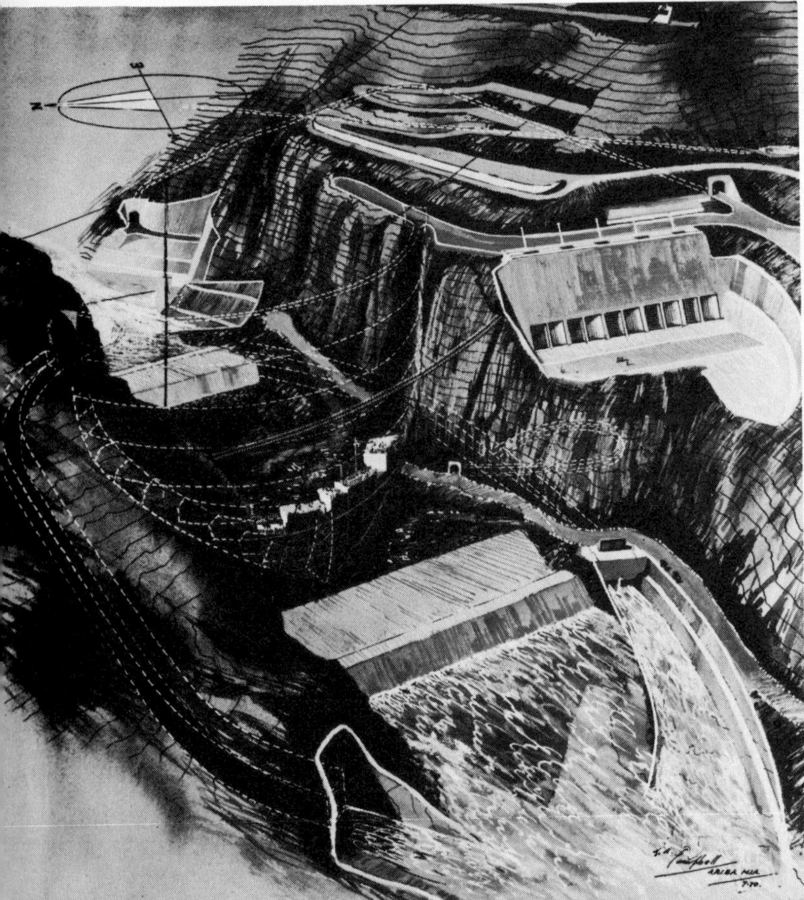

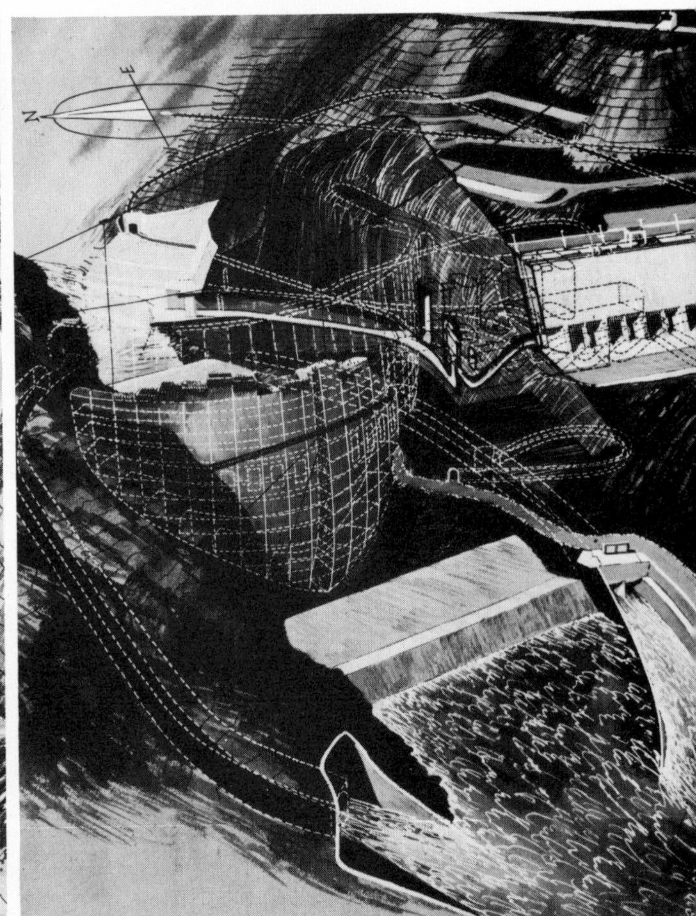

3

5

converter station at Irene, 7 km from Pretoria:

Stage 1 – 1975	. .	800 MW
Stage 2 – 1977	. .	1 250 MW
Stage 3 – 1979	. .	1 750 MW

astruction of Cabora Bassa – Phase 1
astruction of Cabora Bassa – Phase 2
astruction of Cabora Bassa – Phase 3
struction of Cabora Bassa – Phase 4
struction of Cabora Bassa – Phase 5

(*d*) SOUTH BANK UNDERGROUND POWER STATION

The station is equipped with 5 generating units, 5 penstocks, 2 tailrace tunnels, 2 surge vessels and the 220 kV feeds to the surface rectifier station 6 km away.

Turbines

Vertical . . .	Francis Type
Output . . .	415 MW
Speed	107,1 rpm
Effective head .	103,5 m
Consumption .	452 cumecs
Penstock length .	170 m, dia. 9,7 m.

Generators

480 MVA	PF 0,85	50 Hz
16kV + 5%		56 pole
Xd = 14,5%		107,1 rpm

Static excitation through thyristors
Rotor diameter 13 m weight 920 tons.
Flywheel effect (GD^2) 115 000 tm^2
A 2 × 500 ton crane is provided for installation of rotors etc. The turbine and generator have a common thrust bearing for a load of 2 150 tons and two guide bearings.

	Units in Operation
Stage 1	3
Stage 2	4
Stage 3	5

The generators are connected to the transformer banks accommodated in a separate cavern by isolated phase buses, having a current-carrying capacity of 18 000 amp.

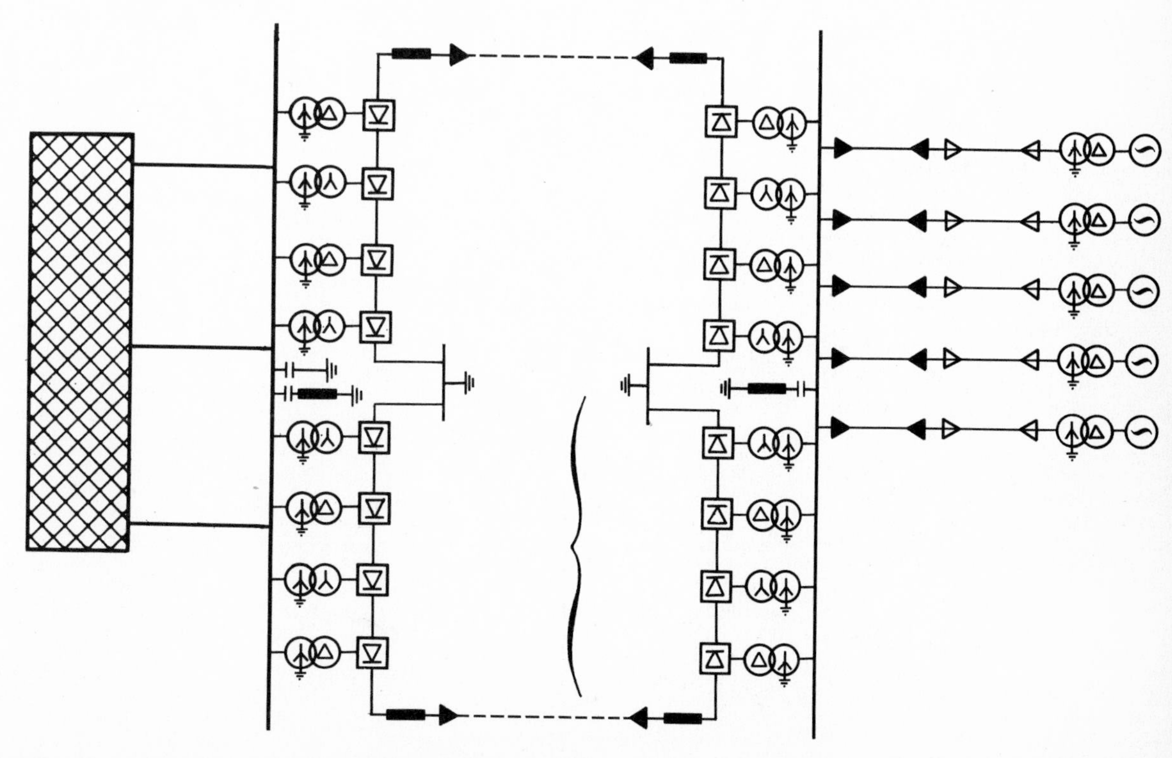

Fundamentals of Transmission Scheme

Opposite page
The Apollo Converter Station at Irene, near Pretoria, where the direct current transmitted from Cabora Bassa is converting to alternating current for distribution into the ESCOM national transmission grid

Generator Transformers

Yd connection single phase units
160 MVA/phase $230/\sqrt{3}$kV $\pm$ 5% 16kV
Impedance 9 – 11%
Wt. without oil 130 tons
Overall dimensions (m) L5, O/W3, O/H4,2

By means of single-core oil-filled cables the power is then transmitted to a platform located 120 m higher and adjacent to the intake works. This is the start of the overhead high voltage line to the Rectifier Station.

TRANSMISSION

CONVERTER STATION EQUIPMENT

A. C. Switchgear

	Cabora Bassa	Apollo
Voltage .	220 kV	275 kV
Interrupting capacity .	24 GVA*	15 GVA
B.I.L. . .		1 130 kV

*Allowance for construction of similar capacity hydro station on northern side in the future.

A. C. Filter Circuits

Tuned banks for 5th, 7th, 11th and 13th harmonics, high pass filter for higher orders.

Converter Transformers

Single phase units
Yd5 for bridges 3, 4, 7 and 8
Yyo for bridges 1, 2, 5 and 6
Tap changer on neutral end with 27 positions
Details of Single Phase Units:

	Cabora Bassa	Apollo
Size . .	96,7 MVA	90,8 MVA

Weight of largest unit (Bridge 7 and 8) without oil – 163 tons
Overall dimensions of largest unit:
Length 9,5 m; Width 3,7 m;
Height 4,3 m.

Converter Bridges

Semiconductors – disc type Thyristor cells
Bridge details: 240 MW, 133 kV, 1 800 A

Converter Bridge

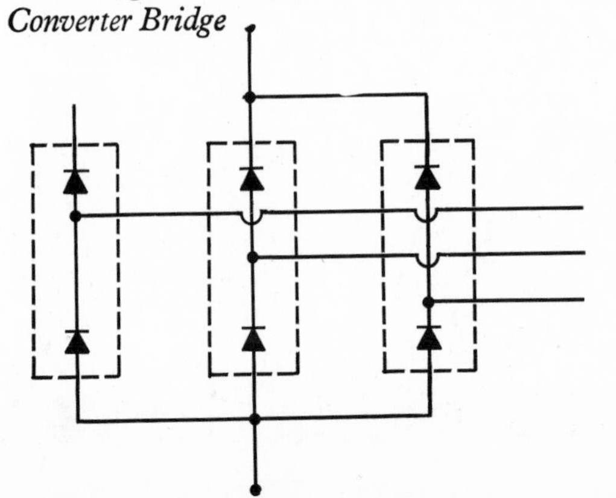

Overvoltage Protection

Lightning
(i) Direct strikes in yard – current limited to 5kA by efficient overhead screening.
(ii) Impulse coming in from line – limited to 1 200 kV by surge diverter.

DC Transmission Line

Tower configuration

2 monopolar lines approx. 1 km apart. Length 1 414 km. 514 km in Republic. Main conductor 42 × 4, 14 mm Al + 7 × 2, 32 mm St. Earth wire 12 × 3, 52 mm + 7 × 3, 52 mm St. Copper equivalent of each conductor 343 mm^2. Cover angle: 15 degrees. Wt. of conductor: Main 1 800 kg/km.
Earthwire 860 kg/km.
Spacing of quad conductors: 45 cm.
No. of towers for whole route approx. 7 000.
Tower height 40 m normal with extensions of –6m, –3m, + 3m, + 6m.
Average span length: 425 m.
Maximum span 700 m using reinforced towers.

Line Voltage	
Stage 1	±266 kV
Stage 2	±400 kV
Stage 3	±533 kV

CONSTRUCTION

The diversion of the Zambezi to permit the construction of the lower parts of the dam in the river bed in the dry was planned so that two large D-shaped diversion tunnels each 16 metres by 16 metres were driven through the left and right banks of the gorge, curving round the dam line. A cofferdam was constructed of rockfill about 40 metres high upstream of the dam to raise the water level and push or divert it into the two tunnels. Another cofferdam was constructed downstream of the dam to prevent the diverted waters emerging from the tunnels from re-entering the dewatered dam site. It was then possible to pump out the water in the area between the two temporary cofferdams and so begin the excavation of foundations in the river bed and the concreting of the permanent dam.

It was realised that during the flood season the flow of water, even with the help of Kariba

upstream holding back some of their discharges, would be such that the cofferdams would have to be designed to be overtopped or else constructed to a height that would be uneconomic.

Model tests carried out in Lisbon and Paris confirmed that this was feasible.

The cofferdams, constructed in accordance with this philosophy, were overtopped on February 22, 1972, and remained submerged for about four weeks. At the peak of the flood the flow was 4 metres deep over the upstream cofferdam. After the flood had receded detailed observations revealed that behaviour had been highly satisfactory. The theory and the model tests had been vindicated!

De-watering between the cofferdams commenced as soon as the cofferdams were exposed again on the falling flood and excavation for the dam was commenced on schedule.

The cofferdams were overtopped again the following year without damage. After that it

Twin direct current transmission lines from Cabora Bassa to Irene, near Pretoria, a distance of 1 400 km.

did not matter any more because the dam concrete had advanced to a height beyond the fury of the Zambezi.

Considering the remote and difficult site conditions and the quality of labour available, the accident rate at Cabora Bassa compares very favourably with other similar projects. The biggest single accident occurred on October 15, 1973, when some 5 000 tons of rock collapsed in the north surge chamber in the underground station, entombing 8 workers. In major hydro-electric projects it is usually necessary to construct either on the upstream or downstream side of the power station a "surge" chamber, sometimes called a chamber of equilibrium. We all know the bang that results if a water tap is suddenly shut. The effect of such a "hammer blow" is magnified a million times when one of these giant turbines in the water circuit is suddenly shut off. The instantaneous pressure blows, or suction vacuums, created could cause catastrophic damage to the power house and tunnels feeding and draining it. A chamber is, therefore, provided to absorb the shock harmlessly. At Cabora Bassa complex mathematical equations indicated that two chambers were necessary on the downstream side of the power station to serve the five tunnels conveying the exit water from the turbines. These rectangular chambers, excavated in solid rock, are 72 metres high, 76 metres long and 20 metres wide. A rockfall of such dimensions with no warning gave little chance for the plant and workers trapped in its path.

One knows the exact properties of copper and steel. How can anyone ever really know the properties of rock, which is not a homogeneous material, from one metre to another? Nowhere is there such "judgement" geology and engineering as in underground mining operations. Technical knowledge has to be reinforced by cumulative experience and precedents.

An immense clearing up and strengthening operation was launched and an accelerated programme was accepted by the Portuguese authorities. As a result the delays to the programme were minimised. Intermittent deliveries of electricity to the Republic were achieved in June 1976 even though tests were still in progress with respect to mechanical and electrical equipment at Cabora Bassa site.

At the peak of construction operations some 7 000 people of many nationalities were employed on the site. Zamco had arranged for an extensive training school at site to train completely unskilled workers in the basic skills of carpentry, steelfixing, safety measures etc. This proved highly successful.

Between 1970 and 25th April 1974, when the revolution took place in Portugal, the Frelimo harassed the road and rail supply lines to the site, and caused considerable

Layout of coffer-dams

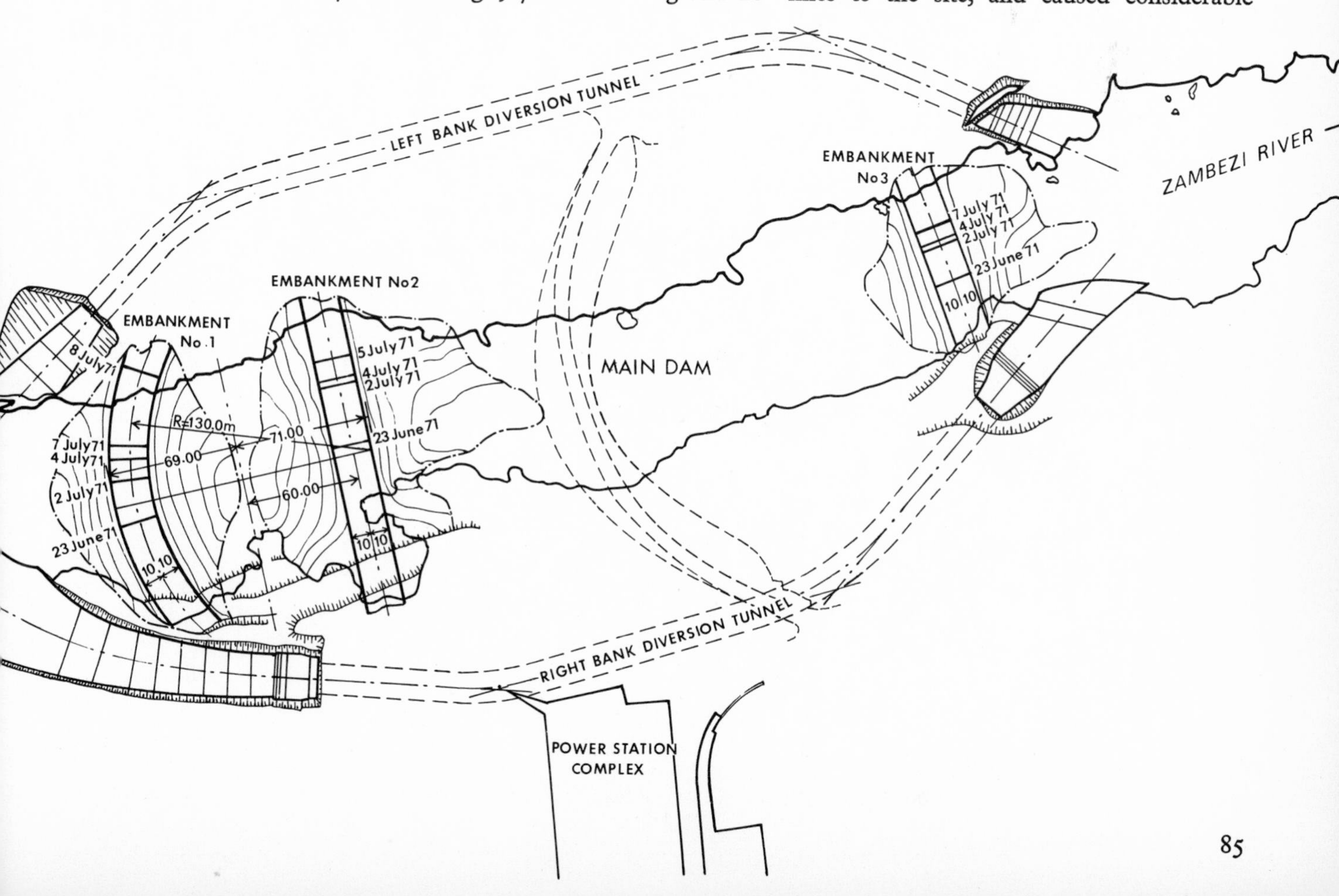

difficulties in the transport of vital supplies, particularly cement, which had to be transported from a factory at Dondo, near Beira. At the site the Portuguese authorities erected a double row of barbed wire fences with a diameter of about 30 kilometres. Between the wire fences the area was extensively mined. Portuguese troops were stationed inside the fenced area which caused additional logistic problems in supplying foodstuffs.

No incident was recorded at the site.

The prominent dates for the contract were:-

Contract awarded to Zamco . . .	17th September 1969
Closure of the river by two gates . .	5th December 1974
Completion of twin transmission lines (More than a year ahead of schedule)	January 1974
First turbo-generator tested . .	25th March 1975
First test power transmitted to South Africa . .	19th May 1975
Intermittent power to South Africa .	June 1976

After Moçambique became independent in 1975 a joint Portuguese-Moçambique Company – Hidro Electrica do Cabora Bassa, was set up to operate and administer the project and discharge the contractual obligations to Escom.

CONSULTING ENGINEERS AND CONTRACTORS – SUPPLIERS

Allgemeine Elektricitäts – Gesellschaft A.E.G. – Telefunken . .	(AEG)	(G)
Brown, Boveri & Cie Aktiengesellschaft	(BBC)	(G)
Compagnie de Constructions Internationales for itself and on behalf of the following companies:		

- Compagnie Industrielle de Travaux
- Entreprises Campenon-Bernard
- Société des Grands Travaux de Marseille
- Société Francaise d'Entreprises de Dragages et de Travaux Publics
- Société Générales d'Entreprises

The complex international and functional linkage is perhaps best illustrated by the following organigram:-

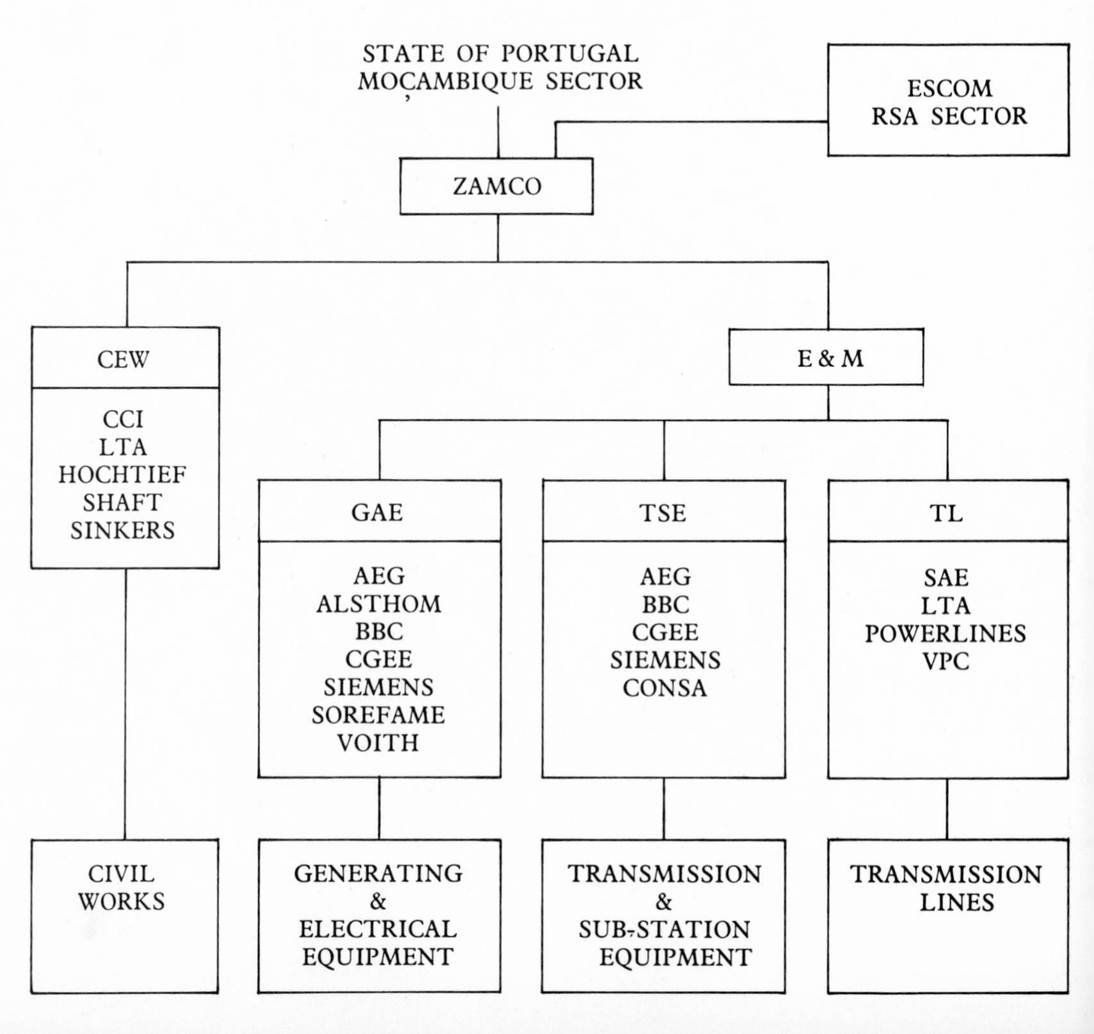

Cabora Bassa upstream rockfill coffer-dam overtopped on February 22, 1972.

Compagnie Générale d'Entreprises Electriques . .	(CGEE-COGELEX)	(F)
Entreprise Fougerolle-Limousin .	(FOUGEROLLE)	(F)
Hochtief Aktiengesellschaft . . .	(HOCHTIEF)	(G)
J. M. Voith GmbH .	(VOITH)	(G)
L T A Limited . .	(LTA)	(SA)
Powerlines Limited	(POWERLINES)	
Siemens Aktiengesellschaft . .	(SIEMENS)	(G)
Shaft Sinkers (Pty) Limited . . .	(SHAFT SINKERS)	(SA)
Sociedades Reunidas de Fabriçacôes Mestalicas – Sorefame, S.A.R.L. .	(SOREFAME)	(P)
Societa Anonima Elettrificazione, S.p.a. . . .	(SAE)	(I)
Societe Générale de Constructions Electriques et Mécaniques Alsthom	(ALSTHOM)	(F)
Vecor Projects and Constructions Limited Consa .	(V.P.C.)	(SA)

Note: G denotes . . West Germany
F denotes . . France
SA denotes . South Africa
P denotes . . Portugal
I denotes . . Italy

The major transport contractor was United Transport Holdings (Pty) Limited.

CHAPTER 7 Rhodesia

KARIBA HYDRO-ELECTRIC SCHEME

Area of Lake Kariba superimposed on map of southern England at the same scale

Rhodesia is a land-locked country situated in the tropical zone between 16° and 23° south having a total area of 382 000 sq. km. It has a central watershed running from southwest to northeast at an altitude varying from 1 200 m to 1 500 m above sea-level from where the rivers flow northward towards the Zambezi and southward to the Limpopo. These two rivers form the northern and southern boundaries of the country. The total population is 6,1 million, of which 5,8 million are African with an annual growth rate of 3,6 which is one of the highest in the world today.

The country's water resources are derived entirely from the annual rainfall which occurs during the summer months from November to March and varies from 400 mm in the south to over 2 000 mm in the east. The average annual rainfall over the whole country is approximately 675 mm. The rainfall also exhibits considerable seasonable variation with periodic regional droughts, particularly in the low rainfall areas. The rainfall is derived from two main sources, the northeast monsoon air flow and the northwest Congo air stream. The former precipitates its moisture in the eastern border region and the latter, intercepted by the central watershed in

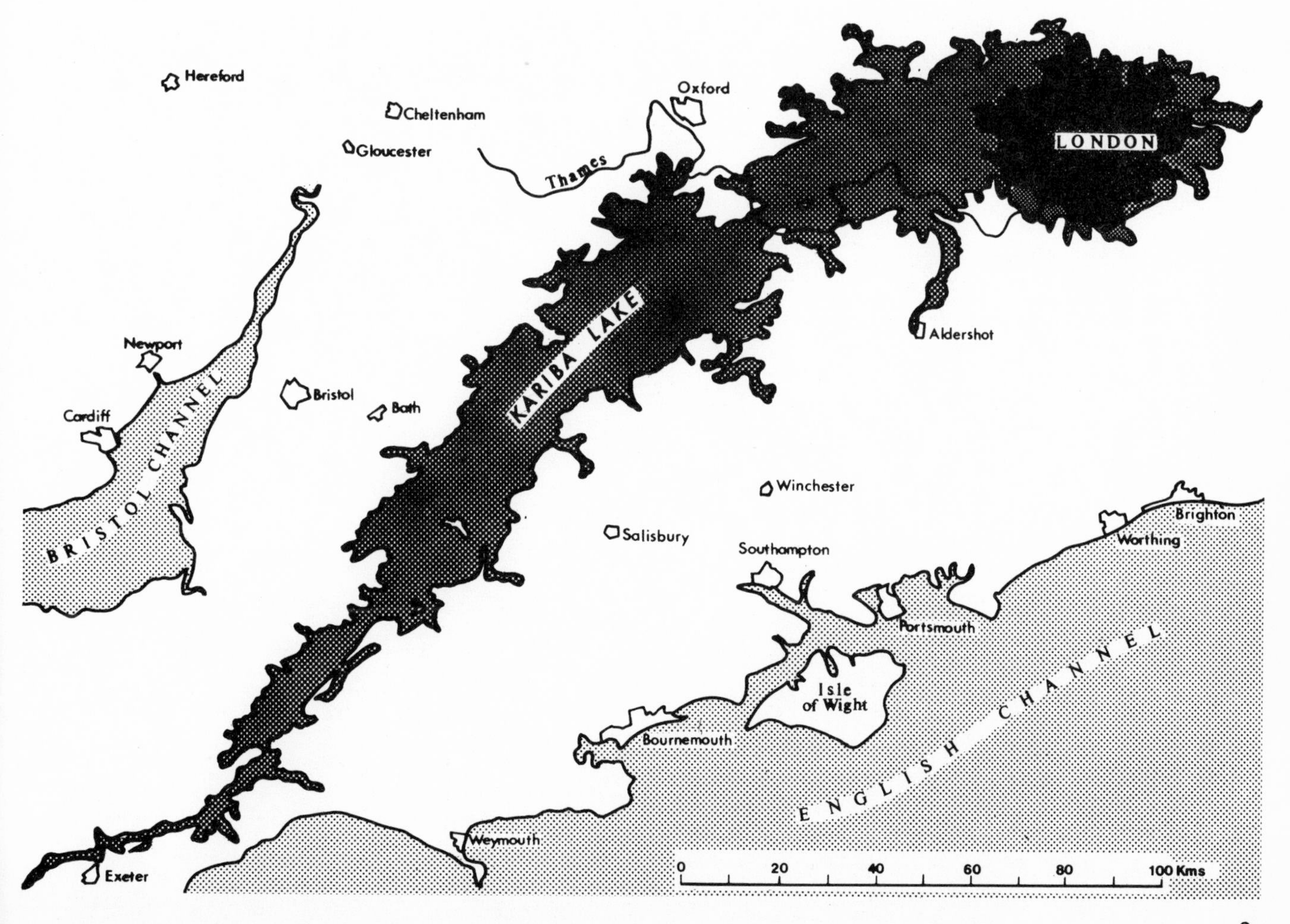

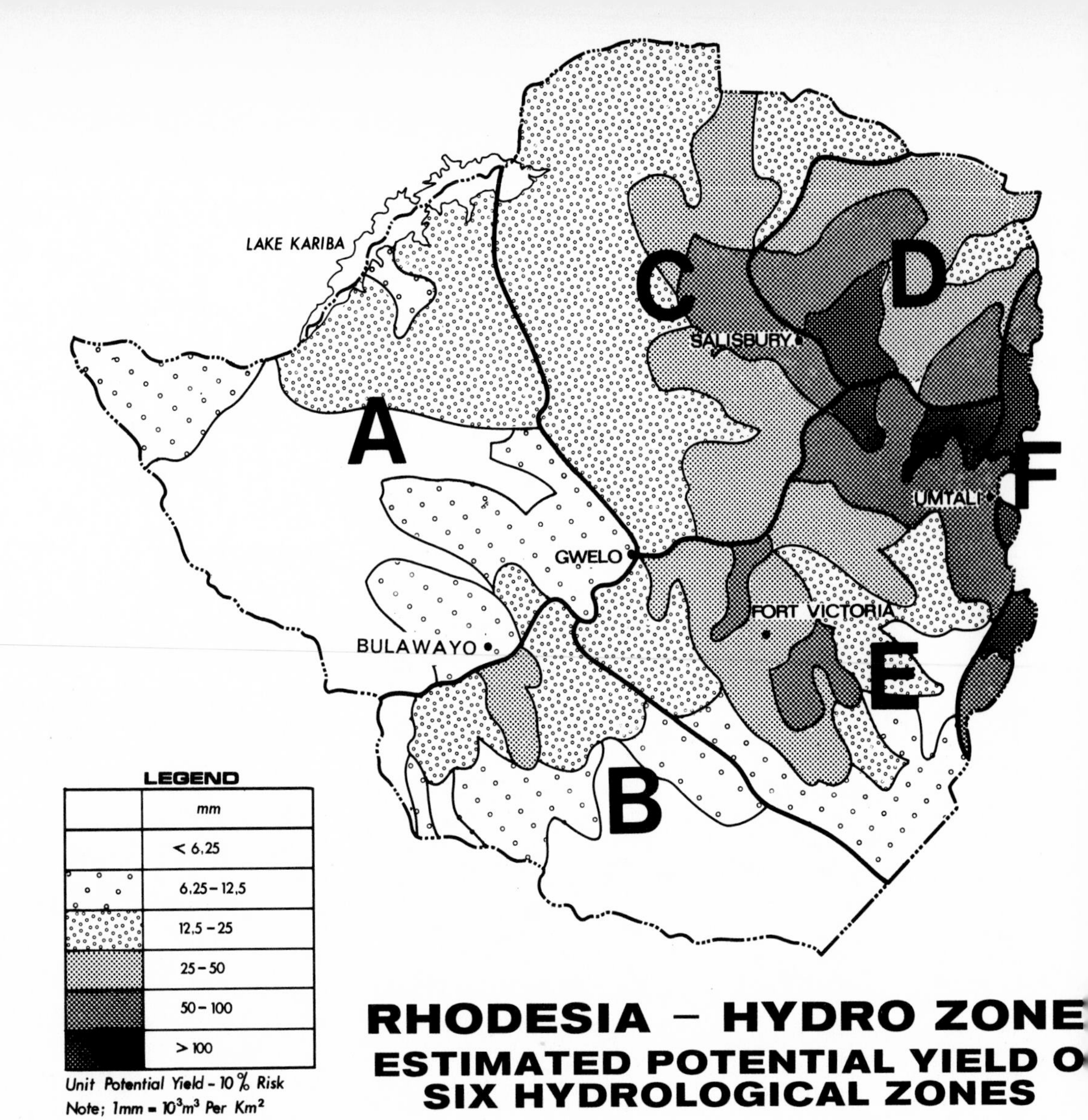

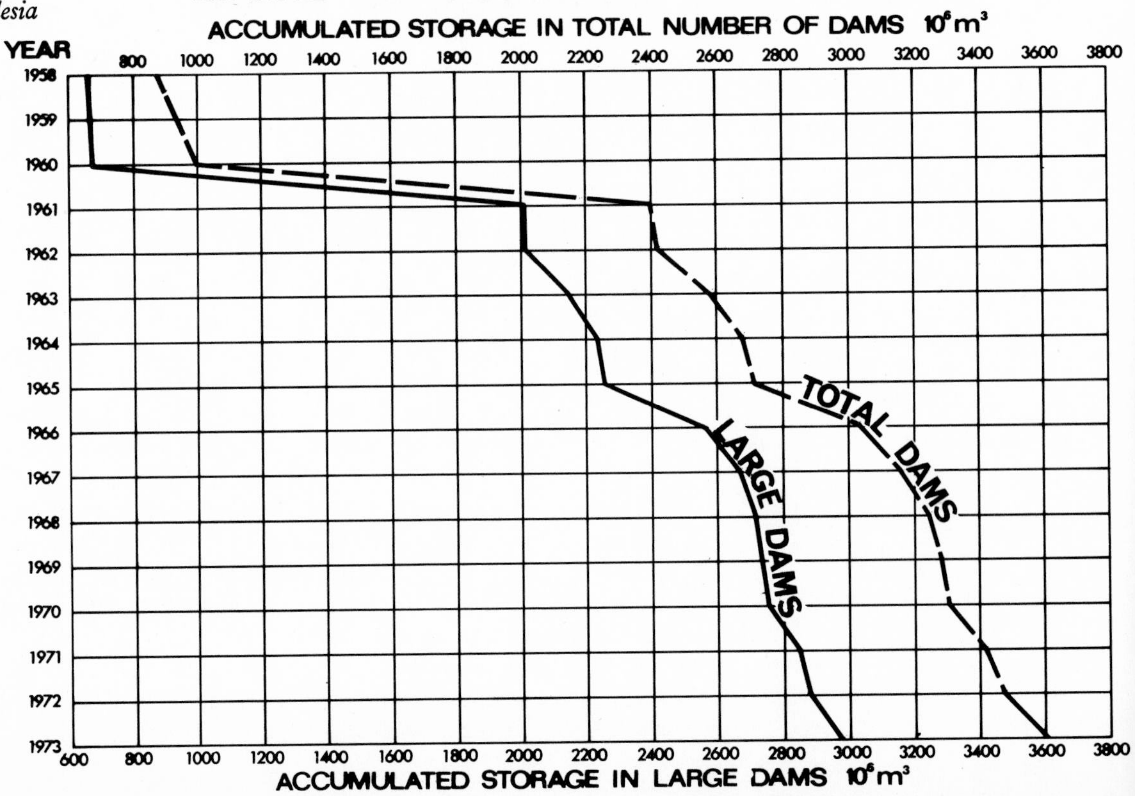

Graph of Dams built in Rhodesia

the northern part of the country. This gives a wet eastern border zone, a fairly well watered northern and central area and a comparatively dry southern area. The rains are confined to the few months from November to March and flood flows have to be stored in dams to satisfy the demand for a sustained reliable water supply throughout the year. A total of 7 200 dams have already been constructed with a total storage of 3 600 $m^3 \times 10^6$, of which 85 are major dams with a total storage of 3 000 $m^3 \times 10^6$.

Rhodesia is reasonably well endowed with water, but the water resources cannot be increased beyond a certain limit and the increasing demand for water will in time make it a limiting factor in the economic development of the country.

Very little information is known about the extent, capacity and re-charge of the country's underground water resources. Some high yielding boreholes have indicated the existence of water aquifers in certain areas and these boreholes are already being used for irrigation.

Surface Water Resources – Rhodesia – Estimated Potential

Hydro Zone	*Area*		*Mean Annual Run-Off (M.A.R.)*	*Optimum Storage*	*Potential Yield*
	km²	*mm*	*10³m³*	*10³m³*	
A	102 980	21	2 208 800	5 340 700	878 900
B	62 060	24	1 467 540	3 246 600	624 900
C	89 980	58	5 390 500	9 372 400	2 688 900
D	37 660	98	3 694 300	4 954 500	2 097 400
E	85 050	71	5 799 200	8 696 700	3 140 400
F	7 080	212	1 504 500	1 175 400	1 047 400
TOTALS	381 810	52	20 064 840	32 786 300	10 477 900

HISTORY

The double-curved concrete arch dam Kariba, is situated on the Zambezi River some 385 kilometres downstream from the Victoria Falls, or the "Smoke that Thunders", as the local inhabitants refer to it. Its left abutment is in Zambia and the right one in Rhodesia.

It was built to regulate what was called at the time the greatest man-made lake, with a surface area of 5 180 square kilometres, a usable storage capacity of 44 000 million m³ with a normal operating range of 9,4 metres, and a total storage capacity at maximum flood level of 185 000 million m³.

It was designed primarily as a hydro-electric project to serve the electrical needs of Zambia and Rhodesia. Since it was completed in late 1959 the great storage capacity has made it possible to attenuate the annually recurring floods in the Zambezi and to improve navigation facilities in the lower reaches of the river.

The full conception of the Kariba project involved two stages. The first stage covered the construction of the dam wall, the underground power station on the south bank with six machines each of 100 megawatts capacity and the transmission system to supply electricity in bulk to Lusaka and the Copperbelt of Northern Rhodesia (Zambia), to Bulawayo, Salisbury and the Rhodesian Electricity Supply Commission. The second stage is an underground power station on the north bank with a further capacity of 600 megawatts with a corresponding reinforcement to the transmission system.

The word Ka-riwa from which Kariba is derived denotes a trap which, according to local legend, is the God Nyamanyami. The god is symbolised by a sombre dark rock that emerged from the surly swirling waters downstream of the confluence of the Sanyati and Zambezi Rivers just before the combined waters enter the gorge. This rock is now 130 metres under Lake Kariba, but for years it had been there above water menacing the Batonka tribesmen and all who ventured into the gorge. Some of them thought this rock was Nyamanyami.

Nothing conclusive was known about the Kariba gorge until serious land surveys were undertaken during the Second World War by the veteran and intrepid surveyor Mr Jeffares, who was known and respected throughout Southern Africa. The Southern Rhodesian Irrigation Department began to take a very active interest in these surveys in the mid-forties when the first findings crystallised into what appeared to be a feasible plan for generating electricity at the gorge. Gauging stations

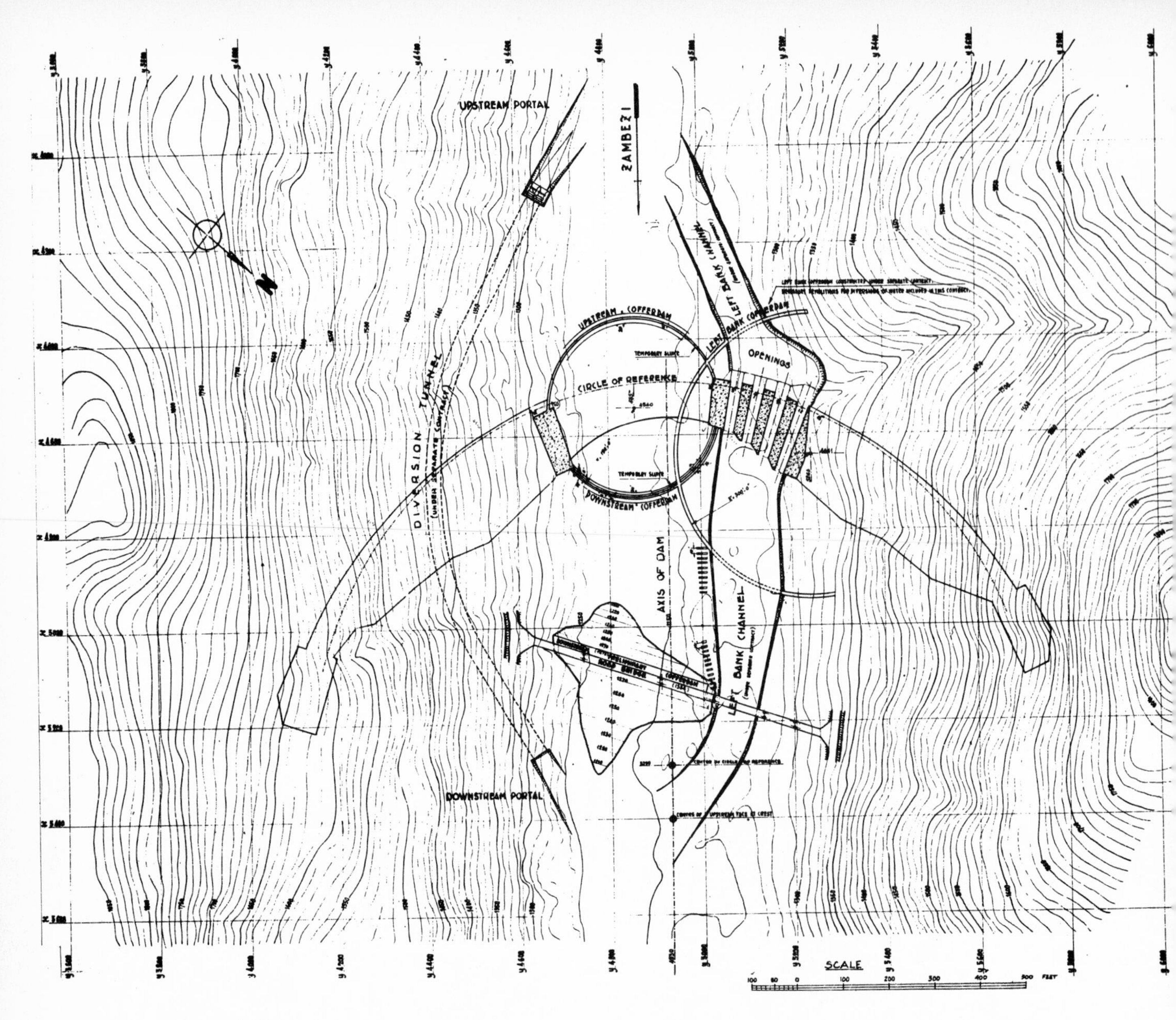

General Plan: Temporary Diversion Structures

LONGITUDINAL SECTION IN THE RIVER BED

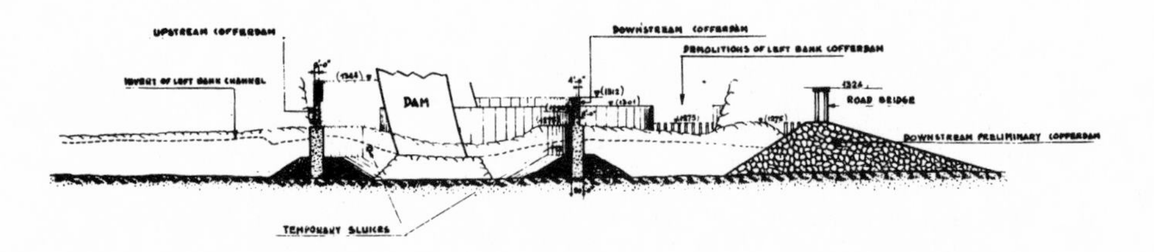

for measuring river flow were set up some 70 kilometres downstream of the gorge and a drilling camp twenty-five kilometres downstream from the present site to test the rock foundations for a major dam.

Early in 1950, Nyamanyami, the River God, showed the Batonka tribe he was against the White man's project. After torrential rains, an avalanche of rock engulfed the hut occupied by a survey party at the lower end of the gorge, killing four men. This was a warning we would be up against him to the end of the job.

When the search for dam sites had narrowed to four, the Southern Rhodesian Irrigation Department engineers were concentrating on site "X", at the Kariba gorge. But some 80 kilometres north of the gorge was another possible hydro-electric site on the Kafue river, a tributary to the Zambezi. The copper boom was starting in the north and the big copper companies favoured Kafue. The

Northern Rhodesian Government appointed the Anglo American Corporation as their consulting engineers on this project.

Meanwhile, the Central African Council had established the Inter-territorial Hydro-electric Commission in 1946. The Commission appointed a panel of well-known British civil and electrical consulting engineers to report on the general prospects of a hydro-electric scheme, and in 1951 they reported in favour of Kariba.

Considerable discussions ensued relating to the merits of Kafue versus Kariba, the northern territory favouring the former, and the southern territory the latter.

Late in 1954 an engineering mission from Electricité de France was asked to study the projects and its reports were reviewed by Monsieur André Coyne who was internationally known as the famous designer of concrete arch dams.

He preferred Kariba and argued that the Kafue project be built later when it could then be operated in conjunction with Kariba with better overall advantage.

The Government accepted his report and a Federal Hydro Electric Board was established in May 1955. Later this became the Federal Power Board.

Consulting Engineers were appointed at the same time and work commenced immediately and finalisation of designs and contract documents and on the mobilisation of a mix of funds for the project. The first contracts were let shortly afterwards for preliminary works followed by the placing of the main civil engineering contract and of orders for power generating and transmitting plant and ancillary equipment.

Prominent dates in the construction of Stage I were:

March 1955	The decision to proceed with Kariba was taken.
July 1955	Contract awarded for preliminary works: Cofferdams, site access, pilot housing scheme.
July 1956	Main Civil Engineering works contract awarded and contracts placed for Mechanical and Electrical plant.

Aerial view of Lake Kariba, the Dam and the township layouts on the Zambian and Rhodesian Banks (Central Office of Information)

July 1957	Diversion of river and start of main cofferdam.
November 1957	Main cofferdam dewatered.
March 1958	Exceptional flood of 16 142 cumecs overtopped cofferdam.
December 1958	River closed – impounding commences.
June 1959	Dam concreting completed.
December 1959	First power sent out to the copperbelt.
May 1960	Project officially inaugurated by Her Majesty the Queen Mother.

In 1963 when the Federation of Rhodesia and Nyasaland was dissolved, the administration of the project in the joint interests of Zambia and Rhodesia was entrusted to the Central African Power Corporation under the chairmanship of Mr. James Ward, C.B.E. Kariba has served Zambia and Rhodesia well in the sixteen years since it was completed. These are two of the few countries in the world where the cost of electricity has been reduced over the past ten years. During this period the cost of electricity in Rhodesia reduced from 0,605 Rhodesian cents to 0,390 Rhodesian cents and by a similar amount in Zambia.

Stage 1 was completed ahead of schedule and within the estimates of £80 million given to the Client, then the Federal Power Board, in 1955, by the Consulting Engineers and the Consulting Accountants. The pattern for loan negotiations for Stage 1 was:

World Bank – An amount in various currencies up to	U.S. $80 million
The Colonial Development Corporation	£15 million
The Commonwealth Finance Corporation	£3 million
The Copper Companies to lend to the Government	£20 million
The British South Africa Company to lend to the Government	£4 million
The Standard Bank of South Africa and Barclays Bank DC & O to lend to the Government	£4 million

DIMENSIONS

(a) POWER GENERATION

Installed capacity of Stage 1 (South Bank Station)	600 MW
Ultimate mean annual power output (Stages 1 and 2)	8 500 million kWh

(b) DIMENSIONS OF UNDERGROUND VAULTS:

Machine Hall:

Length	146 m
Width	23 m
Height	41 m

Transformer Hall:

Length	168 m
Width	17 m
Height	19 m

(c) DETAILS OF MAJOR PLANT

Turbines:

Maximum gross head	110 m
Mean net head	97 m
Corresponding rated flow through each turbine – full load at 88 m net head	134,6 cumecs
Weight of turbine runner	40 tons
Diameter of turbine runner	4,2 metres

Alternators:

Operating speed	167 rpm
Maximum continuous capacity at 0,9 power factor	100 MW
Generator voltage and frequency	18 kV, 50 c/s
Diameter of rotor	8 m
Combined weight of rotating parts of alternator and turbine	475 tons
Overall diameter of alternator	10 m
Total weight of alternator	720 tons

Transformers:

Type and rating – single phase 80 MVA 18/330 kV	
Number installed	10

(d) DAM

Maximum height	131 m
Crest length	633 m
Mean radius of arch	250 m
Number of spillway gates 6 size 9,4 × 9,7 m each	

(e) RESERVOIR

Total capacity at maximum flood level	185 000 million m³

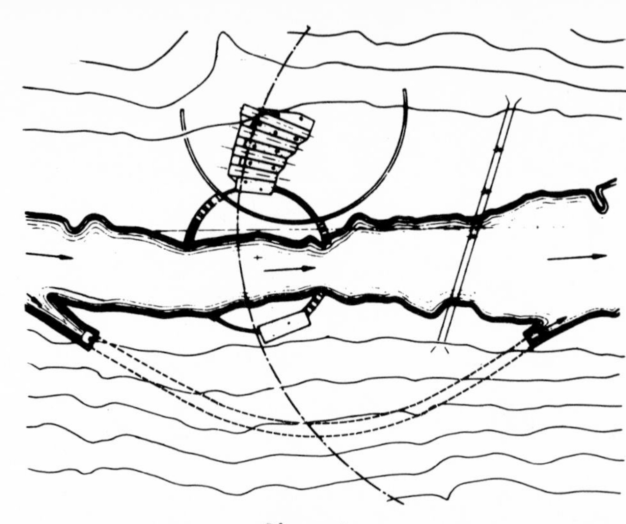

Phase 1

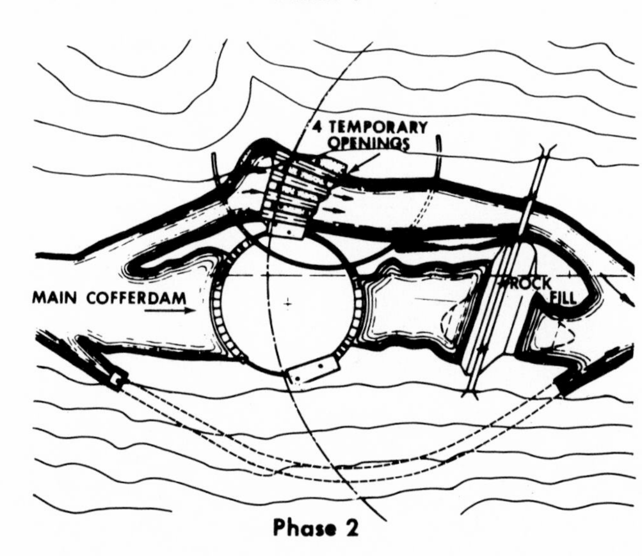

Phase 2

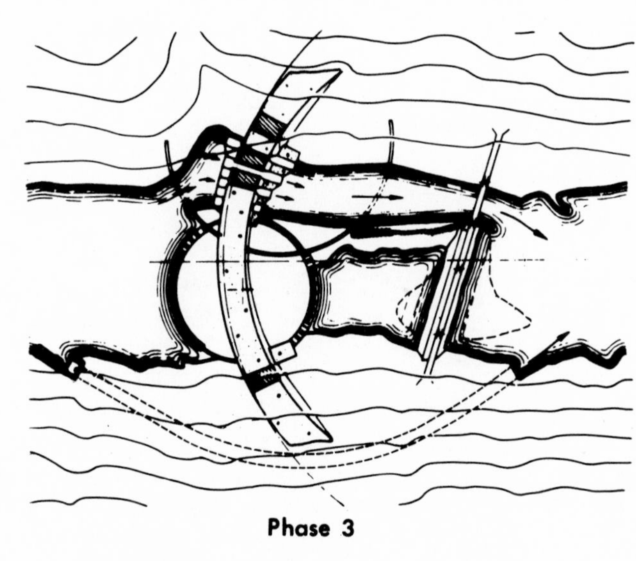

Phase 3

Phases of River Control

	Total capacity at normal top water level	160 000 million m³
	Useful storage within normal operating range of 9,4 metres . . .	44 000 million m³
	Area of reservoir at normal top water level . .	5 180 km²
(*f*)	HYDROLOGY	
	Catchment area .	663 000 km²
	Mean river flow .	1 444 cumecs
	Total mean firm flow per year . .	46 000 million m³
	Mean water balance:	
	Utilisation for power	83%
	Losses by spillage	8%
	Losses by evaporation	9%
	Recorded extremes of flow:	
	Minimum (1949–54)	227 cumecs
	Maximum (1958).	16 142 cumecs
	Capacity of sluiceways:	
	At maximum flood level	9 515 cumecs
	At normal top water level . .	8 665 cumecs
(*g*)	MAJOR ITEMS OF QUANTITIES	
	Volume of concrete in dam . .	975 000 m³
	Peak rate of monthly concreting .	80 400 m³
	Excavation in dam foundations . .	382 300 m³
	Excavation in underground works	577 000 m³
	Concrete placed underground . .	145 000 m³
	Total reinforcing steel in dam and underground works	10 950 tons
	Total cement delivered . . .	360 000 tons
	Peak rate of consumption of diesel oil per month .	2 million litres
	Peak rate of consumption of petrol per month . .	227 000 litres
	Total amount of petrol and oil consumed . . .	54,5 million litres

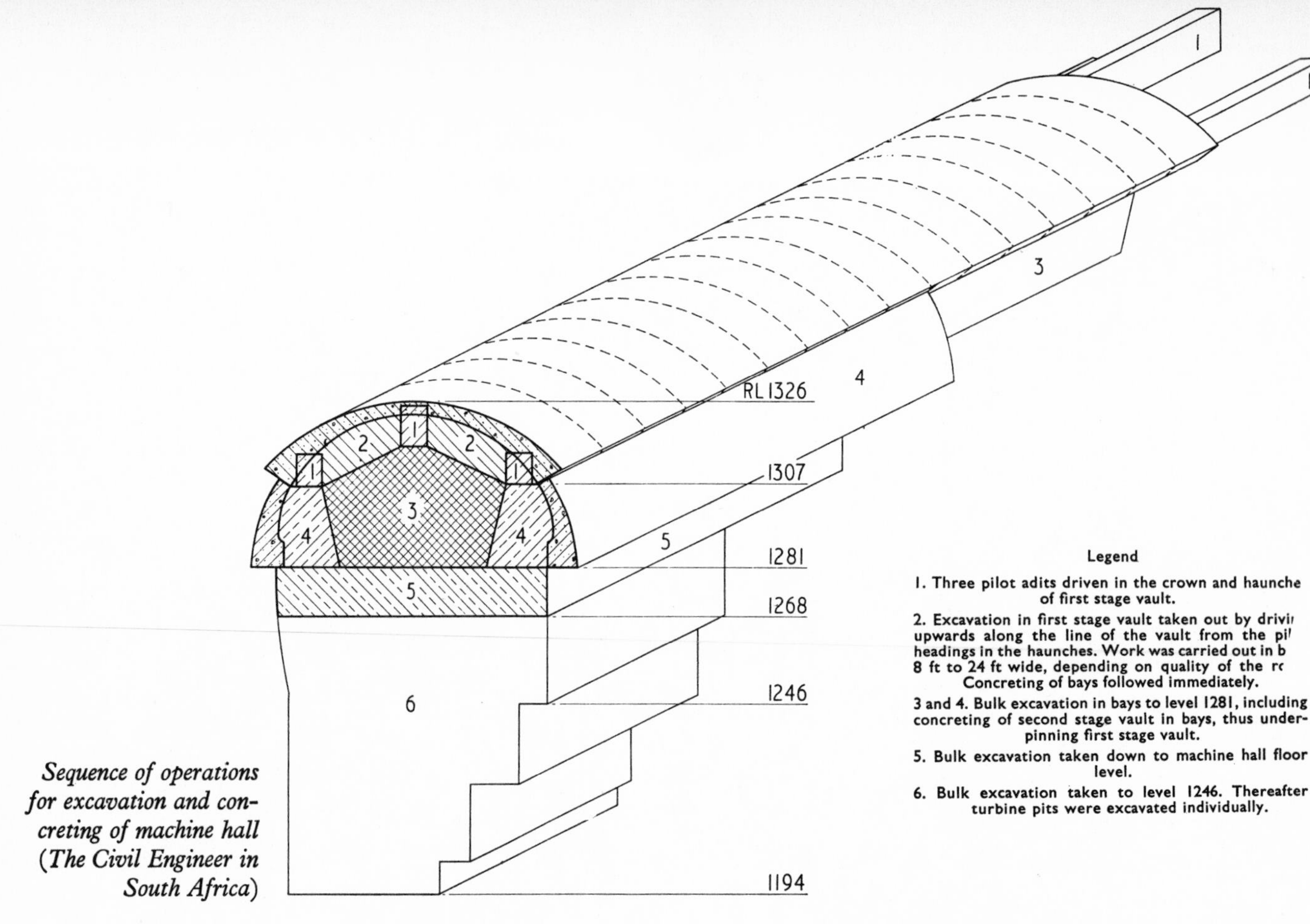

Sequence of operations for excavation and concreting of machine hall (The Civil Engineer in South Africa)

Total amount of materials moved approx. . . .	700 000 tons
Installed capacity of temporary diesel station . .	8 400 kW

CONSTRUCTION

Major problems during construction of Stage I were the diversion and closure of the river, and the extraordinary flood of 1958.

Diversion was carried out in three phases.

In phase I the low water levels before the 1955–56 rainy season were used to excavate a canal on the north bank and to construct a large semicircular cofferdam. Simultaneously, it was necessary to drive a large diversion tunnel through the south bank to handle the expected flood of early 1956 and to construct a light bridge and a suspension foot-bridge over the Zambezi. Segments of the dam were built inside the north bank cofferdam, leaving four temporary openings to accommodate the Zambezi flow during Phase 2 operations.

In Phase 2 a cofferdam was constructed by tipping rock in flowing water from the bridge constructed during Phase I. At this stage sections of the north bank cofferdam were blasted and removed, and the rising wall of the rockfill cofferdam pushed the Zambezi into the north bank canal and through the temporary openings left for this purpose in the section of the main dam. In the still water conditions created upstream of the rock cofferdam and in the middle of the river it was then possible to construct a circular concrete cofferdam and, when pumped dry, the deepest sections of the main dam could be constructed. The flank sections of the main dam could be constructed irrespective of river levels.

Phase 3 consisted of demolishing the downstream half-circle of the central cofferdam and closing the temporary openings in the main dam on the north bank. This moment heralded the birth of the new lake – Lake Kariba.

Closure of the river was effected in two successive stages. First the diversion tunnel was sealed off and thereafter the temporary openings were closed. In the first phase coarse rockfill was dumped at the downstream portal (exit) of the diversion tunnel and, as soon as relatively steady flow conditions had been achieved, sheet piles were driven at the upstream portal (entrance). These were sealed by rockfill placed on the outside of the piles in order to ensure that no breakthrough of water should occur at this point, once the lake started filling and before the tunnel could be plugged with a solid concrete "cork". A concrete semicircular arched dome was placed over the entire upstream portal. The downstream portal was then finally sealed by dumping fine materials over the coarse rock and the

Cabora Bassa – Artist's impression looking upstream – 'A Thing of Beauty'

Aerial view of Lake Kariba looking upstream

Sunset on Lake Kariba

Kariba South bank underground power station

Night-shift – Hendrik Verwoerd Dam (Ministry of the Interior)

Recreational facilities on Lake Kariba after the lake had filled

Vanderkloof underground power station showing two penstocks coming in from the right

Vanderkloof underground power station showing one spiral casing under erection

Below: *Wemmershoek Dam, Cape. Spillway control in relation to reservoir (Douglas)* Above: *Steenbras D*

tunnel pumped dry. A strong concrete plug inside the tunnel completed its final sealing.

The second stage, which involved the closure of the four temporary openings in the dam, is illustrated in the drawing below.

The method was highly effective, and flow through the openings was reduced from 500 cumecs, before the placing of coarse aggregate, to zero in 11 days, with the water level of the lake thereafter rising by 0,67 metres per day.

Work on all parts of the dam proceeded speedily until early in 1958 the warning system brought news that the Zambezi was in wholly unprecedented flood.

The effect of the volume and spate of the flood when it arrived at Kariba in late February – early March was quite devastating. The central cofferdam was flooded, the road-bridge washed away, the foot-bridge seriously damaged, the banks eroded and for a short period the underground works had to be sealed off. A record discharge of 16 000 cumecs passed through the gorge on March 5, 1958. A large quantity of debris was deposited downstream of the dam and across the tailrace outlets.

An accelerated programme was agreed to between the Client and the Contractor, which not only recouped the time lost while repairing or minimising the damage, but ensured that power would still be available on or before 1st January 1960 as contracted.

The method adopted to construct the underground power station is illustrated in the drawing opposite as also the orientation of the underground powerstation relative to the reservoir. Geology influenced the final location and orientation of the underground power station. In a sense it was a unique technical concept in that the intakes to the turbines and the outlets to the river, i.e. the tailraces, were on the same side of the power station.

A major concern in design of the dam was the dissipation of energy of the water discharged through the six sluice gates which have a capacity of 9 515 cumecs at maximum flood level.

A series of model tests indicated that a scour hole of some 55 metres depth would be formed in the gneiss rock. By 1966 this depth was approached and the scour hole appeared to be stabilised as predicted by the model tests and the rock bar formed on the downstream lip of the crater assists materially in the dissipation of energy.

Discharge through all the gates is an impressive sight – involving the dissipation of millions of horsepower in energy.

CONSULTING ENGINEERS AND CONTRACTORS

Consulting Engineers:

Civil Engineering: Sir Alexander Gibb and Partners, London

A. Coyne & J. Bellier, Paris

Société Générale de Exploitations Industrielles, Paris represented in the Federation by Gibb Coyne Sogei (Kariba) (Pvt) Ltd.

Co-consulting Engineers with Gibb Coyne Sogei (Kariba) (Pvt) Ltd.

Jeffares & Green, Salisbury, for South and North access roads and

Sir Alexander Gibb and Partners (Africa),

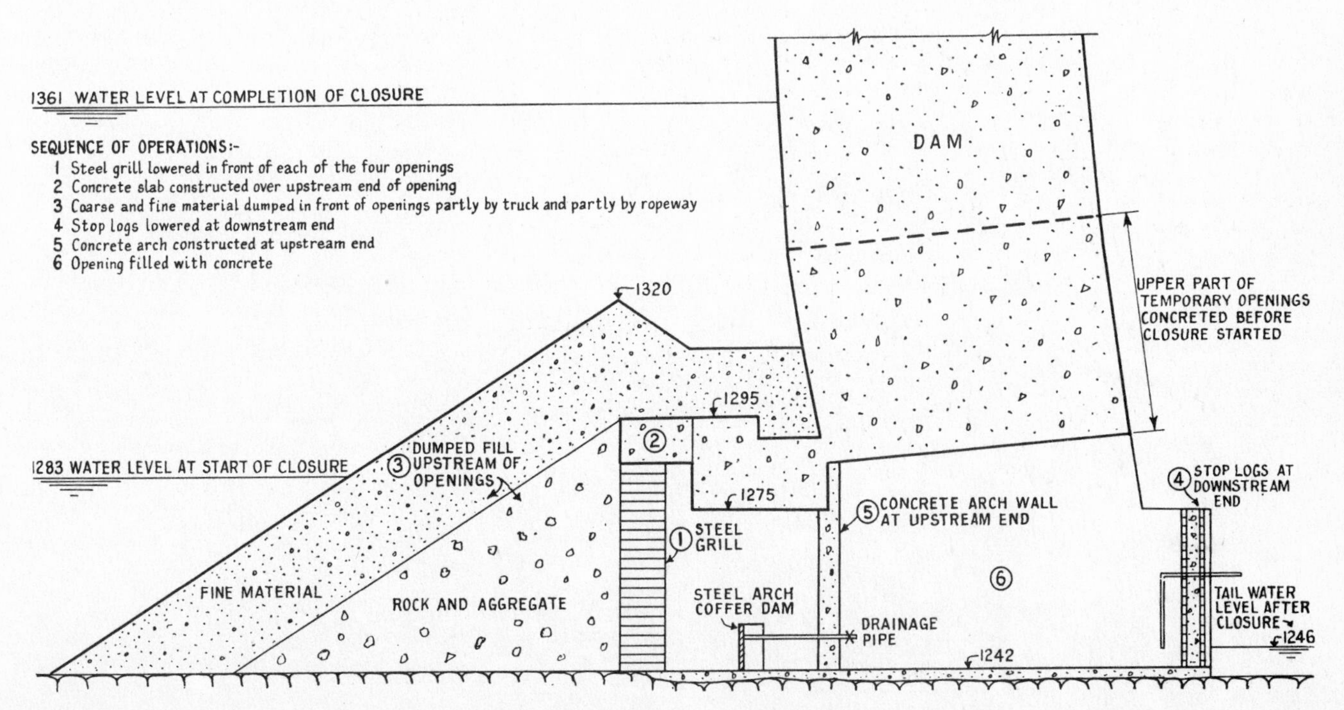

Method of closure of temporary openings in main dam (The Civil Engineer in South Africa)

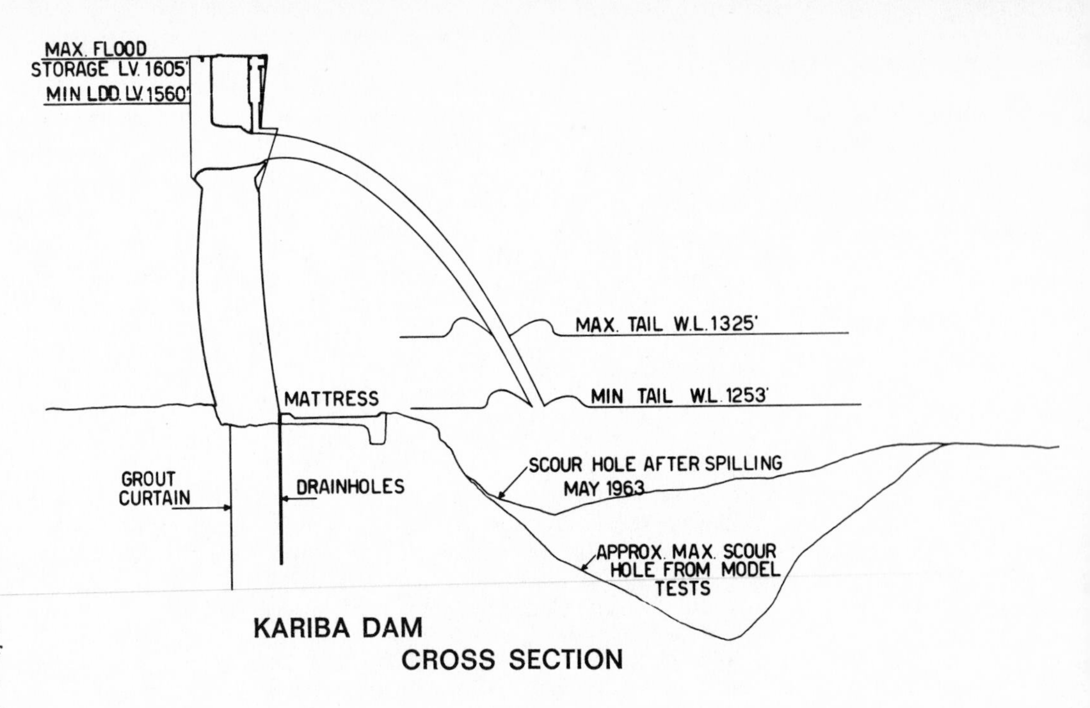

Cross section of dam showing formation of scour hole with spilling (Gibb Coyne Sogei (Pvt.) Ltd.)

for railway sidings, certain access works and the permanent airfield.

Mechanical and Electrical Engineering: Merz & McLellan, London

Chartered Accountants: Coopers & Lybrand, London and Salisbury

Preliminary Works	The Cementation Company Limited
Main Civil Engineering Contract	Impresit Kariba (Pvt) Limited
Intake and Draft Tube Gates, Screens, etc; Flood Gates for Dam	Etablissements Neyrpic
Housing and Associated Works	Richard Costain (Southern Rhodesia) Limited
South Access Road	Division of Irrigation, Southern Rhodesia
South Access Road – High Level Realignment; Railway Sidings, Lion's Den and Kafue; Permanent Aerodrome, Kariba; Salisbury and Bulawayo Substation Buildings	John Laing & Son (Rhodesia) Limited
North Access Road	A. G. Burton Limited
Bridges, Steelwork	Dorman Long (Africa) Limited
Kitwe Substation Buildings; System Control Buildings, Sherwood	Roberts Construction Company (Central Africa) Limited
Norton Substation Buildings	Lewis Construction Company (Rhodesia) Limited
Sherwood and Lusaka Substation Buildings	Sir Alfred McAlpine & Son (Rhodesia) (Pvt) Limited
140 000 b.h.p. Water Turbines	Boving & Company Limited
100 MW Water-Turbine-Driven Generators; 330 kV Switchgear; 220 kV Switchgear at Kitwe	Associated Electrical Industries Export Limited
18 kV Generator and Station Transformer Switchgear	Brown, Boveri & Company Limited
330 kV Generator Transformers	The English Electric Company Limited
330 kV and Ancillary Cables	British Insulated Callender's Cables Limited
Power Station Control and Indicating Equipment; Control and Communications Equipment	Standard Telephones & Cables Limited
Power Station Cables; Power Station Lighting	Clough, Smith & Company (Central Africa) (Pvt) Limited
Auxiliary Transformers	J. Mann & Company (Pvt) Limited
11 kV and Lower Voltage Switchgear; 33 kV Switchgear at	

Model of underground power station showing section through generator and turbines

North underground station under construction. Machine hall looking away from loading bay

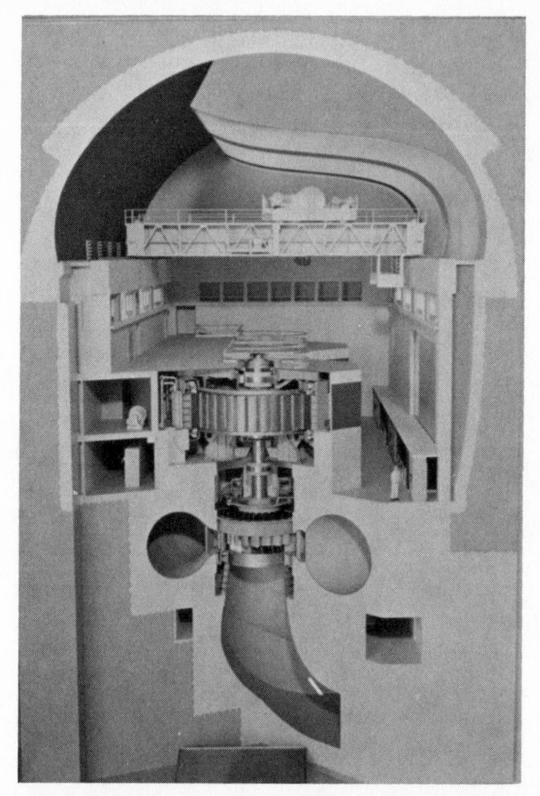

Salisbury; 88 kV Switchgear at Lusaka	A. Reyrolle & Company Limited
Passenger Lifts in Power Station	Schindler Lifts (S.A.) Limited
Power Station Ventilation Equipment	Johnson & Fletcher Limited
200-ton Power Station Cranes; 100-ton "Goliath" Crane, Lion's Den	Babcock & Wilcox Limited
330 kV Substation Transformers; Metering equipment	Ferranti Limited
Shunt Compensating Reactors	C.A. Parsons & Company Limited
11 kV Airblast Switchgear for Shunt Compensating Reactor Control	Allmanna Svenska -Elektriska Aktiebolaget
88 kV Switchgear at Norton, Bulawayo and Sherwood	The Southern Rhodesia Electricity Supply Commission as Agents for the Board
88 kV, 33 kV, 11 kV and Auxiliary Cables	British General Electric Company of Central Africa (Pvt) Limited
330 kV Transmission Lines	Rhodesian Power Lines (Pvt) Limited
Transport	Kariba Transport Limited
Road Transport, Heavy Lifts	Thornton's Transportation Rhodesia (Pvt) Limited
Petrol and Diesel Fuel	The Shell Company of Rhodesia Limited
Cement	Chilanga Cement Limited
Transmission Line Surveys	Swart and Hancock

STAGE 2

HISTORY

The financing arrangements were such that from the revenues of Stage 1 a Development Fund was created to finance partially the cost of Stage 2 development.

By 1973 Rhod. $43,5 million or Kwacha 49,5 million had accrued to this Fund.

Work on Stage 2, the North Bank, did not commence until 14th December 1970. After some contractual problems in the early stages work progressed well and the North Bank Station was commissioned during 1976.

DIMENSIONS

Machine hall	130 m long × 24 m wide × 48 m high
Transformers	On surface bench
Access Tunnel	315 m long × 8 m wide × 6,4 m high
Lift Shaft	152 m long × 4,5 m dia.
Intakes	4 No. 12 m wide × 15 m high with vertical lift gates
Penstocks	4 No. 250 m long × 6,75 m dia.
Tailraces	4 No. 131 m long × 9,0 m equivalent dia.

Turbines (4 No. Vertical Francis)	Max. Gross Head – 107,6 m Max. Net Head – 104 m Mean Head – 92 m Flow 186,79 m³/s at 92 m head Weight of runner – 65 t Dia. of runner – 5,05 m
Alternators (4 No.)	Operating speed – 136,4 r.p.m. Max. continuous capacity at 0,9 power factor – 150 Mw Generator voltage – 18 kV at 50 c/s Dia. of rotor – 9,428 m Combined weight of rotating parts of alternator and turbine (including hydraulic thrust) 1 070 t. Generator Housing (octagonal) outside dimension – 15,0 m. Total weight of each alternator 860 t.
Transformers	Type and rating – core type 167 MVA Number installed – 4

CONSTRUCTION

Induced seismicity from the vast volume of water stored was expected and experienced. A comprehensive network of seismic gauges was set up before the lake started to fill.

Seismic activity in 1973 was the highest on record since 1963. In an area about 10 kilometres radius some 65 kilometres west of the dam, a large number of tremors was recorded during December ranging in magnitude up to 5,4 on the Richter scale. The activity gradually reduced over the first quarter of 1973. In another area, approximately 20 kilometres south of the dam a slight increase in activity was recorded during March, 1973.

CONSULTING ENGINEERS AND CONTRACTORS

Consulting Civil Engineers	Sir Alexander Gibb and Partners
Mechanical and Electrical Consulting Engineers	Merz & McLellan
Chartered Accountants	Coopers & Lybrand
Date of Award of Main Civil Engineering Contract	14th December 1970
Main Civil Engineering Contractor	Mitchell Construction, Kinnear Moody Group Limited
Date contract ceased for first contractor	15th February 1973
Date contract awarded to new contractor	Energoprojekt Engineering & Contracting Company. 28th March 1973.
Noteworthy Sub-contractors	Housing stages I and II – Burton Construction Limited, Zambia (and later Cohen and Company Lusaka). Intake and outfall gates – Voest Austria.
Main mechanical and electrical contractors	Turbines – Voest, Austria Generators – Rade Koncar, Yugoslavia Cranes – Metalna, Yugoslavia Generator Transformers – ASEA, Sweden Generator Connections – Cogelex, France Auxiliary Switchgear etc. – Hawker Siddeley U.K. 330 kV switchgear – Brown Boveri, Switzerland Power Station control equipment – Brown Boveri, Switzerland Passengers Lift – Otis, South Africa Air Conditioning Equipment – Airmec, Zambia 330 kV Cables – Pirelli, Italy Fire Protection Equipment – Mather & Platt, England Power Station Cabling – Balfour Kilpatrick, England Power Station Lighting – Balfour Kilpatrick, England Metering Equipment – Ferranti, England Transport – United Transport Overseas, England

Central circular coffer dam overtopped at peak of the flood on March 5, 1958, when sixteen thousand cubic metres of water per second passed through the gorge

KARIBA

KYLE DAM

HISTORY

Kyle Dam, which is situated a short distance below the confluence of the Mtilikwe and Umshagashe Rivers, is the key structure in the complex of dams and canals that support the irrigation development in the Triangle and Hippo Valley area.

Work on the dam started in 1958 and construction was completed at the beginning of 1961.

DIMENSIONS

Catchment Area	3 989 square kilometres
Yield (10%) .	7 790 litres/sec.
Design Flood .	1 740 m³/sec.
Height of Dam .	63,1 metres
Length of Crest	309 metres
Volume of Concrete . .	54 650 m³
Capacity . .	1 332 million cubic metres
Surface Area .	9 105 hectares

Kyle Dam occupies a key position

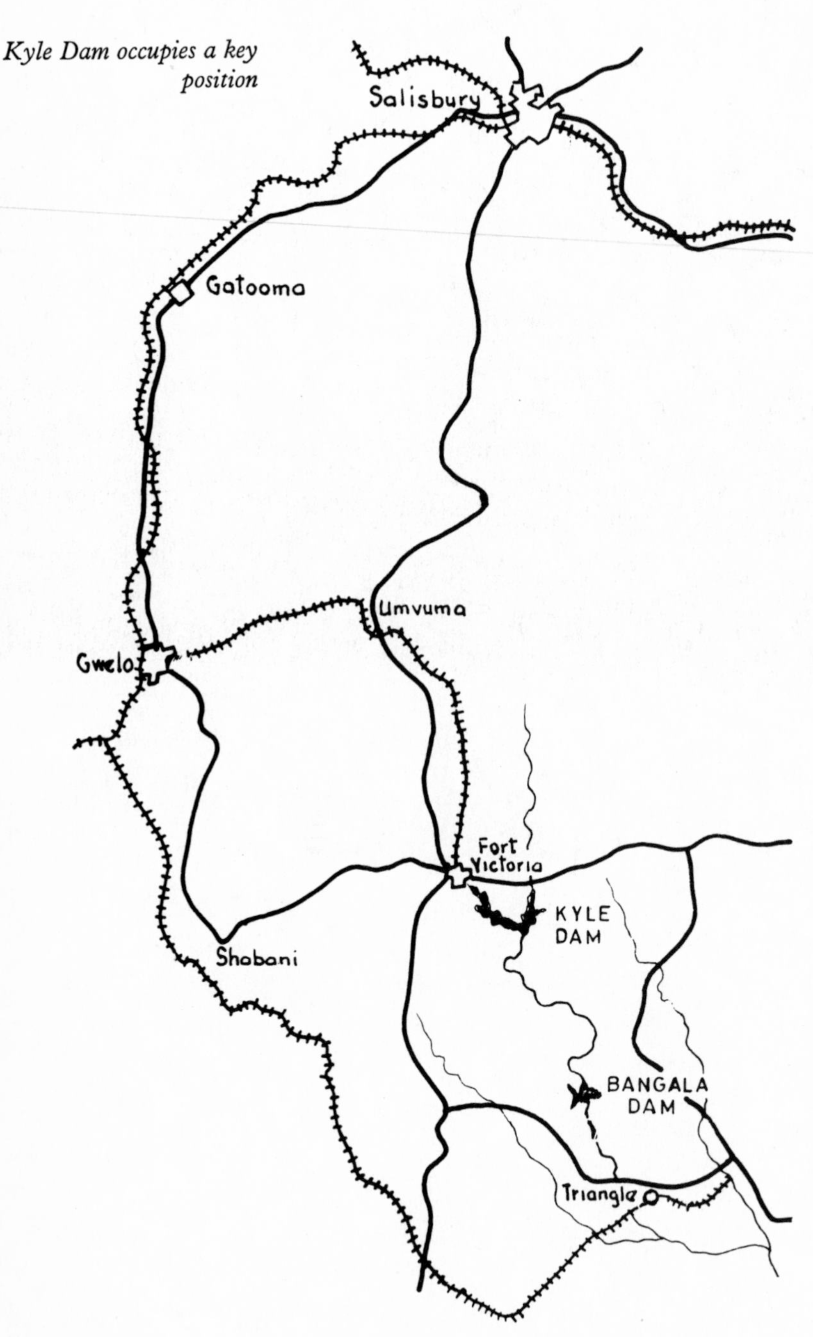

CONSTRUCTION

It is a thin arch of double curvature with a central overflow spillway section discharging on to a reinforced concrete mattress anchored on to solid rock.

The spillway section consists of 5 openings each 12,2 metres wide. The single lane roadway bridge over the spillway is constructed with precast prestressed beams.

The parapets are of precast construction.

The outlets consist of two 1 100 mm dia. pipes controlled on the downstream side by cone valves and on the upstream side by hydraulically operated emergency gates running in a tower structure which is independent from the dam wall.

The dam is divided into blocks approximately 12,2 metres long and the contraction joints are sealed into panels by copper sealing strips on the upstream and mild steel strips on the downstream side. The joints are provided with both primary grouting grooves and a secondary system of reinjectable grout valves.

Deflection targets are inset into the downstream face of the dam and deflections are measured geodetically by observations from beacons on the banks. Within the dam is a system of vibrating wire strain gauges which enable the stresses in the concrete to be calculated.

Water for irrigation is discharged down the river where it travels for a distance of some 97 kilometres before entering the canal system at Esquilingwe Weir.

CONSULTING ENGINEERS AND CONTRACTORS

Design by . . .	Coyne et Bellier
Supervision of Construction . .	Ministry of Water Development
Main Contractor .	McAlpine and Concor (Pvt) Ltd.
Hydraulic Equipment	Sorefame

General view of dam and lake

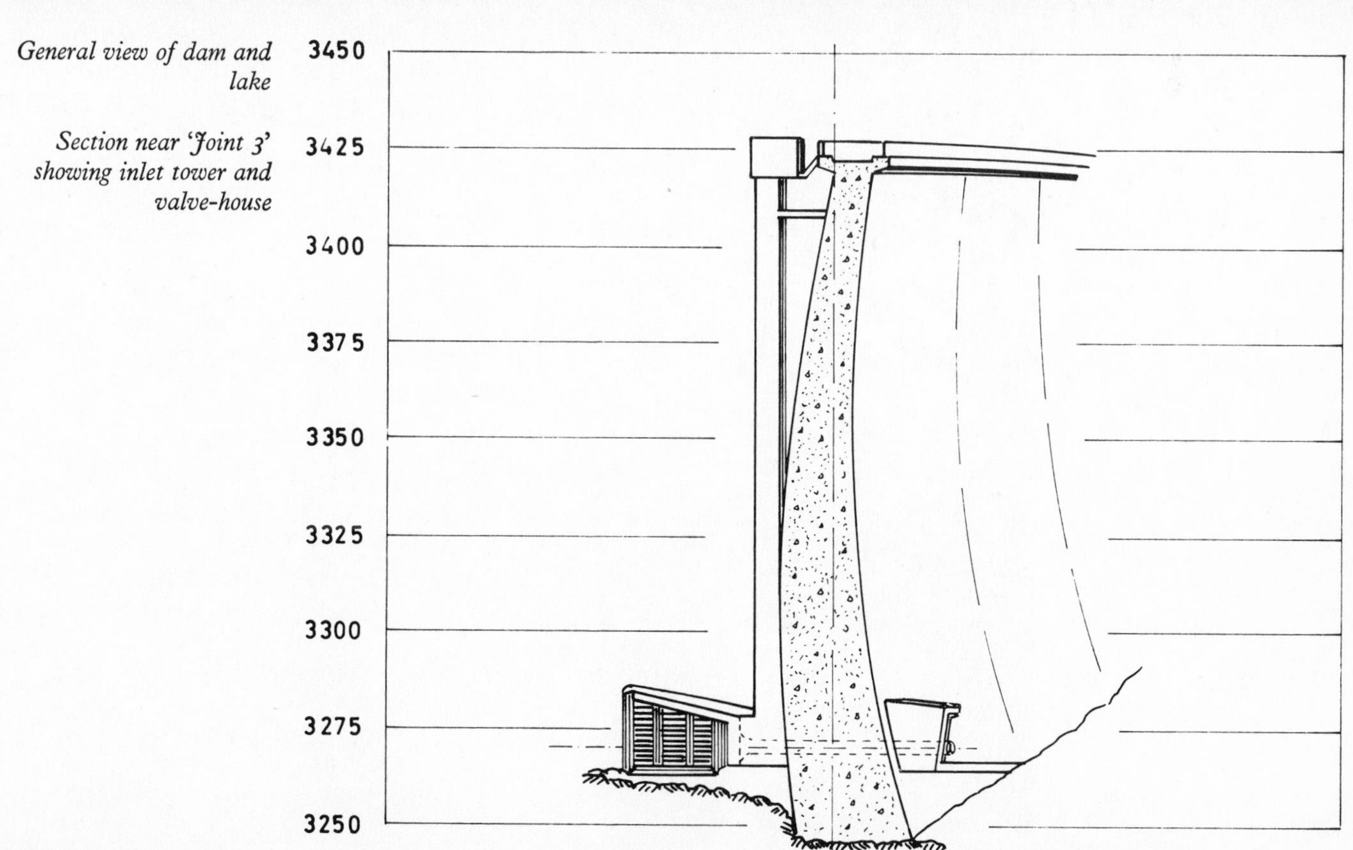

Section near 'Joint 3' showing inlet tower and valve-house

Kyle Dam takes its graceful shape

First spillage at Lake Kyle

LESAPI DAM

HISTORY

Lesapi Dam is part of the development of the great irrigation potential of the Sabi Valley. Stored water from the dam will be passed downstream to the pump station at the Middle Sabi Irrigation Scheme some 200 kilometres downstream. As the Sabi River irrigation complex is developed, larger dams will be built and water from Lesapi Dam will be made available to highveld irrigators. Rusape town will also draw their requirements from Lesapi Dam.

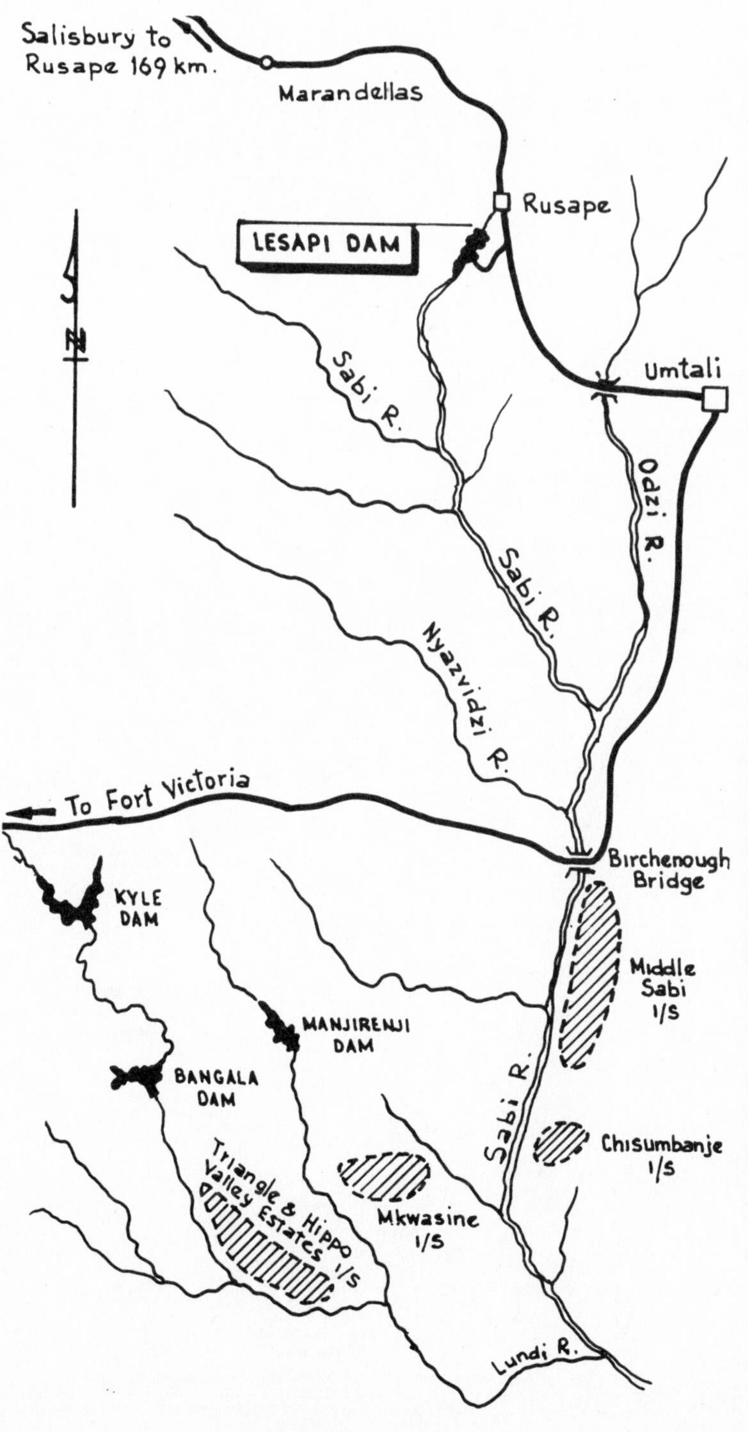

Lesapi Dam – construction method of rockfill cofferdam designed for overtopping.

Lesapi rockfill cofferdam successfully overtopped during construction.

Lesapi Dam, showing spillway gates.

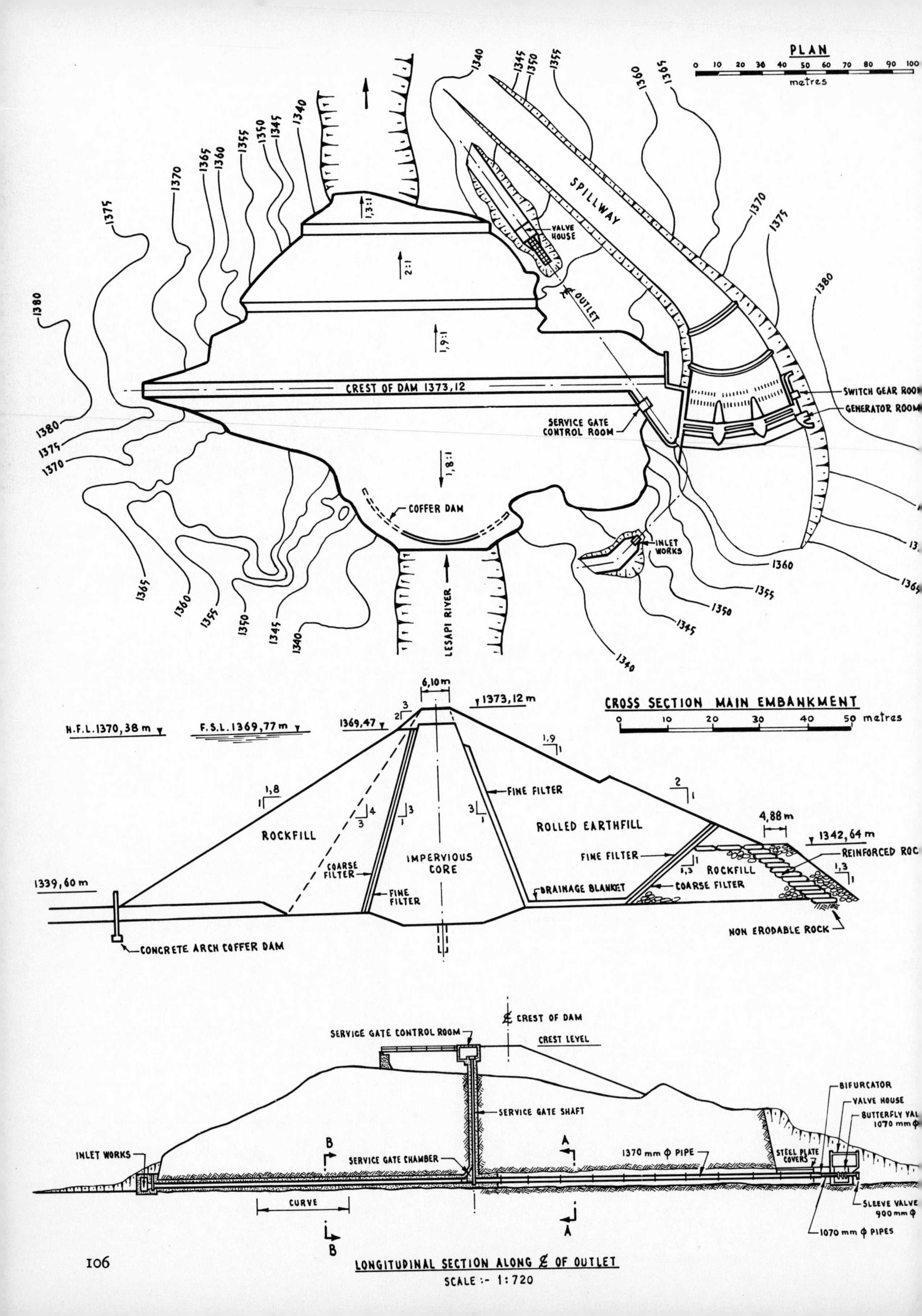
PLAN
0 10 20 30 40 50 60 70 80 90 100
metres
SPILLWAY
VALVE HOUSE
OUTLET
CREST OF DAM 1373,12
SWITCH GEAR ROOM
GENERATOR ROOM
SERVICE GATE CONTROL ROOM
COFFER DAM
INLET WORKS
LESAPI RIVER
1,3:1
2:1
1,9:1
1,8:1
CROSS SECTION MAIN EMBANKMENT
0 10 20 30 40 50 metres
H.F.L.1370,38 m
F.S.L.1369,77 m
1369,47
6,10 m
1373,12 m
ROCKFILL
COARSE FILTER
FINE FILTER
IMPERVIOUS CORE
ROLLED EARTHFILL
DRAINAGE BLANKET
ROCKFILL
4,88 m
1342,64 m
REINFORCED ROCK
NON ERODABLE ROCK
1339,60 m
CONCRETE ARCH COFFER DAM
CREST OF DAM
CREST LEVEL
SERVICE GATE CONTROL ROOM
SERVICE GATE SHAFT
INLET WORKS
SERVICE GATE CHAMBER
1370 mm ϕ PIPE
STEEL PLATE COVERS
BIFURCATOR
VALVE HOUSE
BUTTERFLY VALVE 1070 mm ϕ
SLEEVE VALVE 900 mm ϕ
1070 mm ϕ PIPES
CURVE

LONGITUDINAL SECTION ALONG ℄ OF OUTLET
SCALE :- 1:720

The dam has a central core of impervious soil supported by rockfill excavated from the spillway channel on the upstream side and rolled earthfill on the downstream side. The design incorporated a reinforced rockfill toe on the downstream side so that floods could be passed over the partially completed dam during construction. This was the first time this method of diversion had been used in Rhodesia.

Lesapi Dam and reservoir.

DIMENSIONS

Catchment area .	674 square kilometres
Length of wall .	240 metres
Maximum height of wall . .	41,2 metres
Maximum depth of water . .	36 metres
Capacity . .	68 million cubic metres
Surface area .	615 hectares
Design flood . .	1 470 cumecs

CONSTRUCTION

The blockey granite foundations necessitated extensive underground works. Cut-off works included core trench excavations up to 20 metres deep, adits, shafts and a heavy grouting pattern.

Irrigation demands require that all stored water can be discharged during the months August to November inclusive and the outlet works are designed to pass a maximum flow of 16 cubic metres per second. The outlets comprise a 160 m long tunnel through the right abutment with a 27,5 m deep service gate shaft. Flow is controlled by 2 No. 900 mm sleeve valves protected by 2 No. 1 050 mm butterfly valves and a service gate.

The chute spillway is designed to take 1 470 cumecs. The three automatic operating radial gates, 13,72 m wide by 6,10 m high, are powered by electricity with a generating set and also diesel operation as standby equipment. Because of the high base flow in the river it is possible to have the full supply level during the rains at a level of 0,6 metres below that for the dry season. This eases the problem of passing the maximum flood.

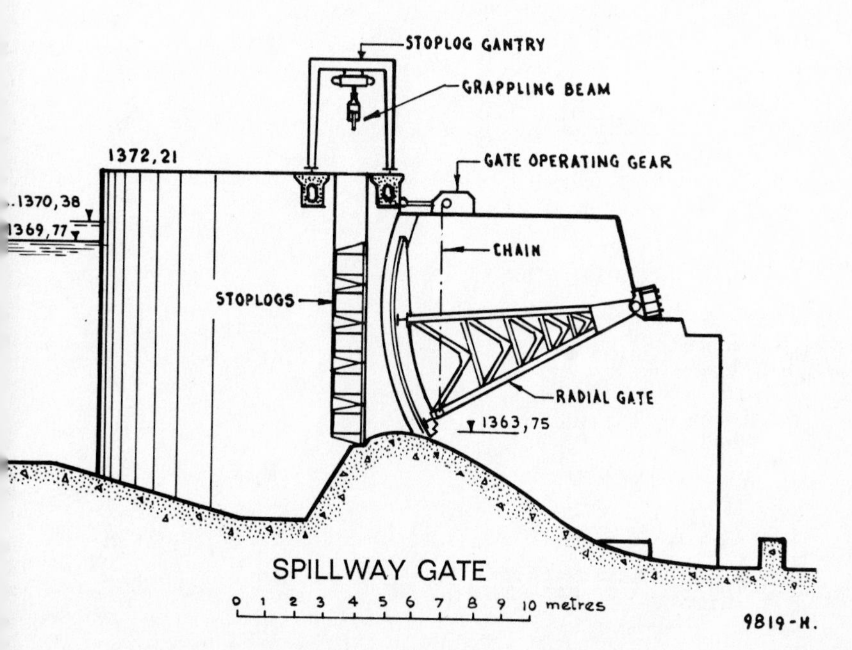

CONSULTING ENGINEERS AND CONTRACTORS

Main contract .	W. J. & R. L. Gulliver (Pvt) Ltd.
Outlet works, Foundation treatment, Exploratory work . . .	Cementation Company (Rhod.) Ltd.
Outlet pipes . .	O. Connolly and Co. (Byo.) (Pvt) Ltd.
Valves . . .	Steelmetals Ltd., Johannesburg
Supply and installation of Spillway Radial Gates . .	Sorefame

The design and supervision of construction of the dam was carried out by the Ministry of Water Development.

UPPER NCEMA DAM

HISTORY

Upper Ncema Dam was built by the City of Bulawayo to supplement its water augmentation scheme. Construction began in August 1971 and was completed in January 1974 at a cost of some S2 500 000. It is close to three other dams owned by the City of Bulawayo – the Lower Ncema Dam, Inyankuni Dam and Umzingwane Dam, all situated in the Essexvale district some 50 km from Bulawayo.

It was originally planned to raise the Inyankuni full supply level by 9 m to bring it onto the same full supply level as the Lower Ncema Dam, and to connect the two dams by means of a tunnel, thereby storing the excess Ncema yield of $45 \times 10^6 \times m^3$ in the Inyankuni basin. However, when the estimate of tunnel costs was made, it was found to be more economical to construct the Upper Ncema Dam, and store the excess catchment yield there.

Upper Ncema Dam is sited 3 km upstream of the Lower Ncema Dam – its prime function being to regulate the stream flow into the lower dam, from where the water is pumped to the Council's purification works closer to Bulawayo.

Upper Ncema Dam stores excess catchment yield

The capacity of the lower dam is $18 \times 10^6 \times m^3$, but the basin provides a very low evaporation rate; at full supply, the lower dam would have a similar evaporation rate to the upper dam at 1/5th its full capacity.

DIMENSIONS

Catchment Area	620 km^2
Capacity	$45 \times 10^6 m^3$
M.A.R.	$52,7 \times 10^6 m^3$
Type of dam	Mass concrete gravity dam
Crest length	232 m
Depth of water	32 m
Maximum height of structure	38 m
Maximum incoming flood	2 850 m^3/s
Maximum outgoing flood	1 550 m^3/s
Flood rise	4 m
Surface Area at f.s.l.	800 hectare
Volume Concrete	68 000 m^3

CONSULTING ENGINEER AND CONTRACTOR

Civil Contractor Andrews & Kidd

The design and supervision of construction of the dam was carried out by Messrs. Watermeyer, Legge, Piesold and Uhlmann.

BANGALA DAM

Bangala Dam has a remarkably high crest length/height ratio

HISTORY

Bangala Dam was constructed to store water from the intermediate catchment below Kyle Dam. Although not of large capacity, it has a high yield when operated in conjunction with Kyle Dam.

The dam was built in record time, design being started in about April 1961, and the construction was complete by December 1962, with the dam full, and passing heavy floods at that time.

DIMENSIONS

Catchment Area .	5 828 sq. kilometres
Yield (10%) . .	4 134 litres/sec.
Design flood . .	4 670 cubic metres/sec.
Height of Dam . .	50,7 metres
Length of Crest .	396 metres
Volume of Concrete .	87 900 cubic metres
Capacity . . .	130 million m³
Surface Area . .	1 133 hectares

CONSTRUCTION

The design is simple, being a constant thickness, constant radius cylindrical arch. The upstream radius is 167 metres and the wall apart from the crest is 6,70 metres thick.

The spillway is approximately 140 metres long and consists of an ogee crest with a heavily cantilevered ski jump discharge.

The dam is remarkable for its very high crest length/height ratio of about 7,7:1.

There is no bridge over the spillway but a gallery runs the whole length within the cantilevered overhang to provide access between the banks.

Forming the massive spillway overhang was a difficult work achieved by a system of suspended precast beams with precast slabs spanning between them.

The outlets and hydraulic equipment are identical to those on Kyle Dam except in the details of arrangement. They consist of two

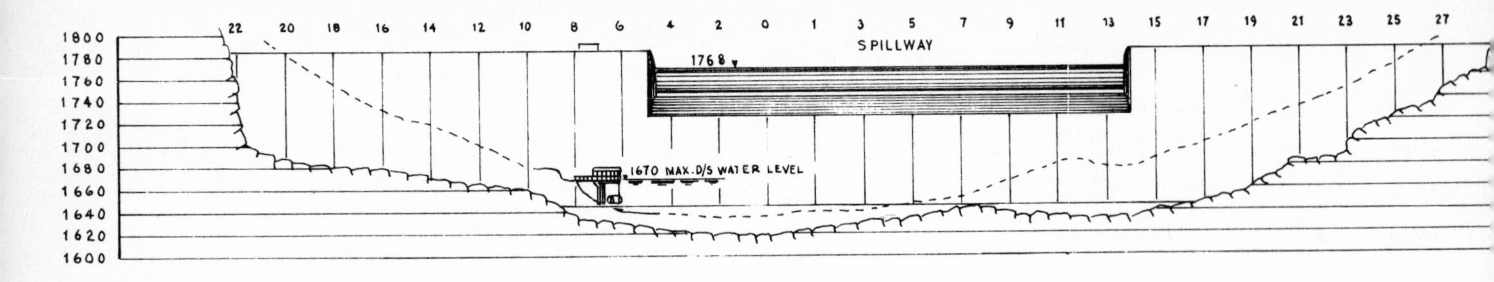

DOWNSTREAM ELEVATION
SCALE 1:1500

Salisbury
Gatooma
Umvuma
Gwelo
Fort Victoria
KYLE DAM
Shabani
BANGALA DAM
Triangle

Scale 1:2,500,000

0 10 20 30 40 50 60 70 80 90 100 Miles

1 100 mm pipes controlled by cone valves and emergency gates.

The dam is divided into blocks approximately 14 metres long and contraction joints are sealed into panels with copper sealing strips on the upstream side and mild steel strips on the downstream side. The joints are provided with both primary grouting grooves and a secondary system of reinjectable grout valves.

Deflection targets are inset into the downstream face of the dam and deflections are measured geodetically by observations from beacons on the banks. Within the dam is a system of vibrating wire strain gauges which enable the stresses in the dam to be calculated.

CONSULTING ENGINEERS AND CONTRACTORS

Design by . .	Coyne et Bellier (Paris)
Supervision of construction .	Ministry of Water Development
Main Contractor	McAlpine and Concor (Pvt) Ltd.
Hydraulic equipment .	Sorefame

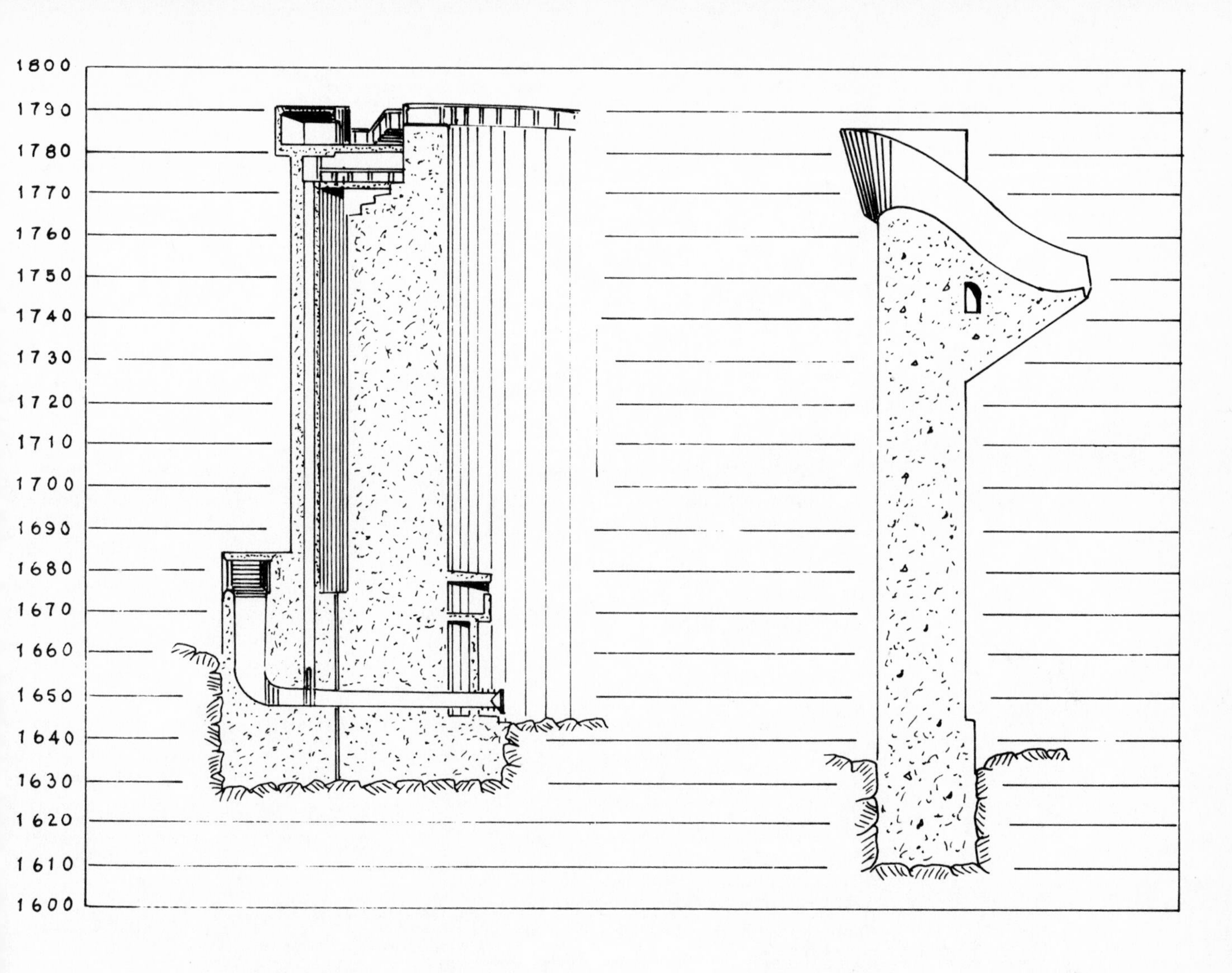

SECTION THROUGH OUTLET

SECTION THROUGH SPILLWAY

SCALE 1 INCH TO 40 FEET

DARWENDALE DAM

HISTORY

Darwendale Dam, which was constructed to satisfy the increasing demands of Greater Salisbury, is situated 56 kilometres west of Salisbury where the Hunyani River cuts through the Great Dyke. It is thirty kilometres downstream from Lake McIlwaine on a direct line, but when it fills, the stored water will back up under the Hunyani bridge on the main Bulawayo Road.

For a dam with a height of only 27,5 metres, it has a fantastic capacity of 490 million cubic metres and a surface area of 8 100 hectares. The excellence of its basin is well illustrated by comparing it with its sister dam Lake McIlwaine which is 36,5 metres high, has a capacity of 250 million cubic metres and a surface area of 2 630 hectares.

DIMENSIONS

Crest level	1 346,0 m
High Flood level	1 342,5 m
Full supply level	1 341,1 m
Maximum height of dam	27,5 m
Maximum depth of water	22,6 m
Length of wall at crest level	1 200 m
Crest width	6,55 m
Maximum inflow flood	4 200 m³/s
Maximum outflow flood	2 190 m³/s
Catchment area . Total	3 792 sq. km.
Catchment area – below McIlwaine	1 590 sq. km.
Capacity	490 million m³
Surface area	8 100 hectares
Yield	3 400 litres/s
Spillway 3 Gates . Each	13,71 m wide 6,10 m high

CONSTRUCTION

To pass the floods three automatically operating spillway radial gates were installed. Each of these is 13,7 m wide and 6,1 m high. The Contractor started work in 1973 and the dam was substantially completed during 1976.

CONSULTING ENGINEERS AND CONTRACTORS

Design and Supervision of Construction	Ministry of Water Development
Earthworks Construction	W. J. & R. L. Gulliver (Pvt) Ltd. (Main Contractor)
Concrete	Grinaker Construction (Rhod.) (Pvt) Ltd. (Sub-Contractor)
Gates	Steel Metals (Pvt) Ltd.

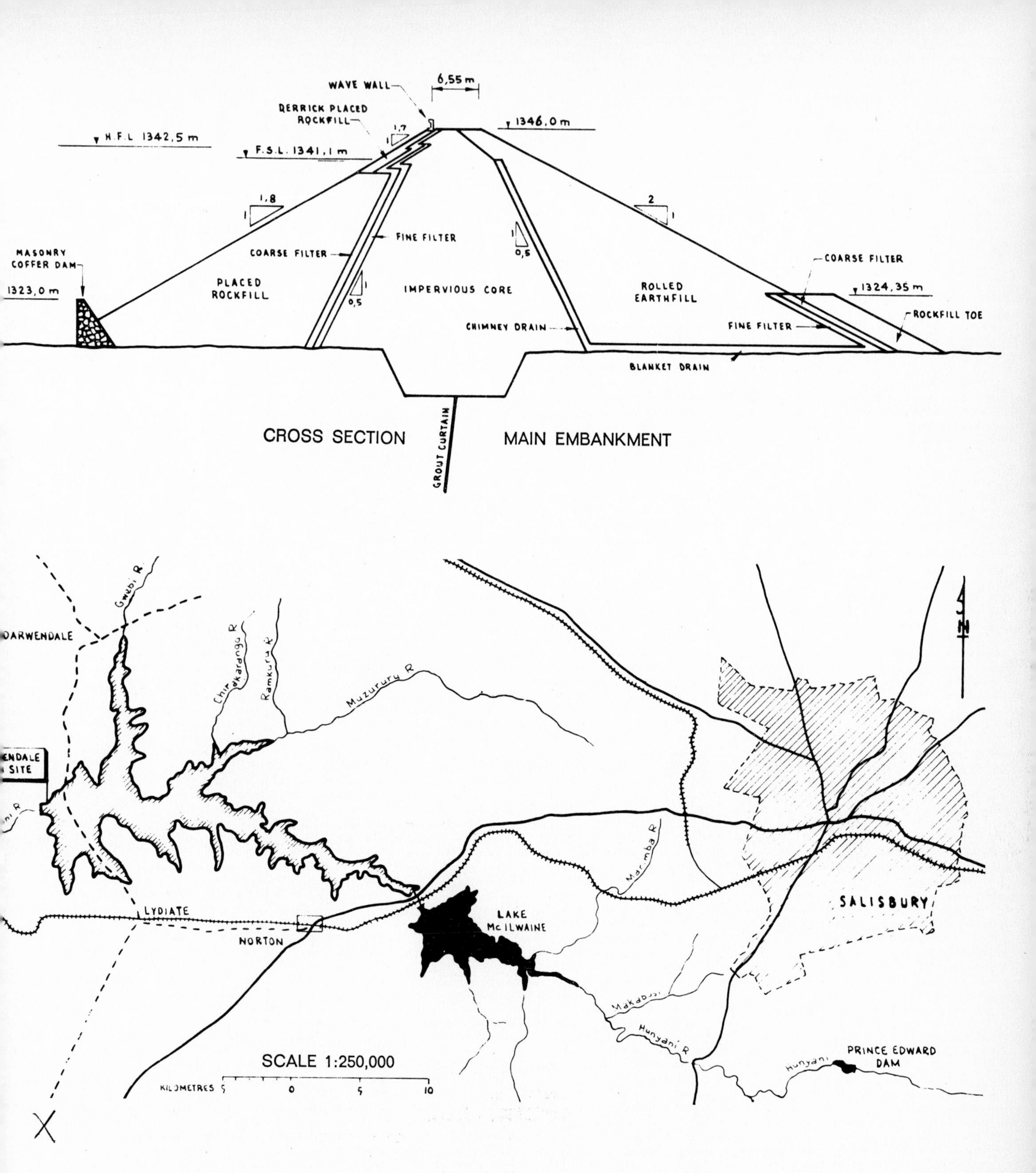
WAVE WALL
6,55 m
DERRICK PLACED ROCKFILL
1346,0 m
H.F.L 1342,5 m
F.S.L. 1341,1 m
1,7
1,8
2
MASONRY COFFER DAM
COARSE FILTER
FINE FILTER
0,5
COARSE FILTER
1323,0 m
PLACED ROCKFILL
IMPERVIOUS CORE
ROLLED EARTHFILL
1324,35 m
CHIMNEY DRAIN
FINE FILTER
ROCKFILL TOE
BLANKET DRAIN
CROSS SECTION
MAIN EMBANKMENT
GROUT CURTAIN
DARWENDALE
Gwebi R
Chirakaranga R
Ramkuru R
Muzururu R
LYDIATE
NORTON
LAKE McILWAINE
Marimba R
SALISBURY
Makabusi
Hunyani R
Hunyani
PRINCE EDWARD DAM
SCALE 1:250,000
KILOMETRES 5 0 5 10
N

HUNYANI POORT DAM (LAKE McILWAINE)

This dam on the Hunyani River is the major source of water for the city of Salisbury. Originally a smaller dam was proposed on this site for this purpose only, and so the Rhodesian Government took over the site and built a larger multi-purpose dam for both urban use and irrigation. Another dam, Darwendale Dam, has been built some 25 km downstream for Salisbury. Advantage will be taken of the more favourable geographic position of Hunyani Poort Dam to Salisbury to supply additional water from this dam for the city, the equivalent irrigation commitment being supplied from Darwendale Dam.

Initially the Salisbury City Council were unhappy with the proposal to build an earth dam rather than a concrete dam. Accordingly an engineer from the U.S. Bureau of Reclamation was commissioned to examine the design proposed by the Ministry for Water Development. This engineer subsequently reported favourably on the proposals and the construction of Hunyani Poort Dam as an earth dam went ahead and was completed in 1952.

DIMENSIONS

Catchment area .	2 220 km^2
Height above lowest foundation .	40 m
Length of crest .	341 m
Volume . . .	780 $\times$ 10^{3}m^3
Gross capacity .	250 $\times$ 10^{6}m^3
Spillway maximum discharge capacity . .	2 150 m^3/sec.
Type of spillway .	Uncontrolled

CONSULTING ENGINEERS AND CONTRACTOR

Engineering by .	Ministry of Water Development, Rhodesian Government
Construction by .	Clifford Harris Limited.

Water hyacinth infestation on Lake McIlwaine

MANJIRENJI DAM

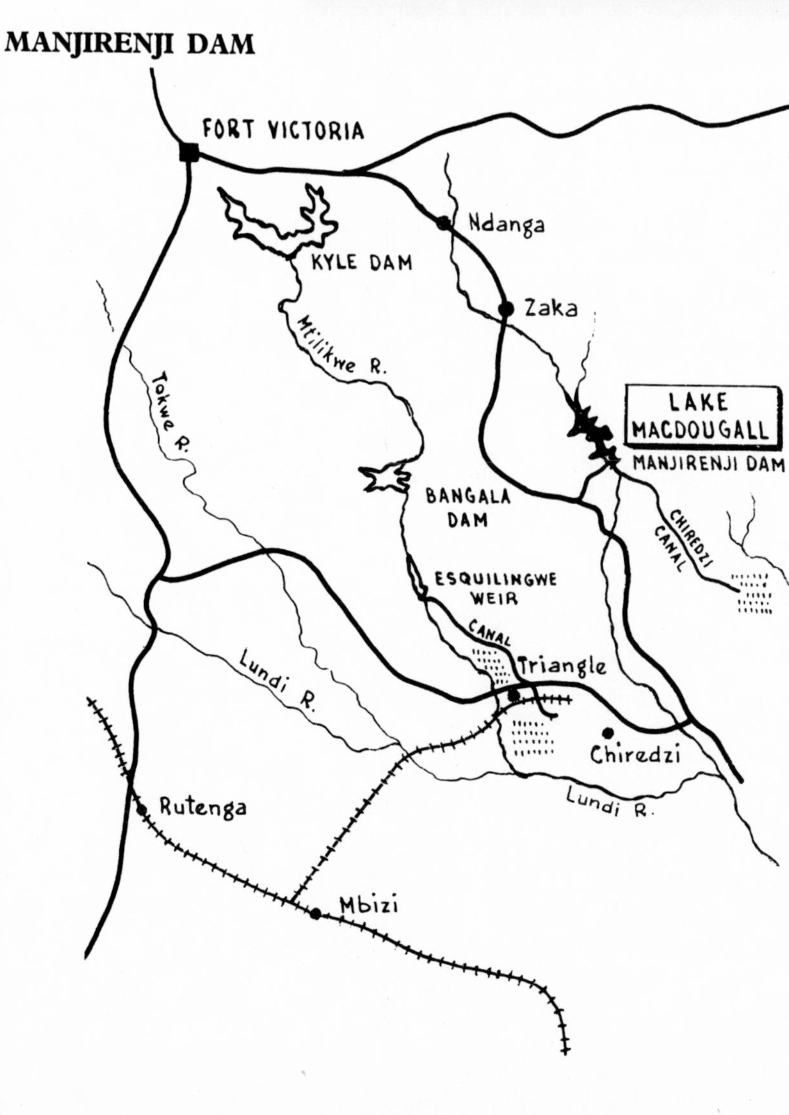

LOCALITY PLAN

HISTORY

Construction of the Manjirenji Dam and Chiredzi Canal was started early in 1964. It was sufficiently far advanced to impound the 1965–1966 flood flow and water was first delivered down the canal in April 1966. The spillway construction and finishing works were completed later in 1966.

The scheme was originally designed for the irrigation of 15 000 acres (6 069 ha) of cane on the proposed Nandi Sugar Estates. With the fall in world prices of sugar the concept of the scheme was changed and under the direction of the Sabi-Limpopo authority the first stage of the Mkwasine wheat scheme was developed early in 1966 to make use of the water provided by the dam.

DIMENSIONS

Lake MacDougall

Catchment area . 593 sq. miles (1 536 sq. km.)

Capacity . 230 000 acre ft. = 62 400 m.g. (283 m³ × 10⁶)

Surface Area . 5 000 acres (2 023 ha)

Chiredzi Canal

Capacity . 210 cusecs (6 m³/s)

Length . 23 miles (37 km)

Manjirenji Dam

Yield . . 145 cusecs (4 m³/s)

Design Flood . 110 000 cusecs (4 119 m³/s)

Height of . dam. . 170 feet (53 m)

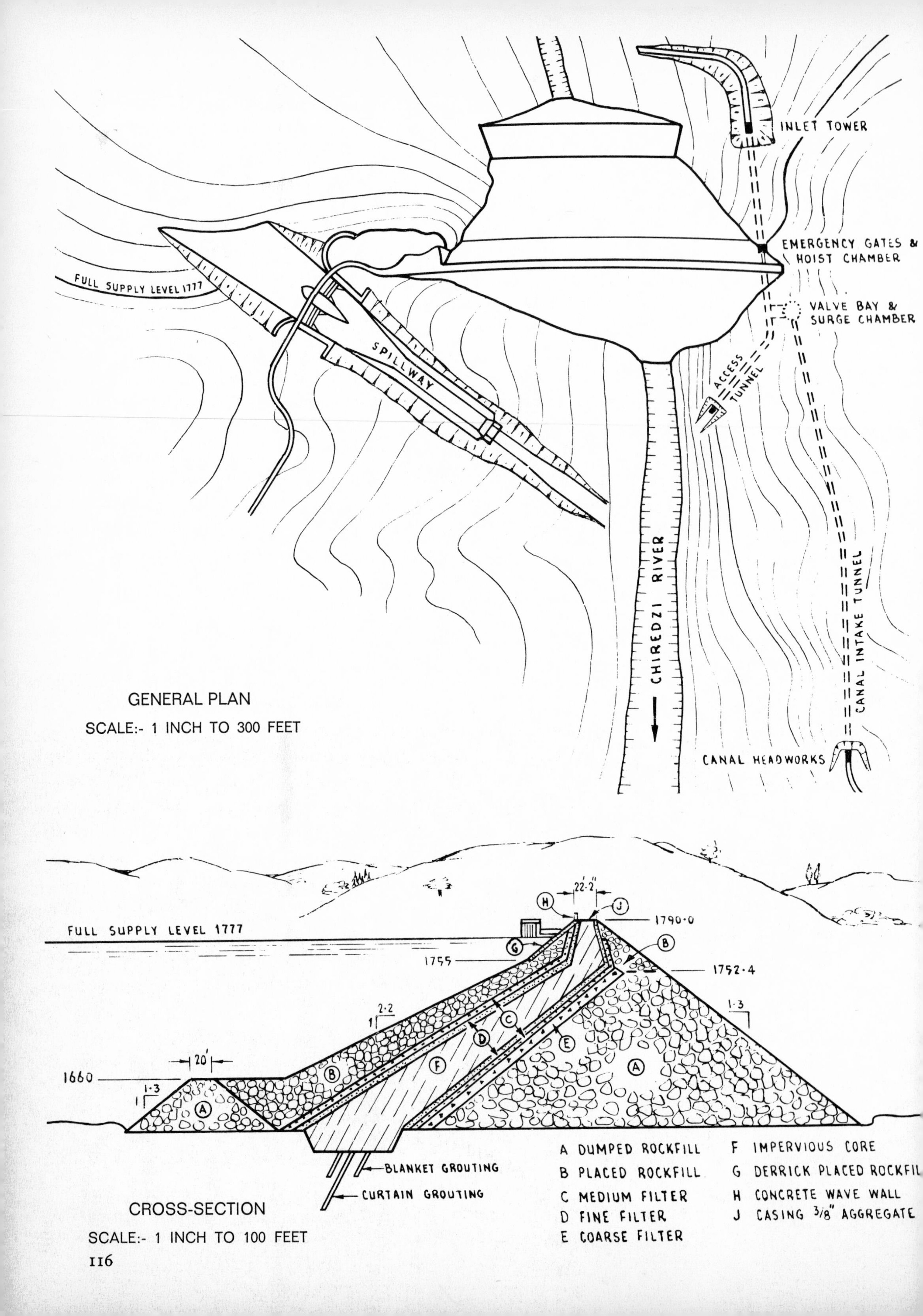
INLET TOWER
EMERGENCY GATES & HOIST CHAMBER
VALVE BAY & SURGE CHAMBER
ACCESS TUNNEL
FULL SUPPLY LEVEL 1777
SPILLWAY
CHIREDZI RIVER
CANAL INTAKE TUNNEL
CANAL HEADWORKS
GENERAL PLAN
SCALE:- 1 INCH TO 300 FEET
22·2"
1790·0
FULL SUPPLY LEVEL 1777
1755
1752·4
2·2
1·3
20'
1660
A DUMPED ROCKFILL
B PLACED ROCKFILL
C MEDIUM FILTER
D FINE FILTER
E COARSE FILTER
F IMPERVIOUS CORE
G DERRICK PLACED ROCKFILL
H CONCRETE WAVE WALL
J CASING 3/8" AGGREGATE
BLANKET GROUTING
CURTAIN GROUTING
CROSS-SECTION
SCALE:- 1 INCH TO 100 FEET

Length of Embankment .	1 035 feet (323 m)
Volume of Embankment .	862 000 cu. yds. (659 000 m³)

CONSTRUCTION

The dam is the largest rockfill dam in Rhodesia. The rockfill, consisting of large quarried boulders dumped into position, is sluiced by a high powered jet of water to consolidate it. Lying against the rock is the sloping clay core which forms an impervious membrane which in turn is covered by a further layer of placed vibrated rock.

The spillway with its concrete lined chute controlled by two large steel radial gates, each 50 foot (15,6 m) wide and 34 foot (10,6 m) high, is the first of its type in Rhodesia.

The gates are automatically controlled by floats housed in the gate piers. As the level in the dam rises due to a flood so the gates open automatically and water is discharged down the chute. The bigger the inflow into the dam the greater the opening of the gates, which, at full opening, will discharge about 90 000 cubic feet of water per second. The Chiredzi Canal is aligned to quickly reach the watershed between the Chiredzi and the Mkwasine Rivers. Designed to carry a flow of up to 210 cubic feet of water per second (6 m³/s) it follows the watershed to the irrigation lands, a distance of about 23 miles (37 km.)

CONSULTING ENGINEERS AND CONTRACTORS

Main contract for dam . . .	Sir Alfred MacAlpine & Son (Rhod.) Pvt. Ltd.
Preliminary contract and grouting . . .	The Cementation Co. (Rhod.) Ltd.
Spillway concrete sub-contract .	Roberts Construction Co. (Rhod.) Ltd.
Spillway gates .	Rheinstahl Union Bruckenbau A. G.
Outlet emergency gates . . .	Edward Bateman Ltd.
Painting of gates	Thermal & Corrosion Contractors (Pvt.) Ltd.
Tunnel piping .	Hume Pipe Co. (Rhod.) Ltd.
Butterfly valves & recorders .	Stewarts & Lloyds of Rhodesia Ltd.
Generating plant	Hawker Siddley Brush (Pvt.) Ltd.
Transport . .	Hooper & Stopforth (Pvt.) Ltd. Dessingtons

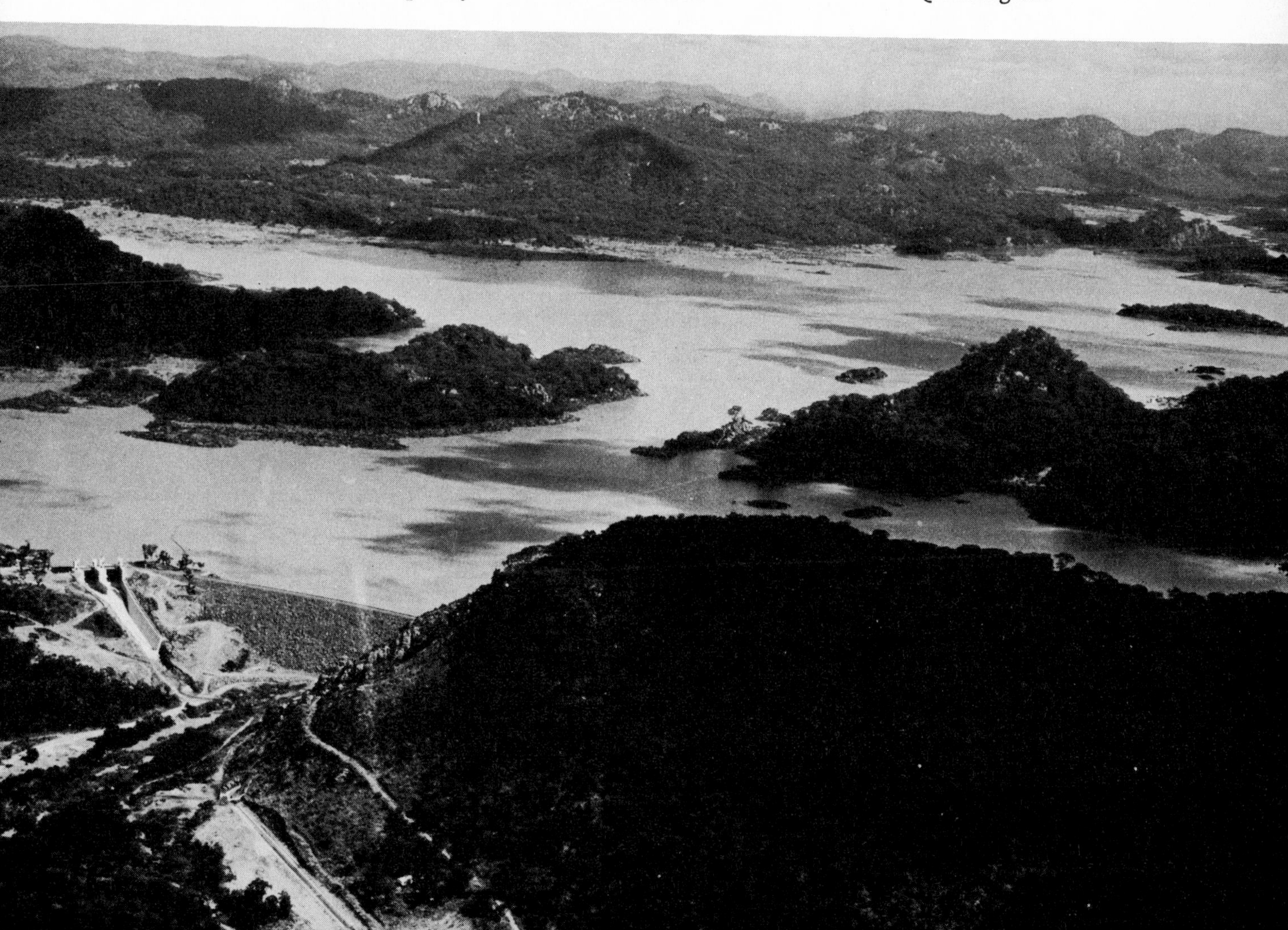

Manjirenji Dam, showing the rockfill wall and spillway gates (Ministry of Information, Salisbury, Rhodesia)

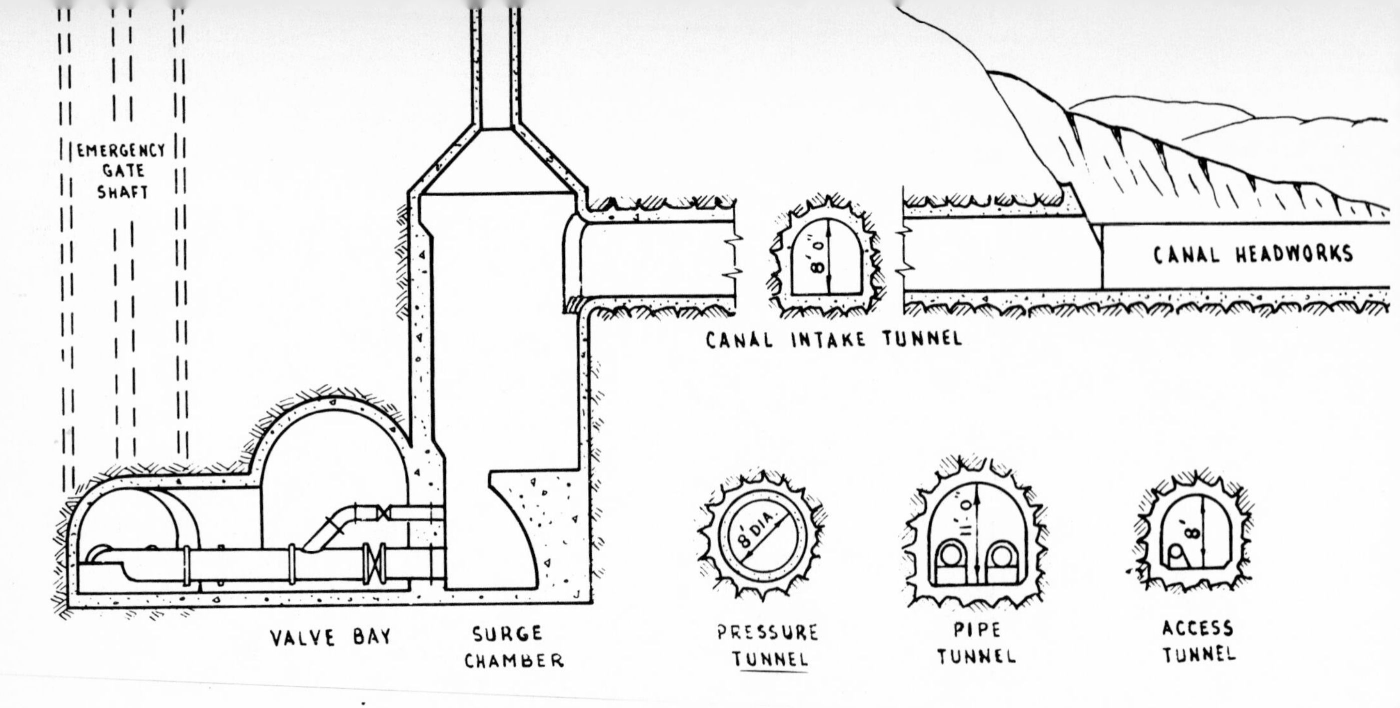

CANAL INTAKE WORKS
SCALE:- 1 INCH TO 20 FEET

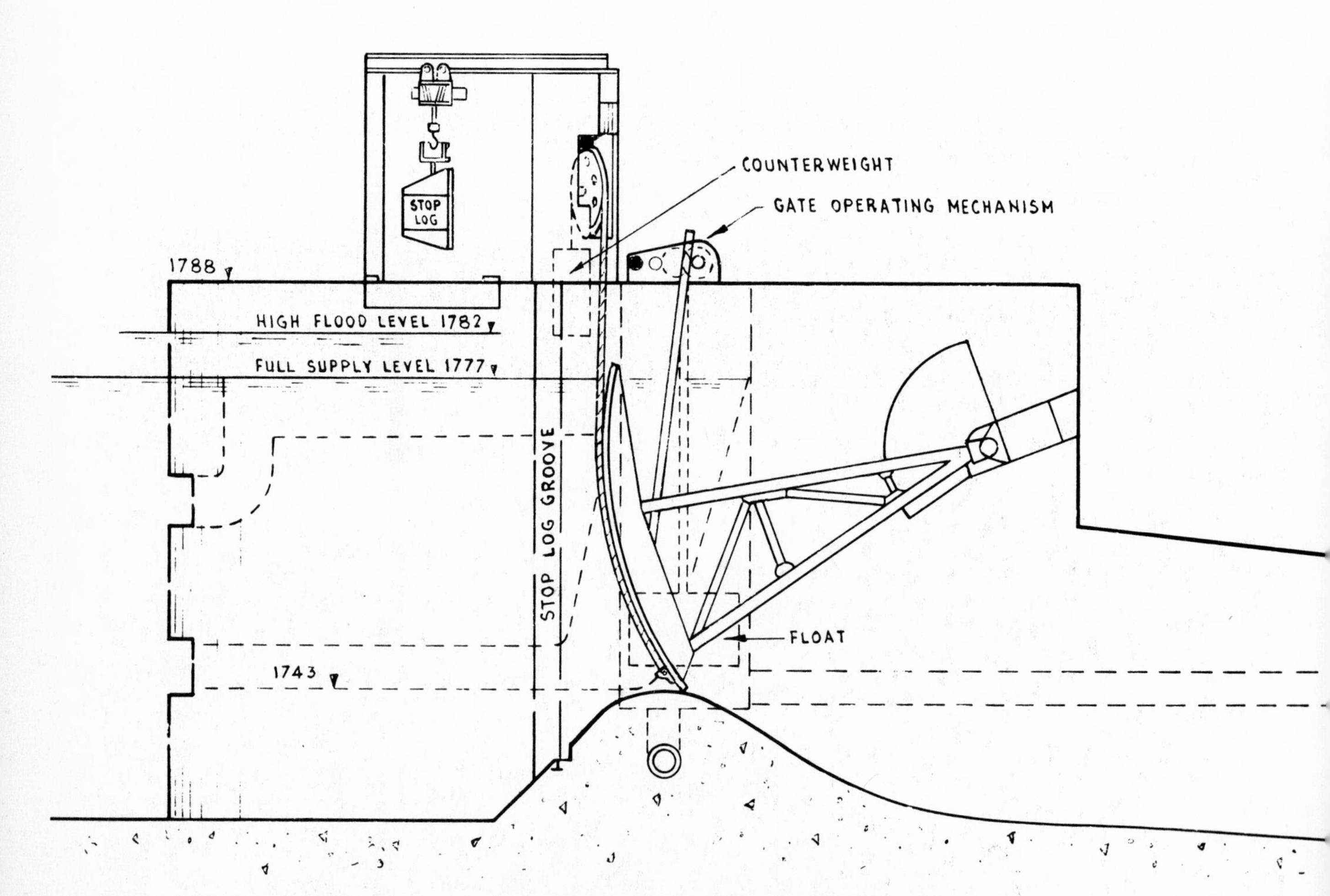

SPILLWAY HEADWORKS
SHOWING RADIAL GATES & STOPLOGS
SCALE:- 1 INCH TO 20 FEET

MANJIRENJI DAM

TYPICAL CANAL CROSS-SECTION
ON IDEAL GROUND
SCALE:- 1 INCH TO 20 FEET

Fuels . . .	Shell (Rhod.) (Pvt.) Ltd.	*Quarrying & crushing contract* . . .	Bulawayo Blasting Co. Ltd.
Spillway Bridge & Steelwork .	Stevens & Dawnays Ltd.		
Main contract Chiredzi Canal	Roberts Construction Co. (Rhod.) Ltd.	*Canal earthworks sub-contract* .	W. J. & R. L. Gulliver (Pvt.) Ltd.

GWENORO DAM

Gwenoro Dam – fascinating geological formations

HISTORY

Gwenoro Dam is located on the headwaters of the Lundi River and is the principal source of water for the City of Gwelo. The dam was completed in 1958, and in the years before Gwelo required the total yield from the dam it provided emergency relief supplies in times of drought to Shabani town and other desperate needs as far as 250 kilometres downstream.

DIMENSIONS

Catchment area . .	411 km^2
Capacity . . .	32 × $10^6 m^3$
Surface Area . .	465 ha
Yield (4%) . . .	13,7 × $10^6 m^3$/annum
Length of wall . .	1 380 m
Maximum height of wall	30 m
Volume of embankments	643 000 m^3
Spillway capacity .	1 130 m^3/s

CONSTRUCTION

The dam is an earth-rock structure with a central impervious earth core. The dominant feature of the dam is the spillway which has been cut through a bald granite kopje on the right bank. The rock is very competent and the spillway is unlined, energy being dissipated in a hydraulic jump stilling post before water is returned to the river.

The fact that the engineer considered that it was unnecessary to line the spillway cutting, has provided the geologist with a fascinating 3 dimensional section in which he can recognise no less than 8 different geological formations.

The outlet works consist of an intake tower with a 92 m long reinforced concrete conduit housing a 533 m and a 610 m steel outlet pipe. The valve house is at the downstream toe of the dam.

CONSULTING ENGINEERS AND CONTRACTORS

Main civil engineering contract .	Impresit (Pvt) Ltd.
Structural steel work . . .	Dorman Long (Pty) Ltd.
Outlet pipes . .	Hume (Rhodesia) Ltd.
Outlet valves . .	Stewarts & Lloyds of Rhodesia Ltd.

The design and supervision of construction of the dam was carried out by the Ministry of Water Development.

INGWEZI IRRIGATION PROJECT

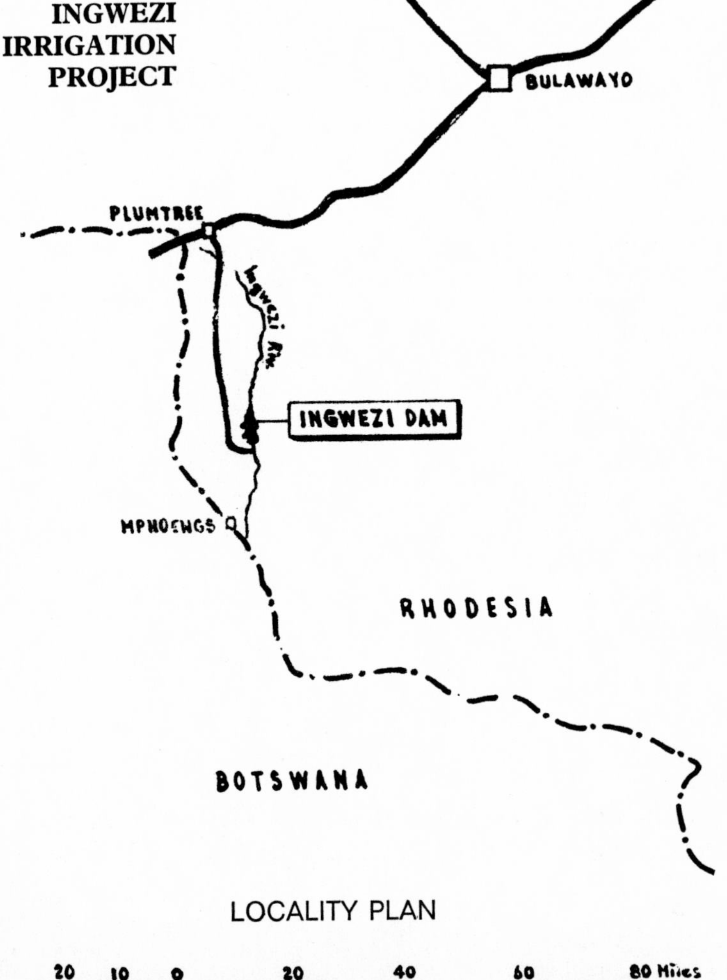

LOCALITY PLAN

HISTORY

The Ingwezi Irrigation Project, which is the largest work to be implemented by the Matabeleland Development Council under the £3 million drought relief scheme, comprises the Ingwesi Dam and Canal for the irrigation of 1 600 acres (647 ha) in the M'Phoengs tribal trust area.

The dam and canal were designed by members of the designs branch of the Ministry of Water Development, whose staff also supervised the construction of the works.

DIMENSIONS

Catchment Area	327 sq. miles (847 km^2)
Capacity . . .	56 500 acre feet (70 m^3 × 10^6)
Surface area . .	2 100 acres (850 ha)
Yield	12,6 cusecs (0,35 m^3/s)
Design Flood . .	60 000 cusecs (1 680 m^3/s)
Max. Height of Dam	130 feet (40 m)
Max. Depth of Water	117,5 feet (36 m)
Length of Embankment . .	900 feet (274 m)
Volume of Embankment . .	938 000 cubic yards (718 000 m^3)
Canal capacity . .	30 cusecs (0,85 m^3/s)
Canal length . . .	12,5 miles (20 km)

Canal grades vary from 1:500 to 1:4 000.

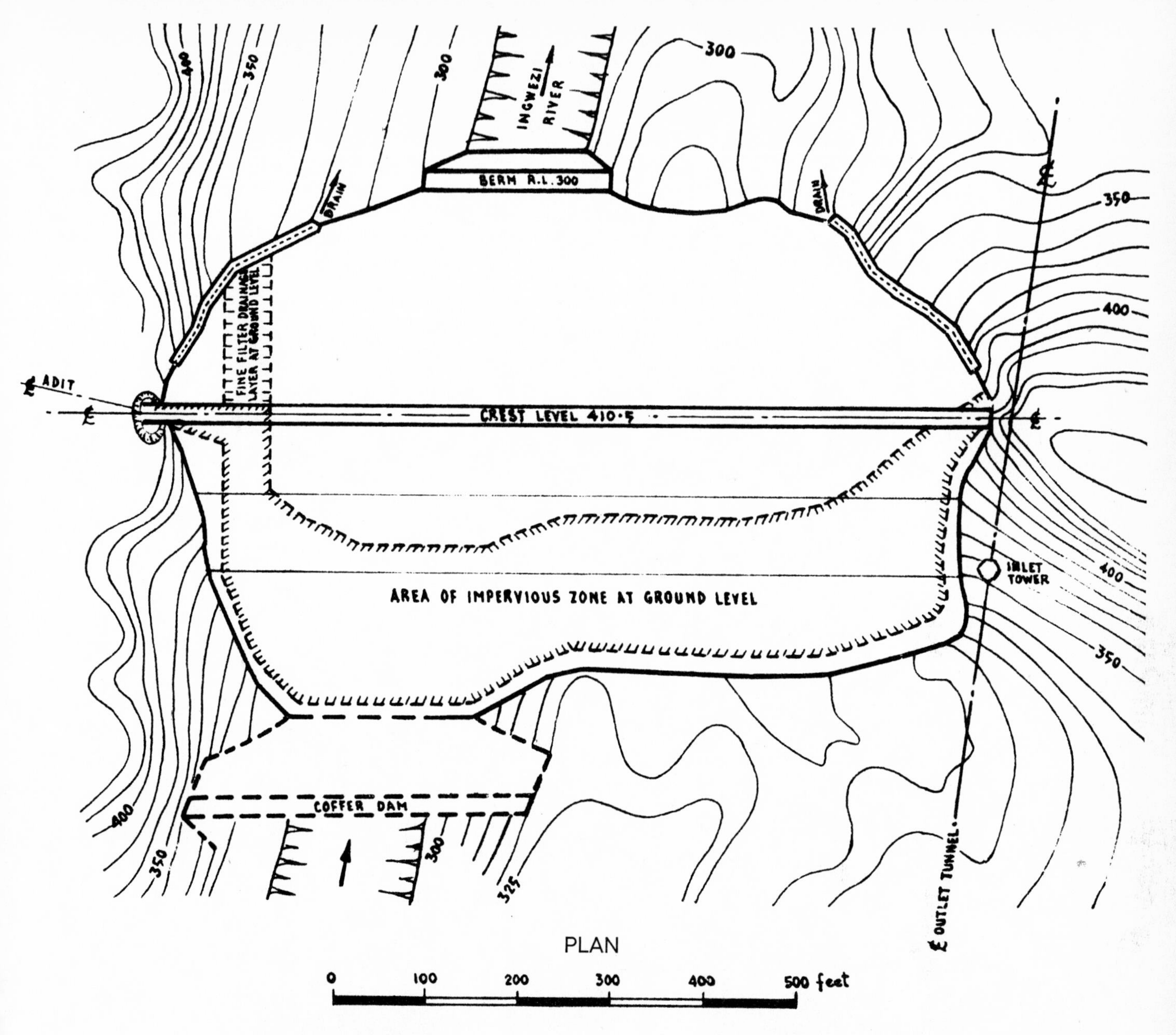

CONSTRUCTION

The dam is the largest in terms of capacity in Matabeleland. It is an earthfill structure with a sloping upstream impervious core resting against a sand filter-drain which is supported by the downstream shell. The embankment is protected from erosion by a thin layer of crushed rock on the downstream face and quarried rock and filters on the upstream side. The spillway is a gabion-protected structure on the left bank.

Water is discharged into the canal through a right bank tunnel with control valves on the upstream side in a dry tower. The valves are protected by service gates.

The parabolic concrete lined canal, some 12½ miles (20 km) long, has a capacity of 30 cusecs (0,85 m^3/s). There are three inverted syphons totalling 1 900 feet (579 m) which consist of 33 inch (838 mm) diameter concrete pipes. A 70 acre-ft (86 000 m^3) earthfill balancing dam has been constructed at the tail of the canal to simplify control of the irrigation water.

Construction work on the dam commenced in September 1966, and the structure was complete by March 1968. The canal contract was let in June 1967 and the first water was discharged down the canal in August, 1968.

CONSULTING ENGINEERS AND CONTRACTORS

DAM

Main contract	W. J. & R. L. Gulliver (Pvt) Ltd.
Foundation treatment and outlet works.	Cementation Company (Rhod.) Ltd.
	Founders Steel Industries

FULL SUPPLY LEVEL 398·0
410·5
20'
380·0
350·0
2·3
2·7
3
2
1·3
3" CRUSHER RUN AGGREGATE
PLACED ROCKFILL
COARSE FILTER
FINE FILTER
ROLLED EARTHFILL ZONE
PLACED ROCKFILL TOE
COFFER DAM
IMPERVIOUS ZONE
FINE FILTER
RIVER SAND COMPACTED IN PLACE

MAIN EMBANKMENT CROSS-SECTION
SCALE:- 1 INCH TO 100 FEET

F. Issels & Sons Ltd.
Taylor, Lourens & Haigh (Pvt) Ltd.
Shell (Rhod.) (Pvt.) Ltd.

CANAL
Main contract

Lewis Construction Co. (Rhod.) Ltd.
East Hunyani Earthmovers (Rhod.) Ltd.
McCullough & Naisbitt
Hume (Rhod.) Ltd.
African Gate & Fenceworks (Rhod.) (Pvt) Ltd.
J. B. Grist (Pvt) Ltd.
Modern Engineering (Pvt) Ltd.

Ingwezi Dam – largest capacity in Matabeleland

CHAPTER 8 Angola/S.W.Africa

KUNENE RIVER PROJECTS

HISTORY

The territory of South West Africa is a country of extremes. It covers more than 80×10^6 hectares and since the rainfall is generally low and inconsistent there are no perennial rivers flowing through this country. However, South West Africa does share two rivers as a common northern boundary with Angola, namely the Kunene River and the Okavango River. The Kunene River rises in the central highlands of Angola, near Nova Lisboa where the annual rainfall exceeds one and a half metres resulting in an estimated annual run-off of $5 \times 10^9 m^3$.

The Odendaal Commission, which was set up in 1962 to investigate the report upon the potential of South West Africa and the measures which could be taken to stimulate its rate of development to the advantage of all population groups, issued its report in 1964. Not only did the Commission emphasise the need to increase greatly the efforts already being made in the direction of conserving water supplies, building road links etc., but it also came to the important conclusion that the utilisation of the waters of the Kunene River for the generation of electric power could provide a substantial and economic contribution towards the increased and accelerated development of South West Africa.

The South African Government accepted practically all the recommendations of this Commission and, in particular, agreed to the Industrial Development Corporation of South Africa Limited undertaking the formation of a private company formed in South West Africa under the title of "S.W.A. Water and Electricity Corporation (Pty) Limited"-SWAWEK for short – with the object of developing the power and water potential of the Kunene River.

In 1963 the Governments of Portugal and South Africa agreed on the procedures to be followed in achieving the best joint utilisation of rivers of mutual interest to both Angola and South West Africa. In 1969 it was jointly announced that agreement had been finally reached concerning the following works to be established in the development of the Kunene River Basin:

A dam at Gove, near Nova Lisboa in Angola, in the Kunene River with a capacity comparable to South Africa's Vaal Dam for the purpose of regulating the Kunene River.

A dam at Calueque, some forty kilometres upstream from the Ruacana Falls, for the further regulation of the Kunene River in accordance with the requirements of the power station to be built at Ruacana.

A hydro-electric power station at the Ruacana Falls, and the associated diversion works.

A pumping scheme at Calueque for abstracting up to six cubic metres per second of water from the Kunene River for human and animal consumption in South West Africa, and initial irrigation in Owambo.

Up to the late fifties and early sixties the level and distribution of economic activity in this territory did not justify the establishment of a country-wide network of electric power transmission lines fed from one or more large generating stations. Each local authority, mine and/or other user had, therefore, to provide its own power supply. Small-scale power generation, coupled with high fuel costs, mainly due to transport, resulted in high power costs.

However, the demand for power was growing rapidly, particularly in Windhoek, in the coastal centres and the mines in and near Tsumeb. It was realised that the availability of abundant power would create new possibilities for mines and industries.

DIMENSIONS

Gove Dam

Earthfill dam of maximum height	58 m
Length	1 111 m
Volume of earthfill	$4 \times 10^6 m^3$
Reservoir capacity	$2\,574 \times 10^6 m^3$
Designed by	Hidrotechnica Portuguesa

Construction supervised by Gabinete do Plano do Cunene

Contractor	A. Campos

Artist's impression of the Kunene Scheme

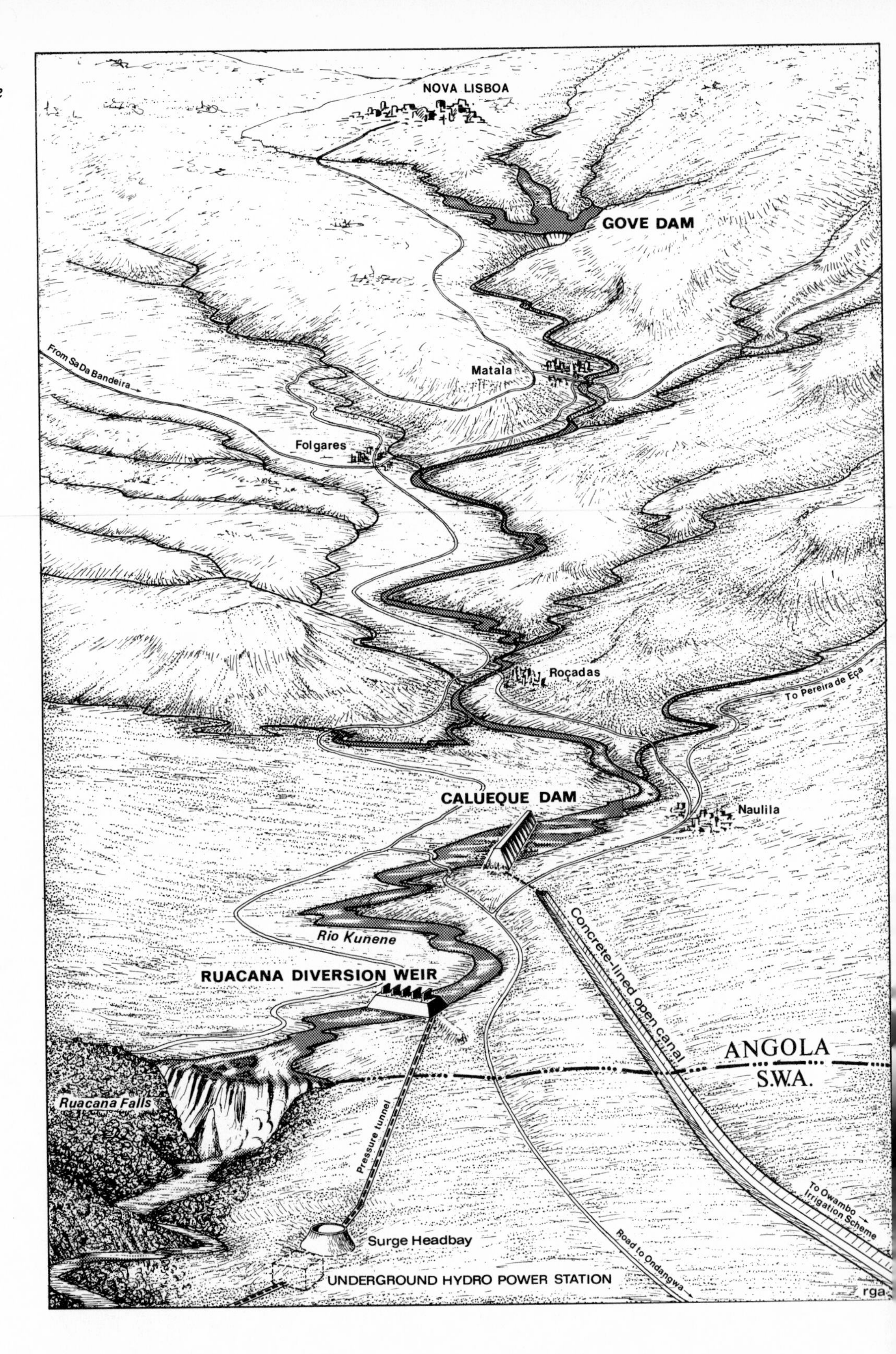

CONSTRUCTION

Thermal Power prior to Hydro-electric Power

Owing to the unavoidable delay in finalising the above-mentioned agreement and the equally unavoidable need to ensure that an adequate market existed for the power that SWAWEK was proposing, SWAWEK found itself in a position during 1967 where it felt obliged to recommend to the South African Government that the hydro-electric scheme on the Kunene River should be preceded by a thermal power station near Windhoek and an associated power distribution network, initially only in the area bounded by Windhoek, Walvis Bay and Tsumeb. This modification to the original scheme envisaged by the Odendaal Commission soon received official blessing. SWAWEK immediately set in motion the machinery required for the building of a 90 MW coal-fired power station, today known as the Van Eck Power Station, in honour of SWAWEK's first chairman, the late Dr. H. J. van Eck.

Some 800 kilometres of 220 kV overhead transmission line forms the backbone of the present SWAWEK transmission system. Each of the three phases comprise a single steel-cored aluminium conductor of 300 square mm and twin overhead earth wires. The three phases, in triangular configuration, are carried on 33 metre high steel lattice masts, with a maximum span length of almost 1 200 metres and a ruling span length of 430 metres (1 500 ft.). Capacitive current compensation is provided by six reactors of 15 MVA each and adequate protection circuits are incorporated.

An automatic trunk exchange system provides telephone access to and between all power stations and all major distribution and substations. This system which operates via power line carrier circuits, is backed up by a high frequency radio communication network.

The Ruacana – Calueque Scheme

The scheme is being undertaken by the SWA Water and Electricity Corporation (SWAWEK) and the objective is to generate hydropower up to a capacity of 320 MW with an annual output of some 1 200 GWh that will be transmitted into the SWA grid reaching to the main centres of Windhoek, Tsumeb and Walvis Bay where support will be given by existing thermal stations.

Water released from the storage regulation dam at Calueque is to be diverted at a weir about 1 km above the Ruacana Falls which provides a direct 100 m drop in the Kunene River. From the intake a pressure tunnel carries flow under the Palmwash ravine and releases it into a 27 m deep oval shaped surge headbay whereafter four vertical steel-lined penstocks supply the underground powerhouse comprising three parallel caverns. Downstream of the draft tubes the water is directed back into the river through a free-flowing tailrace tunnel.

Tunnelling operations at Ruacana comprised:

1. A 360 m long access gallery for vehicular traffic from the hillside to the underground caverns, excavated in modified horseshoe shape with an equivalent finished diameter of 8,5 m. Some 26 500 m³ of excavation including a side trench for services and an 18 m extension between caverns were required.
2. A 1 500 m long pressure tunnel of which some 1 300 m required underground excavation to a nominal diameter of 9,5 m for the main concrete lined reaches of 8,3 m internal size and varied to suit short steel-lined sections of 7,4 diameter either side of the Palmwash ravine where rock conditions are severely disturbed by a wide fault. Portions of the concrete lining over minor fault or shear zones are heavily reinforced against internal pressure which attains a maximum 62,5 m head under surge conditions. Other sections where the rock cover is inadequate are also reinforced but over a considerable length the water pressure is carried largely by the rock with only nominal mesh included in the circular concrete lining. Altogether rock excavation amounted to 97 000 m³.
3. Three parallel powerhouse caverns for:
   the transformer hall, 126 m long × 15,5 m wide × 15 m deep at the crown
   the machine cavern, 141,5 m long × 16 m wide × 36,5 m deep at turbine basement pits, and
   the surge chamber 70 m long × 11,5 m wide × 28 m deep
   along with eleven horizontal or steeply inclined galleries for the busbars, draft tubes, drainage and other duct facilities. These all required some 135 000 m³ of intricately shaped excavation.
4. Seven vertical shafts each about 130 m deep for the four penstocks of 4,8 m excavated diameter, two cable shafts of 3,9 m diameter and the 7,6 m diameter elevator shaft. These involved a total of 20 000 m³ of excavation.
5. A 590 m long tailrace tunnel in semi-Dee shape with an upper diameter of 11,0 m by 13,9 m deep and a tapered bottom width of 8,9 m. Some 83 000 m³ of excavation in top and bottom headings for the 140 m² face area were required.
6. An aggregate of 180 m length in three special construction adits generally of 25 m² cross-section but widened for mucking bays amounting to nearly 9 000 m³ of excavation.

Underground excavations totalling nearly 370 000 m³ were needed for the various tunnels, caverns, shafts, special galleries and construction adits for the hydro-power station at Ruacana. In addition surface excavations for portals and the large surge headbay amounted

RUACANA HYDRO-POWER

Following the report in the July issue of the Division's site visit to the Cunene Scheme, a photographic presentation illustrates the work being undertaken.

The Cunene river rises in Angola and from Ruacana becomes the international boundary to SWA. The mean flow is 190 cumecs – about 80 per cent of that of the Orange river. Ruacana Falls has a sudden drop of 100 m and the photograph shows the magnificent sight of a 1 200 cumec flood.

Here the river is down to a flow of only 50 cumecs. The 2,5-MW interim power station, built in 1970 for construction supplies, appears in the foreground and about 1 km upstream of the falls concreting is in progress on the diversion weir.

Basic regulation for the Cunene has been provided well upstream at the Gove Dam near Nova Lisboa and the SWA Water and Electricity Corporation is proceeding with construction of the Calueque Dam about 40 km upstream of Ruacana. This is a composite earthfill and concrete dam with 10 large radial gates providing spillway capacity for extreme floods up to 8000 cumecs. Piers under construction are shown above. A pump station to supply 6 cumecs of water to Ovambo is already in operation and is being integrated into the dam structure.

The diversion weir at Ruacana consists mainly of a concrete gravity wall topped by five fish-belly flap gates, each 56 m long by 3,5 m high, between trunnion control piers.

Intake works on the left bank of the weir carry water into a pressure tunnel of 8,3-m diameter. The photograph shows the north portal and start of tunnel concrete lining.

Over certain sections of the tunnel which is 1 500 m long the use of steel ribs was required but otherwise for basic support during excavation rock-bolting was applied.

The tunnel drops under a deep ravine known as the Palmwash where there is a fault zone. Over this section steel lining will be employed and the photograph shows the portal face taken beyond the fault gouge with appropriate hood protection. Higher up, one of the original exploratory adits can be seen.

The tunnel discharges water into an oval surge headbay 80 m by 55 m in plan and 27 m deep. From here water is taken underground through individual penstocks of 3,6-m diameter to develop a gross head of 140 m in 4 by 80 MW turbines.

The penstocks emerge at a lower bend below the transformer hall and pass through a gallery into the parallel machine cavern.

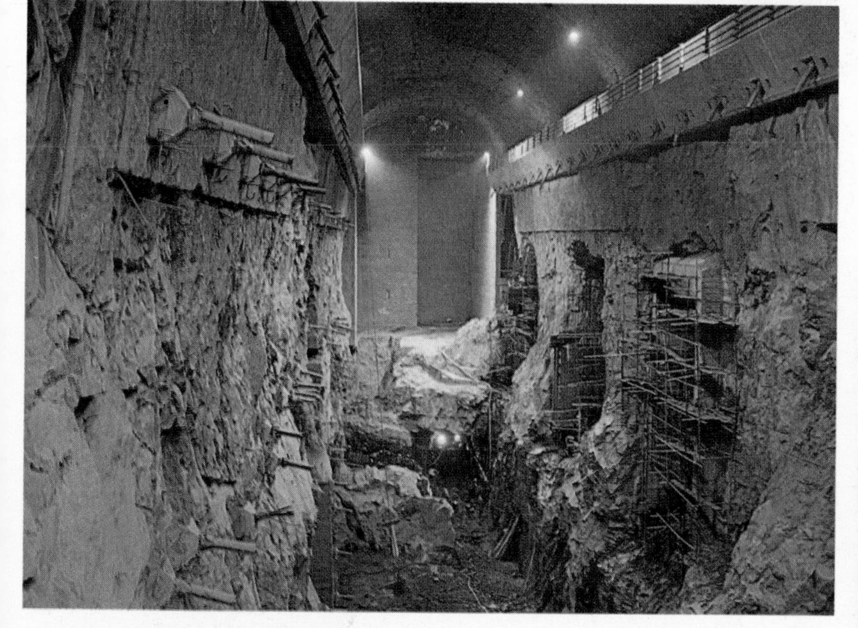

The turbines and generators will be installed in a cavern 160 m long by 16 m average width and 37 m deep at the pits. At one end is the workshop and at the other the control building shown above.

Alongside is the third parallel cavern known as the surge chamber and serving as discharge manifold from draft tubes to the tailrace tunnel which leads the water back over 590 m to the Cunene river.

Photographs by:
R. Immelman
W.C.S. Legge
J.B. Richard

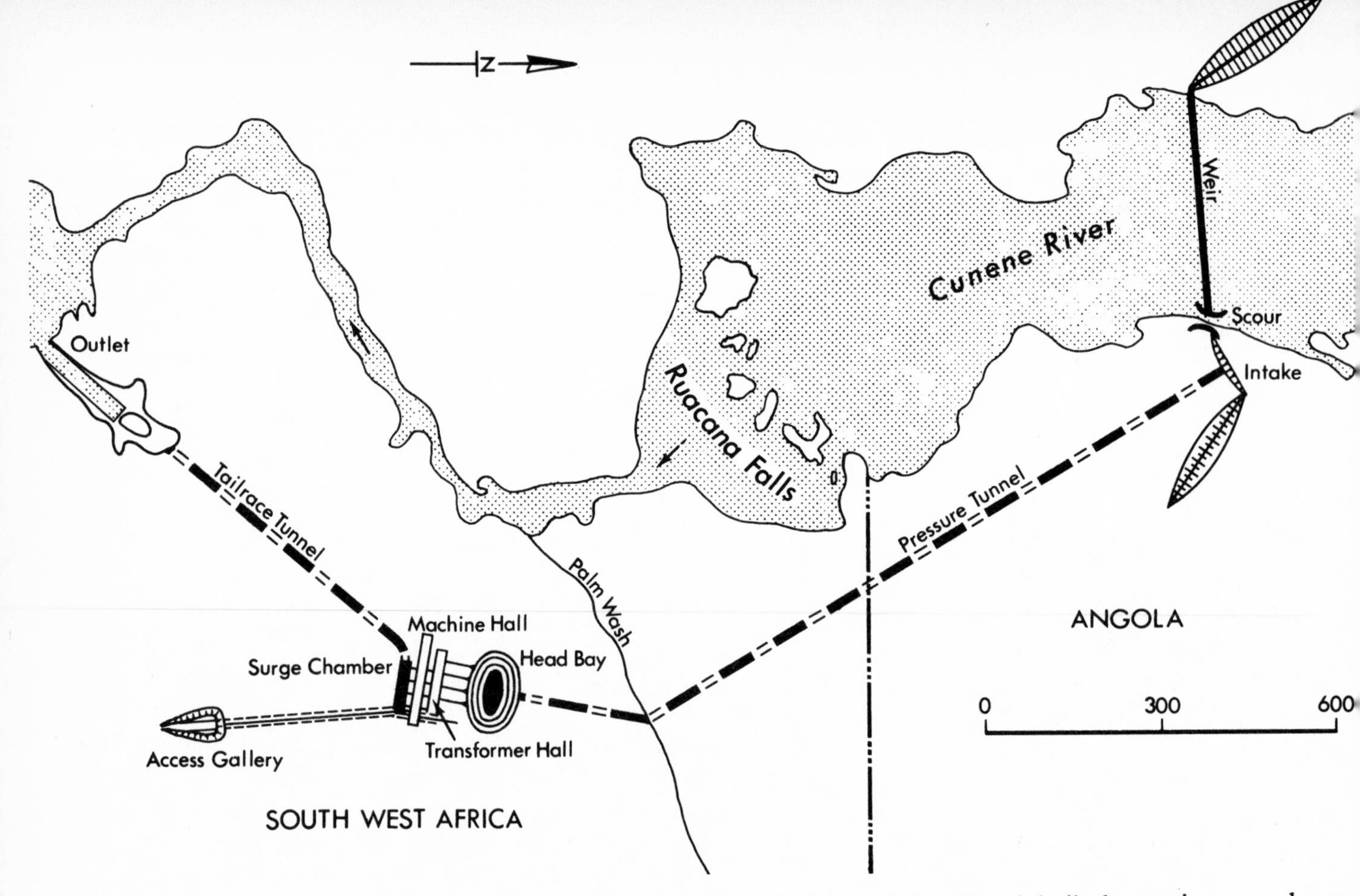

to almost 300 000 m^3, apart from a further 120 000 m^3 involved in foundations for the diversion weir.

Calueque Pumping Scheme

The first stage of the pumping scheme which has been operating for several years, is housed in an 18 metre concrete section which forms part of the central section of the final Calueque Dam. The two pumps, which can operate in parallel consist of vertical spindle, variable speed, direct driven pumps with a delivery capacity of 3 m^3 per second each. The pumped water passes through a steel-lined rising main approximately 2,5 kilometres long, to a concrete lined open canal. The section of canal in Angola runs due south until it crosses the border between Angola and Owambo, from where it swings south-east via a bifurcation in the canal and past the Olushandja Balancing Dam. The bifurcation will provide the starting point for a future canal system to supply water to places as far south as Swakopmund. The Olushandja Balancing Dam is being used to accept the surplus flow of water and to provide water for Owambo should a breakdown in the pumping scheme occur.

The Gove Dam

The Gove Dam, at the junction of the Cunhangamva and Kunene Rivers just south of Nova Lisboa, is complete. This dam was designed and built by engineers and contractors from Portugal and Angola.

The Calueque Dam

The dam consists of a central concrete section of 250 metres which includes the existing pump station. Furthermore, provision is made for extending the pumping facility at a future date. The central concrete section also holds the two radial gates which will be used to regulate the flow of the Kunene River, and ten spillway gates to handle the release of floodwater. The concrete section will require approximately 60×10^3 of concrete whilst the two earth flanks will require more than $10^6 m^3$ of earthfill. The total length of the embankment is 2,5 kilometres.

The Ruacana Diversion Weir

The diversion weir, situated less than a kilometre upstream from the Ruacana Falls, is intended to divert the flow of the Kunene River via the pressure tunnel to the Ruacana Hydro Power Station.

The diversion weir consists of a central concrete section that houses five steel flood release gates, a scouring gate and the inlet structure to the pressure tunnel. Two earth flanks and two auxiliary earth embankments complete the weir. The entire weir and a section of pressure tunnel is constructed within Angola

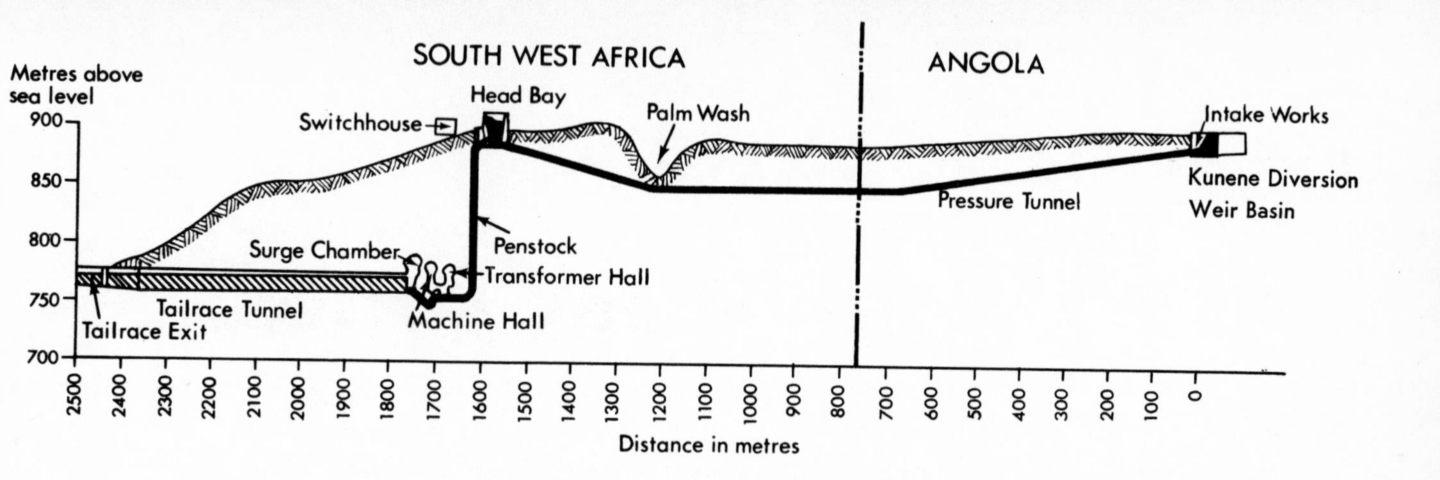

The Untapped Further Potential of the Kunene River

As the waters of the Kunene rush towards the Atlantic Ocean, several cataracts and falls are encountered. Furthermore, when the Gove and Calueque Dams are complete, the Kunene River will be regulated and hence it will be technically feasible to continue to develop the power potential of the Kunene River downstream. This further potential has been estimated by SWAWEK at 1 560 MW at 8 sites.

CONSULTING ENGINEERS AND CONTRACTORS

Ruacana – Calueque Projects

Client

South West African Water and Electricity Corporation

Consulting Engineers

Hydroconsults

Main Contractors

COSINT – Civil

Construzioni Interrazionali STA

Generators: Westinghouse U.S.A.

Turbines: Voest – Alpine – Austria

Switchgear: Siemens Ltd.

Transformers: ASEA – Electrical

Penstocks, Gates etc.: B.V.S. Ltd.

CHAPTER 9

South West Africa

SWAKOP RIVER STATE WATER SCHEME

VON BACH DAM

HISTORY

In expectation that the water sources for Windhoek would require supplementation, a river flow gauging station was established near a promising dam site in the Swakop River in 1945, a short distance upstream of the town of Okahandja.

By 1947 it was possible to decide, albeit on the basis of only provisional estimates, that considerable quantities of water could be made available from a dam at the site mentioned. Consequently in September 1947 almost 4 800 ha of land was purchased by the South West Africa Administration.

Various administrative and other problems hampered progress on the project, until the Commission of Enquiry into the Affairs of South West Africa recommended construction of a project to supply Windhoek with water from the Swakop River. Meanwhile the water demand for Windhoek had increased inexorably and by 1965 it was clear that supplementary sources of water would have to be commissioned as a matter of urgency.

By the end of 1965 Consulting Engineers had been appointed to investigate the feasibility and to prepare a planning report for the scheme. By March 1967 it was possible to place before the Administrator-in-Executive Committee proposals for the scheme which were approved. Funds had provisionally been provided on the budget and by the end of 1967 the main contract for the construction of the dam was awarded. The scheme was inaugurated on 21st August 1971 and named the Von Bach dam after Senator Sartorius Von Bach. The total cost was R8 848 000.

DIMENSIONS

Hydrology:	
Catchment area	3 000 km²
Mean annual rainfall	400 mm
Mean annual run-off at dam site	$14 \times 10^6 m^3$
Full storage capacity of dam	$53 \times 10^6 m^3$
Yield on variable draft basis	$6 \times 10^6 m^3$
Surface area at full supply level	500 ha
Reach of water surface up river when dam is full	9 km
Main Embankment:	
Height of crest above river bed	35 m
Height of crest above lowest foundation	40 m
Crest length of embankment	270 m
Crest width of embankment	7 m
Upstream slope of embankment (vert: horiz.)	1:1,7
Downstream slope of embankment (vert:horiz.)	1:1,5
Depth of grout curtain below river bed	40 m
Spillway:	
Number of sluices	2
Size of sluices: Width	10 m
Height	15 m
Size of channel: Width	23 m
Length	212 m
Discharge capacity of spillway	2 200 m³/s
Design flood frequency	1 in 200 years
Secondary embankment:	
Crest length	230 m
Crest width	5 m
Maximum height	4 m

CONSTRUCTION

The water for the Swakop River State Wate Scheme is derived from a dam in the Swako River with a maximum storage capacity o $53 \times 10^6 m^3$. Liberal allowance has also bee made, in constructing the dam, for the accu mulation of silt so as to ensure that the yiel of the dam will not be adversely affected for number of years. The dam has been so de signed that it will be possible, at a later stage to increase the capacity to $64 \times 10^6 m^3$. Fron the dam the water is conveyed by means of gravity pipeline 3 km long to the purificatio works. At the purification works, and afte purification, the water is pumped to a termina reservoir at Okahandja some 5 km away an also to the first of two booster pump station situated along the 62 km pipeline to the termi nal reservoirs at Windhoek.

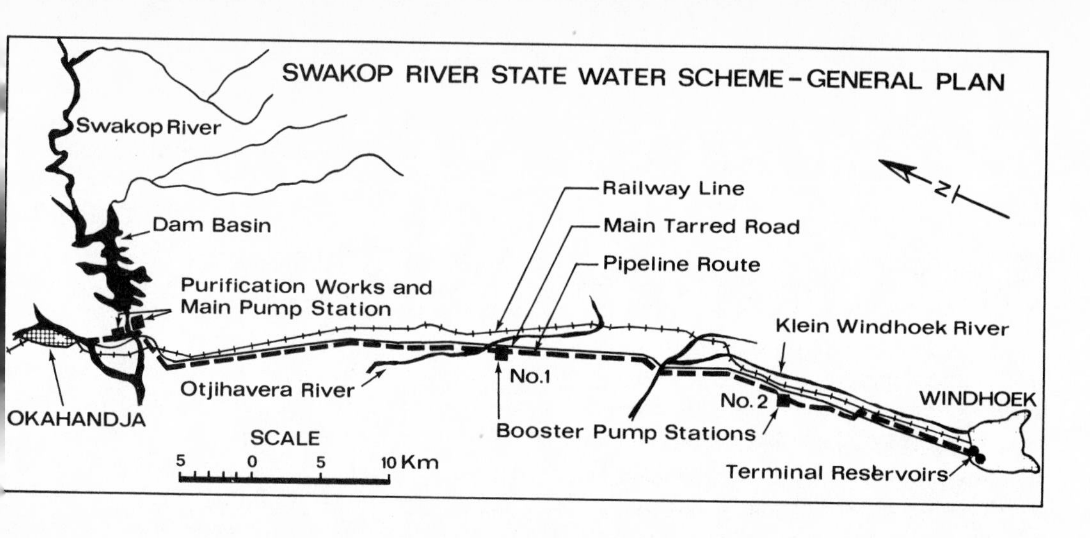
SWAKOP RIVER STATE WATER SCHEME - GENERAL PLAN
Swakop River
Dam Basin
Purification Works and
Main Pump Station
Railway Line
Main Tarred Road
Pipeline Route
Klein Windhoek River
Otjihavera River
No.1
No.2
OKAHANDJA
WINDHOEK
Booster Pump Stations
Terminal Reservoirs
SCALE
5 0 5 10 Km
N

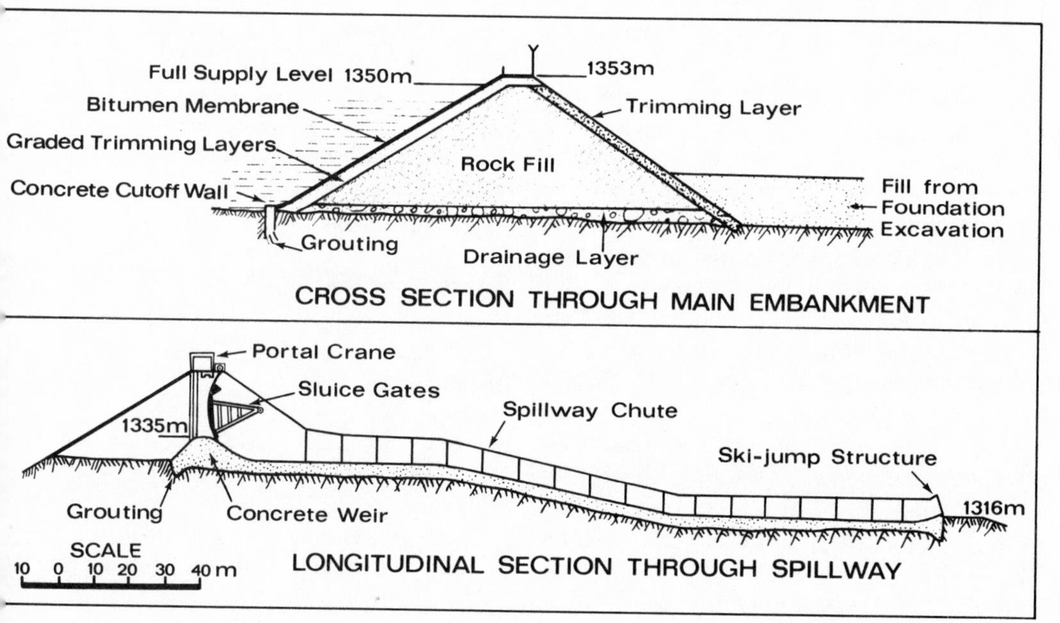
Full Supply Level 1350m
1353m
Bitumen Membrane
Trimming Layer
Graded Trimming Layers
Rock Fill
Concrete Cutoff Wall
Fill from
Foundation
Excavation
Grouting
Drainage Layer
CROSS SECTION THROUGH MAIN EMBANKMENT
Portal Crane
Sluice Gates
Spillway Chute
1335m
Ski-jump Structure
1316m
Grouting
Concrete Weir
SCALE
10 0 10 20 30 40 m
LONGITUDINAL SECTION THROUGH SPILLWAY

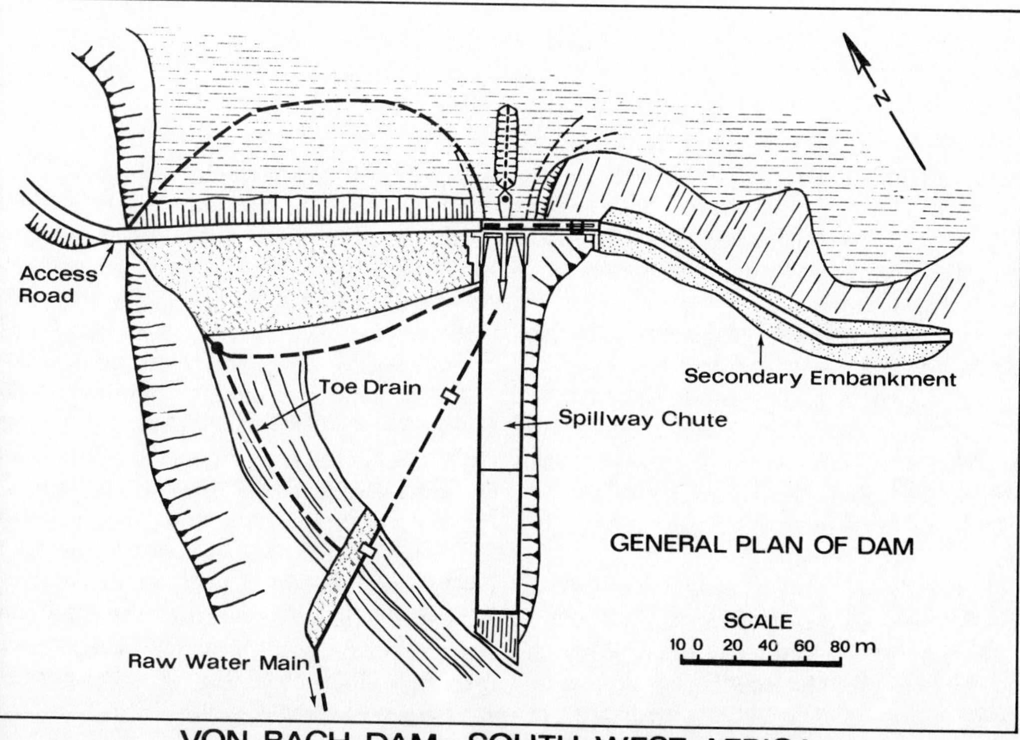
N
Access
Road
Toe Drain
Secondary Embankment
Spillway Chute
GENERAL PLAN OF DAM
SCALE
10 0 20 40 60 80 m
Raw Water Main

VON BACH DAM - SOUTH WEST AFRICA

The dam embankment is constructed of quartzitic micaceous rock rubble which has been laid and compacted in layers. The upstream face of the embankment has been covered with a bitumen layer, of special design, to form a watertight blanket. Quartsitic river sand has been embedded in the upper surface of the bitumen layer in an effort to reflect as much sunlight as possible and so to reduce the effect of increased temperatures on the bitumen blanket. A concrete cut-off wall and grout curtain have been constructed to form a water seal between the embankment and its foundation. The spillway and outlet works are combined in one structure on the left bank of the river. Flood discharge is controlled by means of two sluices mounted over a low concrete weir in the spillway channel. The sluices are completely automatically controlled and electrically operated. The outlet works are accommodated in an outlet tower situated between the two sluices. The outlets are in the form of twin bellmouth inlets each of the four pairs being situated at different levels in the tower. A low secondary rockfill dam straddles a saddle close to the main dam and has been designed so that the crest of the embankment is 1,5 metres lower than that of the main embankment and will breach if catastrophic floods endanger the main embankment in any way. The four pairs of bellmouth inlets in the outlet tower unite in two manifolds 914 mm in diameter which in turn connect to a single raw water main. The raw water main has been constructed so that duplication of the main is impossible. The raw water main pipes have an internal diameter of 610 mm and have been both lined and covered with bitumen and wrapped in fibre-glass fabric. The raw water main is 2,8 km long and the gravity head available enables a maximum flow of 1 425 m^3/h to pass down the line.

CONSULTANTS

Consulting Engineers	Hydro-consults
Basin and dam site survey . . .	Fotogramensura (Pty) Ltd.
Site investigations .	Boart and Hard Metal Products
Access roads . .	Herma Bros. (Pty) Ltd.
Permanent housing .	H. Meyer
Von Bach Dam .	E. Lafrenz (Pty) Ltd.
Spillway gates . .	Consani Engineering Ltd.
Raw water main and Okahandja Pipeline	Suidwes Konstruksie
Purification works .	J. Murphy and Sons (S.A.) (Pty) Ltd.
Windhoek pipeline .	Hume Ltd.
Reservoirs and pump houses . . .	Swemkor (Pty) Ltd.
Main pumping system . . .	Mather and Platt
Minor pumps . .	M & Z Motors and Engineering Ltd.
Cranes . . .	South West Engineering Ltd.
Purification plant .	Jeffrey Manufacturing Co.

NAUTE DAM

HISTORY

The Naute State Water Scheme provides primarily for the supply of potable water to the town of Keetmanshoop. It consists of a storage dam in the Löwen River, 45 km to the south west of Keetmanshoop, a water purification plant near the dam, pumping stations, pipelines and reservoirs.

The construction of a storage dam in the Löwen River as a permanent source of water, primarily for irrigation purposes, was first investigated by the privately owned "Syndicate for Irrigation Schemes in German South West Africa" in conjunction with the German Colonial Government in the period 1897 to 1902.

During the 1950's the need for an additional or completely new water supply scheme for Keetmanshoop became apparent when it was realised that the existing sources consisting of boreholes and the van Rhyn Dam could no longer meet the ever-increasing demand.

In 1961 the Water Affairs Branch of the S.W.A. Administration recommended the construction of a scheme on the Löwen River for the supply of water to the Municipality of Keetmanshoop and for a pilot irrigation scheme on the Seeheim plain. The Executive Committee of the South West African Legislative Assembly accepted this proposal.

In 1964 the Executive Committee of the South West African Legislative Assembly appointed Consulting Engineers to carry out the design work for the proposed scheme. In the same year the first contract was awarded for diamond core drilling and the geological investigation of the foundation materials at the two possible dam sites. In 1966 the construction of the access road, and extension to the Jürgen Railway Siding were commenced. In 1967 the works for the establishment of

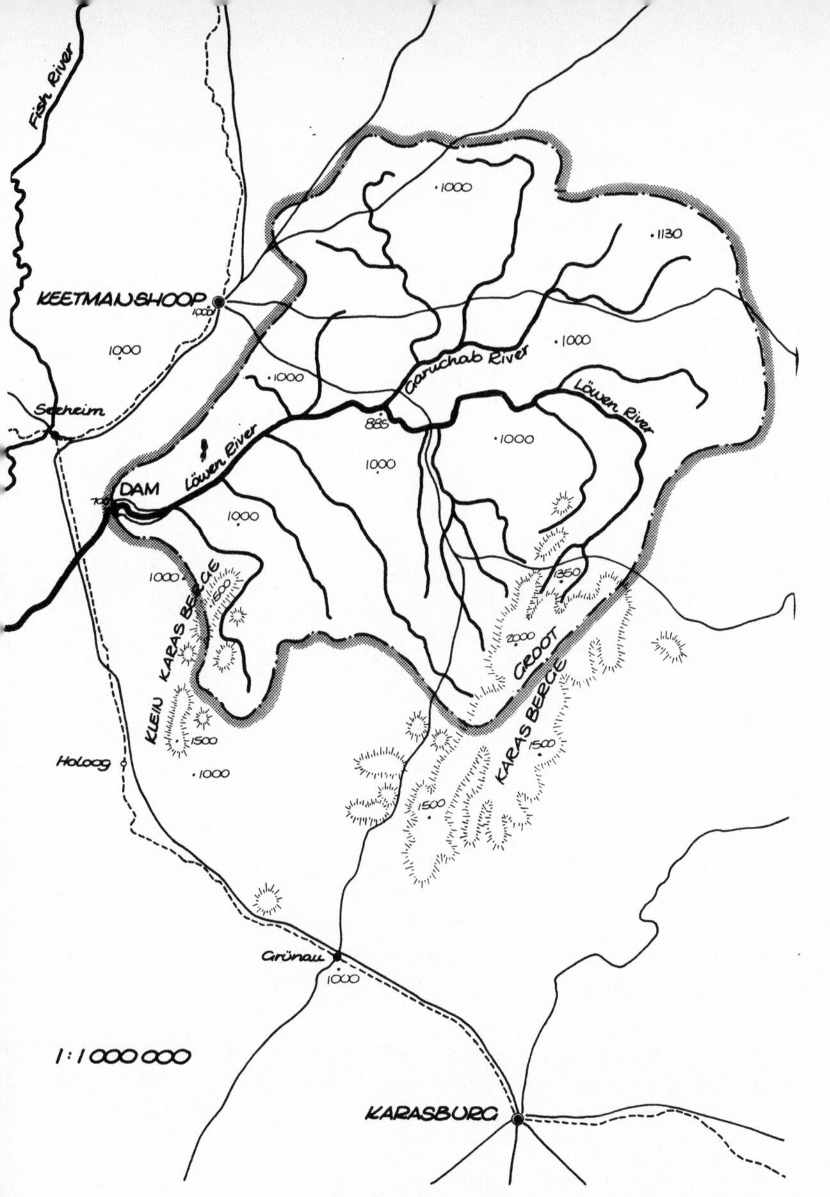

Catchment area

housing facilities, potable and construction water supply and power generation were put in hand. In 1968 the works commenced on the Naute Dam proper and in 1969 work on the purification works, pump stations, pipelines and reservoirs started. The last major contract, awarded in 1970, involved the construction of the powerline and transformers.

The first flood waters were retained in the storage basin during the 1970/1 rainy season.

All components of the scheme were completed by the end of 1971 and the supply of water to Keetmanshoop commenced during the second half of 1972 after the extensions to the Keetmanshoop Municipal Power Station, from where the entire scheme is fed with electricity, had been commissioned.

The Catchment Area:

At the Naute Dam the Löwen River has a catchment area of 8 800 km². The Löwen River and its southern tributaries drain the mountainous and hilly areas on the northern and western slopes of the Grosse and Kleine Karasberge mountain ranges. This area, because of its higher rainfall, steeper topography and geological composition, delivers a high percentage of the yield of the catchment area. Its main northern tributary, the Garuchab, de-waters the gently sloping plains to the east of Keetmanshoop.

The climate in the entire region is semi-arid. The vegetation is generally sparse and consists of a thin grass cover, small scattered bushes and euphorbia. A belt of trees and dense bush lines most of the river banks.

The principal hydrological data are:-

Catchment area	8 800 km²
Mean annual rainfall	150 mm
Mean annual run-off	45 m³ × 10⁶
Maximum flood once in 200 years	2 380 m³/s
Mean yearly evaporation in area	2,65 m
Capacity of the dam basin	70 m³ × 10⁶

DIMENSIONS

Height of crest above riverbed	37 m
Height of crest above lowest foundation	41 m
Total crest length of dam	470 m
Crest length of arch	150 m
Radius of arch	110 m
Crest width (2 sidewalks and 3,5 m wide roadway)	5,5 m
Maximum arch thickness	11 m
Minimum arch thickness	3,5 m
Total concrete volume	111 000 m³
Total reinforcing-steel mass	900 t

CONSTRUCTION

The dam is a composite structure of a central double curvature circular arch abutting against the massive spillway blocks which in turn are flanked by gravity walls to close off the valley. The design allows for raising of the full supply level by 4 m without structural alterations to the dam as presently built.

The dam consists generally of mass concrete; only the spillway wing walls, which also support the radial gate bearings, required large amounts of reinforcing steel. The coarse aggregate for the concrete was crushed from dolerite rock recovered from the dam excavation. For the fine aggregates the natural sands of the Löwen and Gab Rivers were used.

The Dam Equipment

Four radial gates provide for the discharge of flood waters which cannot be stored in the basin. They are located in pairs in each of the two spillway blocks, such that their discharge jets collide at right-angles. This results in very effective energy dissipation from the discharged water so eliminating the need for extensive training work and erosion protection downstream of the dam.

The radial gates are operated by hoist ropes and winders driven either by electric motors or standby petrol engines. The opening and closing of the radial gates is automatically controlled through a time delay mechanism in response to the dam water level fluctuations.

DIMENSIONS

Radial Gates –	4 off
Dimensions of one gate 7 m high × 12,5 m wide	
Discharge capacity of one gate with high flood level in dam	600 m³/s
Minimum time to open (or close) one gate fully	30 min.
Scour outlets	2 off
Diameter of pipes	1 200 mm
Diameter of valves	900 mm
Maximum discharge for one outlet	12,5 m³/s
Irrigation water outlets	2 off
Diameter	900 mm
Raw water outlets	3 off
Diameter	600 mm

All control and maintenance valves are electrically operated and can be controlled from a central switch panel in the raw water pump station.

Raw water pipeline

A raw water pipeline carries water from the dam to the purification works.

DIMENSIONS

Length	1,9 km
Static head	73 m
Diameter	500 mm
Discharge initially	7,46 m³/min.
Pump stages	2 off
Length of the first	34 km
Length of the second	10 km
Pumping head in each stage	200 m
Static head total for both stages	280 m
Diameter	380 mm
Discharge initially	3,39 m³/min.
The initial power requirements are:-	
Raw water pump	155 kW
Clear water pumps, Stages I and II	200 kW each

Purification Works

The raw water from the dam is mixed with lime and alum and led into settling tanks with mechanical stirrers where the chemicals promote the flocculation and subsequent settlement of the suspended particles in the form of sludge. The clear water leaves the settling tanks continuously over outflow notches around the upper edge of the tank whilst

Upstream view of the dam and the spillage openings

sludge is drawn off periodically from the bottom.

Any remaining undissolved impurities are removed in the rapid flow sand filters through which the water is subsequently led. These filters are cleaned at intervals by backwashing with water and air and by floating off the scum and dirt.

Beyond the sand filters the water is sterilised by chlorination and its acidity corrected by the addition of lime.

DIMENSIONS

Settling tanks	2 off
Retention period in settling tanks	2,5 hr
Sand filters, rapid gravity type	4 off
Filter depth	0,86 m
Filter area, each bed	33 m³
Capacity of clear water tank	1 800 m³
Initial maximum flow capacity of works	450 m³/h
Initial power requirement	130 kW

Terminal Reservoirs

The two terminal reservoirs are built on a hill, 2 km to the west of Keetmanshoop at an elevation compatible with the pressure requirements for town reticulation purposes. They are reinforced concrete, cylindrical tanks with roofs, each with a capacity of 3 000 m³. From these, water is fed directly into the town reticulation system.

Operation and Control

The scheme requires for its maintenance and operation four mechanical/electrical artisans together with twelve labourers.

The entire purification and pumping process can be controlled by remote action from the purification plant building with signals being transmitted to the dam, the intermediate pump station and terminal reservoirs over postal telephone lines.

CONSULTING ENGINEERS AND CONTRACTORS

		R
Foundation Investigation Drilling	Rodio SA (Pty) Ltd. Johannesburg	80 000
Access Road	Mariental Transport (Pty) Ltd. Mariental	468 000
Survey	Air Survey Co. (Pty) Ltd., Durban	48 000

State water scheme: Naute Dam, locality plan

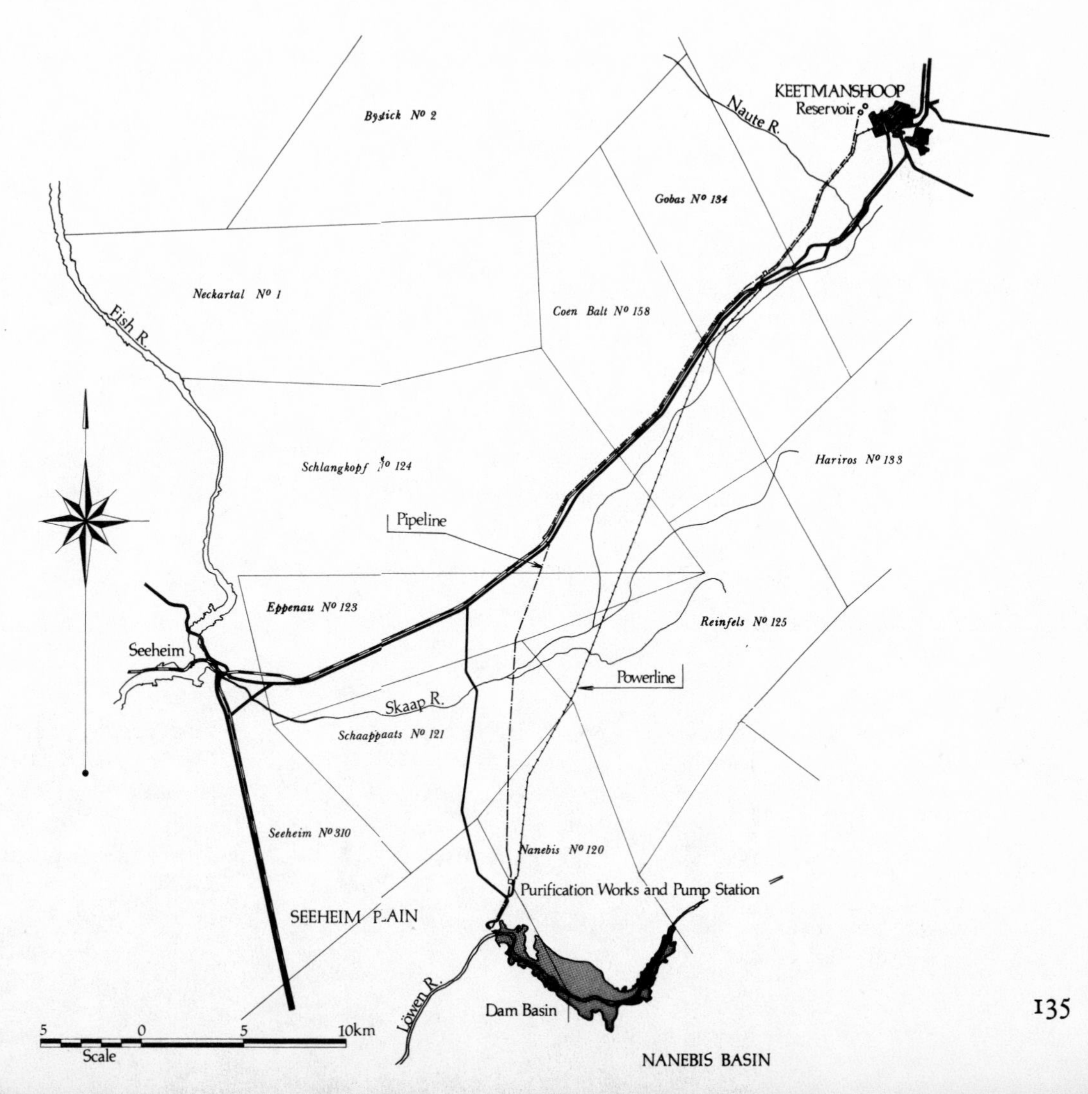

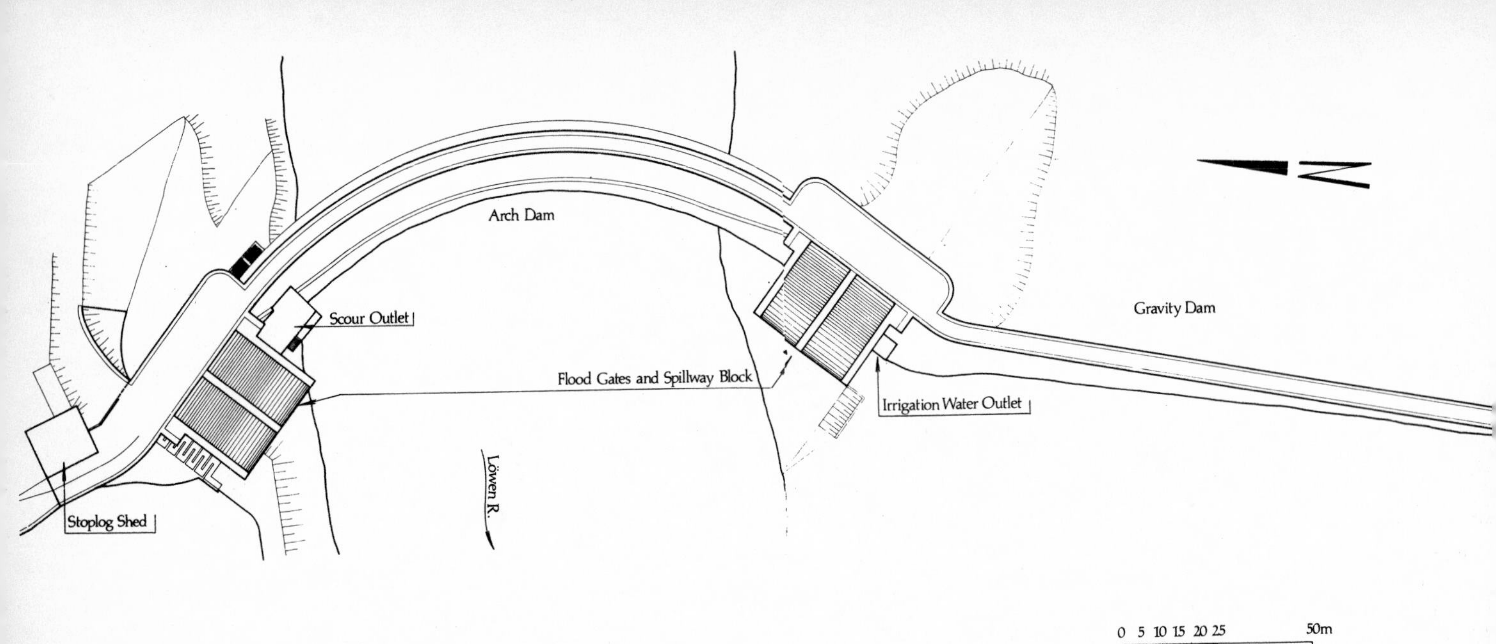

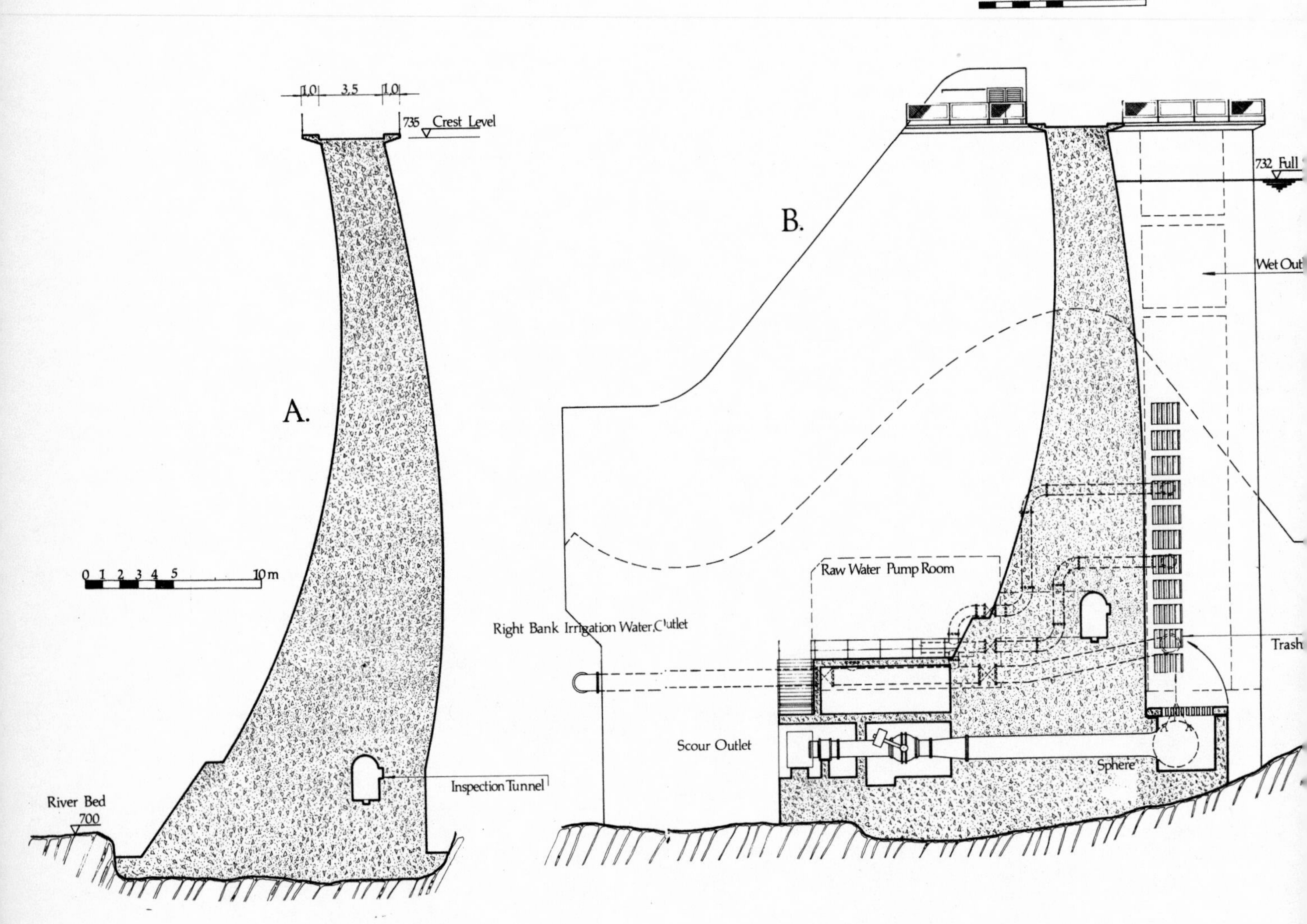

Top: Plan of dam

Bottom: A Typical section through archdam

B *Section of scour and raw-water outlet*

Site Development (Houses, Services)	Van N.E.P. Construction (Pty) Ltd., Keetmanshoop	243 000
Dam	Concor Construction (Pty) Ltd., Johannesburg	4 420 000
Purification Works Building	Concor Construction (Pty) Ltd., Johannesburg	396 00
Purification Plant	Jeffrey Manufacturing Co. (Pty) Ltd., Johannesburg	130 00

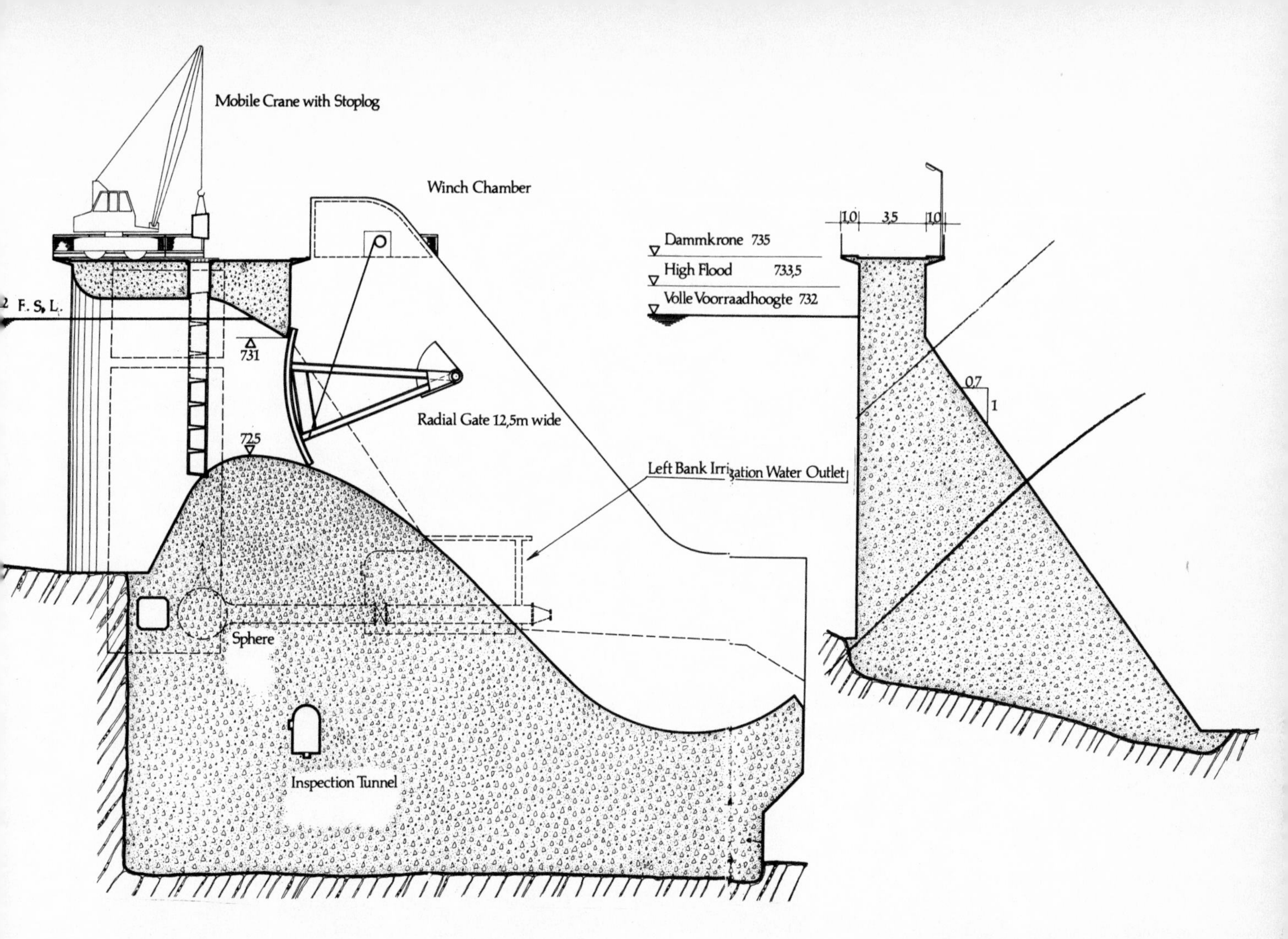

Section through spillway, left irrigation water-outlet

Pumping Plant	M & Z Motors and Engineering Ltd., Windhoek	110 000
Terminal Reservoirs	I. Bonadei Construction (Pty) Ltd., Windhoek	113 000
Pipelines – supply of materials	Hume Ltd., Okahandja/Germiston	662 000
Pipelines – construction	I. Bonadei Construction (Pty) Ltd., Windhoek	234 000
Bushclearing of Dam Basin	Concor Construction (Pty) Ltd., Johannesburg	58 000
Remote Control Equipment	Telewei (Pty) Ltd., Windhoek	30 000
Power transmission line	Hubert Davies (Pty) Ltd., Windhoek	325 000
Professional fees		362 000
Site supervision		117 000
	Total:	R7 796 000

Civil Engineers	J. W. Stein & Partners, Windhoek, Pretoria
Electrical Engineers	G. S. Fainsinger and Associates, Windhoek
Architects	Leon Hurter, Chase & Holmes, Windhoek

CHAPTER 10
Swaziland

KOMATI WEIR AND MHLUME CANAL

HISTORY

At the time of writing there are no major dams in Swaziland. However, there is significant dam and storage potential, especially on the Komati and Lomati Rivers and, having regard to the fertile soils, the climatic conditions, the pastoral habits of the inhabitants, and the geographical location of the country and its rivers, there is little doubt that, within a few decades, the available water storage potential will be used to best advantage of all the inhabitants and its neighbours.

The area of the country is 19 440 square kilometres. It has a population of about 300 000 of which Africans number 289 000, Europeans 7 000 and other nationalities 4 000.

Because of the relatively high rainfall, irrigation was not practised extensively until 1958, when the Colonial Development Corporation (later the Commonwealth Development Corporation) inaugurated its Swaziland Irrigation Scheme (S.I.S.) commanded by waters from the Komati River.

Swaziland may be divided into four regions with distinctive physiographic, climatological and other characteristics affecting the growing of crops and livestock. Three of these, the High, Middle and Lowveld are of major importance. They run from west to east, the elevation falling from a range of 1 829 metres in the Highveld to 305 metres in the Lowveld. To the east of the Lowveld is the fourth region, the Lebombo plateau, which tops an abrupt escarpment dividing Swaziland from Moçambique.

It can be said that the major and best run irrigation scheme is the 12 000 hectare Swaziland Irrigation Scheme, situated between the Komati and Umbuluzi Rivers and fed from a canal taking off from the Komati weir on the Komati River near Border Gate.

The Komati weir is situated on the Komati River in Swaziland at approximate latitude 26°S and longitude 31°E.

In 1950 the Colonial Development Corporation purchased an estate of 42 493 hectares at an altitude of approximately 280 metres above sea-level. The climate is equitable with an average annual rainfall of 730 mm.

A pilot pumped scheme proved that the soils were fertile and particularly suitable for growing crops such as sugar and rice. A pilot survey indicated that at least 12 000 ha of soils were suitable for irrigation.

The Komati flows eastward in the north of Swaziland and is located at a higher altitude than another river to the south – the Black Umbuluzi. It was considered possible to abstract water from the Komati and divert it by a canal about 66 km long through a saddle to cross the watershed between the Komati and Umbuluzi Rivers.

The first stage plans were to irrigate an area of 2 630 hectares between the canal and the Komati River and 6 890 hectares which drains towards the Black Umbuluzi.

The Consulting Engineers appointed by the Colonial Development Corporation in November 1954 were Sir Alexander Gibb and Partners, London, and Sir Alexander Gibb and Partners (Africa) Nairobi.

Their report and recommendations to the C.D.C. dated 28 February 1955 were accepted and work on the project commenced in September 1955.

CONSTRUCTION

In terms of the water award granted by the Government to the C.D.C. in 1950, a maximum of 9,7 cumecs can be abstracted from the Komati. At least 1,3 cumecs have to be passed down the river as compensation water.

It was estimated that an extraordinary flood might be of the order of 5 666 cumecs from a catchment area of 7 252 square kilometres.

The concrete curved weir across the Komati, 248 metres long and 9,0 metres high was designed to pass a flood of 5 666 cumecs, and to limit the upstream water levels during floods. The design helped also to simplify the headworks for diversion into the canal.

The weir includes two sluice gates 1,83 m × 1,22 m for lowering the reservoir if necessary and also for passing sand which might otherwise be deposited. The canal headworks

Komati Weir and the start of the Mhlume Canal

contain two 3 metres by 1,22 metres radial control gates followed by a sand trap and measuring flume. The compensation water is passed back to the river via the sand trap.

Generally the canal follows the contours with a gradient of 1:5 000 but for economy the four principal valleys are crossed by inverted *in situ* siphons, the longest of which is 1,45 km long and is subjected at its lowest point to a head of 27,4 metres. Other minor valleys are crossed by box culverts and in some cases the canal is so arranged that water from the valley can be passed over it in overpass. The main canal was originally 46 km long with a capacity during the first stage of development of 9,7 cumecs for the first 11 km, after which the capacity reduced to 5,6 cumecs and gradually to 0,85 cumecs at the end of the main canal as various turnouts abstract water for irrigation use. Beyond the main canal a further 21 km of canal, called the "Herzov Furrow", was constructed to supply 0,4 cumecs to a private parcel of land.

The first stage was inaugurated by King Sobhuza II on 14th August 1957, and he called the canal "Mhlume Water" which means 'good growth'.

This canal has become the lifeblood of the development which has occurred since then.

The total cost was £1,1 million.

Mhlume Canal flowing through lush irrigated fields

CONSULTING ENGINEERS AND CONTRACTORS

Komati Weir . .	John Laing and Sons
Mhlume Water	Keir and Cawder Ltd.
Canal	Stirling Astaldi Ltd.

Earth dam

In 1964 work started on Stage II with the construction of an earth dam across the Sand River valley.

Low flows of the Komati River in years of drought or when summer rainfall is delayed would have prevented further development of the run-of-river system and the dam was to act as an insurance against drought and to augment the main canal during periods of peak demand. The dam is filled partly by runoff from the catchment area and partly by pumping from the main canal at times of higher river flow.

The outlet canal from the Sand River dam to the main canal is 8 565 metres long, and has a capacity of 4,53 cumecs at a grade of 1:5 000.

Mhlume Canal distributaries in the fields showing automatic offtakes

The main canal was enlarged in 1965 to take advantage of the additional water provided by the Sand River reservoir and its maximum capacity upgraded to 12,18 cumecs. The

Lined section of Mhlume Canal

SAND RIVER RESERVOIR

SAND RIVER DAM

"Herzov Furrow" was also considerably enlarged and now has a capacity at the tail of 2,8 cumecs. The present length of the main canal is thus 46 + 21 = 67 km.

Sand River Dam:

Consulting Engineers:	(*a*)	Dam – Van Niekerk, Klein & Edwards (Pretoria)
	(*b*)	Pump Station – Dommisse & Durham (Pretoria)
Contractors:	(*a*)	Dam – S. Hilton Barber (Pty) Ltd.
	(*b*)	Pump Station – Mather & Platt (S.A.) (Pty) Ltd. H. Incledon & Co. S.A. Ltd. Hume Ltd. Liebherr-Africa (Pty) Ltd. S.A. Liquid Meters (Pty) Ltd.
Cost:	(*a*)	Dam – R1 121 244
	(*b*)	Pumping Installations – R184 514

Pumps:	4 × Lister Blackstone diesels each capable of pumping 0,85 cumecs against static head of 11 metres.	
Dam Wall:	Length . . .	1 478 metres
	Maximum ht.	25,6 metres
	Greatest base width . .	151 metres
	Approx. quantity of core and fill materials .	765 000 cu. metres
	Spillway . .	183 metres long.
Reservoir:	Total capacity	$49{,}94\ m^3 \times 10^6$
	Effective capacity . .	$40{,}69\ m^3 \times 16^6$
	Surface area .	688 ha

Reservoir Outlet Canal and Main Canal Enlargement:

Engineered by S.I.S.

Contractor:	Thompson-Cramond (Pty) Ltd.
Commenced:	8th September 1964
Completed:	23rd December 1965
Cost:	R374 831

Opened by H.M. King Sobhuza II – 5th October 1966.

The growth in development since 1960 may be illustrated by the figures for cultivation given by the C.D.C. for 1976.

Total areas dependent on water from the Komati weir:-

	(*ha*)
Sugar	10 420
Citrus	526,5
Rice	891
Other Crops	904
	12 741,5

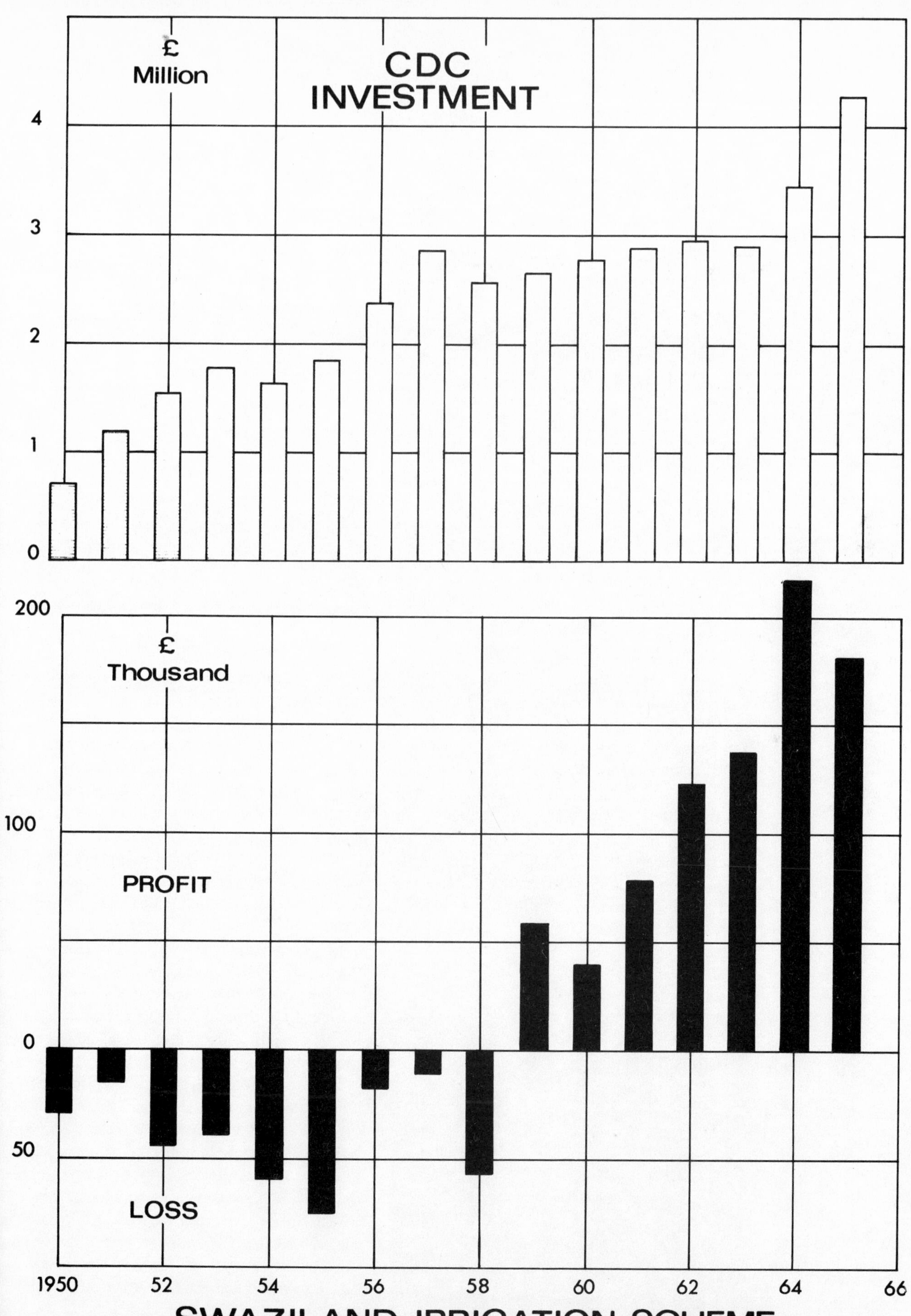

SWAZILAND IRRIGATION SCHEME

INVESTMENT PROFIT AND LOSS FIGURES PUBLISHED BY THE COMMONWEALTH DEVELOPMENT CORPORATION (1966)

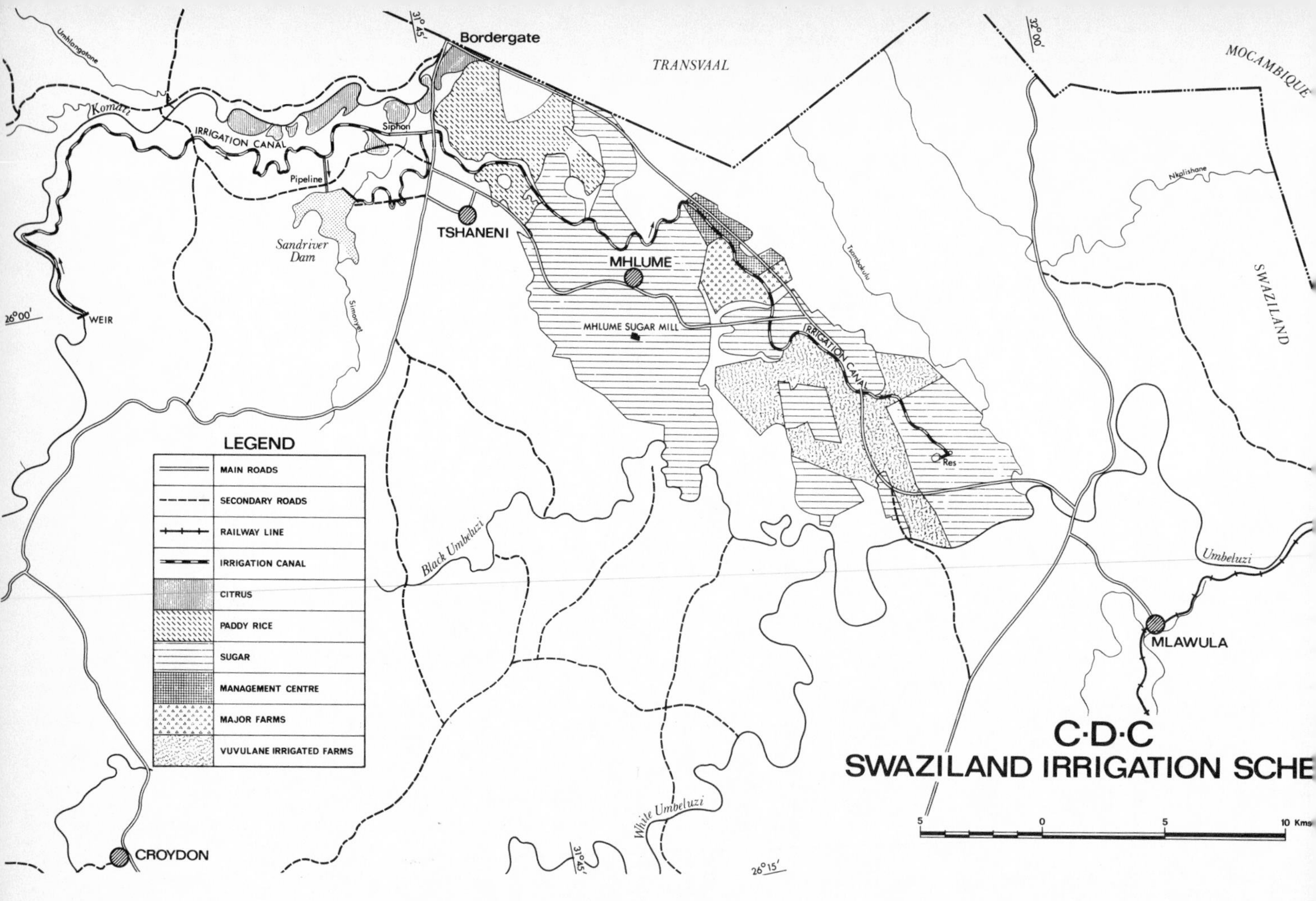

The value of sugar and molasses exports (free at the border) in 1974 amounted to E48 million* or about 43 per cent of Swaziland's total exports for that year.

In 1960 two modern sugar mills, the C.D.C. one at Mhlume and one at Big Bend on the Usutu River, came into production. The Government has under consideration the establishment of a third mill and cane growing area near the Umbuluzi River. This will increase the capacity of the industry by about 50 per cent to more than 300 000 tonnes of sugar per annum.

Overspill section of Komati Weir and Mhlume Canal

The industry employs about 8 000 persons – or about 15 per cent of the total wage earning population. Allowing for dependants and people engaged in associated activities it is estimated that 50 000 persons depend in whole or in part on this industry of which the C.D.C. Mhlume Water Project forms an important part.

The most telling evidence of progress made with such a well organized irrigation project is found in the profit and loss figures for the years 1958 – 1965 published by the C.D.C. in their brochure produced for the opening of the Sand River reservoir in October 1966.

The graph shows that losses were turned into profit within a year after the commissioning of the Mhlume Water Canal in 1957 and the profits have risen steadily since.

It is estimated that the total potential for irrigation in the country, if only good soils on suitable slopes are employed, is 260 000 hectares. Thus even though the major rivers are international in the sense that they rise in the Republic of South Africa and pass through Swaziland before entering Moçambique, it is considered that the water storage potential on the Usutu, Komati and Lomati Rivers is such that an equitable sharing of water will be possible for the foreseeable future.

* Note: The currency of Swaziland is Emalageni.
R1 = 1 Emalageni.

CHAPTER 11 # Botswana

PUTIMOLONWANE DAM (MOPIPI)

HISTORY

The most striking features of the country are its size – about 582 750 square kilometres – its relatively small population – about 450 000 – its arid climate, the predominance of the cattle industry and the general shortage of water with dependence mainly on underground supplies increasingly threatened by depletion.

It has been estimated by the British Mission in its "Report of an Economic Survey Mission" published by Her Majesty's Stationery Office in March 1960 that the total potential arable land, most of which is in the eastern region of the territory, is 3 240 000 ha, of which about 5% or 160 000 ha is under cultivation.

Surface water is available from three major rivers: The Okavango, Chobe and Limpopo.

The Okavango rises in Angola, passes through the Caprivi Strip and brings very large quantities of water into the country on the north west corner of the Okavango Swamp where most of it is lost by evapo-transpiration.

The Chobe forms part of the northern boundary between Botswana and the Caprivi Strip. It is also perennial.

The Limpopo forms the boundary between Botswana and the Transvaal. It carries large flows in the rainy season but it is liable to cease flow in some years during dry weather.

Thus all three major river supplies are international in character. The greatest source of water income is from the Okavango which shows marked regularity of behaviour. The maximum flow occurs during March-April and the lowest during October.

A report by Mr. W. G. Brind, who was Director of Public Works in Botswana (then Bechuanaland), for some years prior to 1951, covering three years observations of gauges,

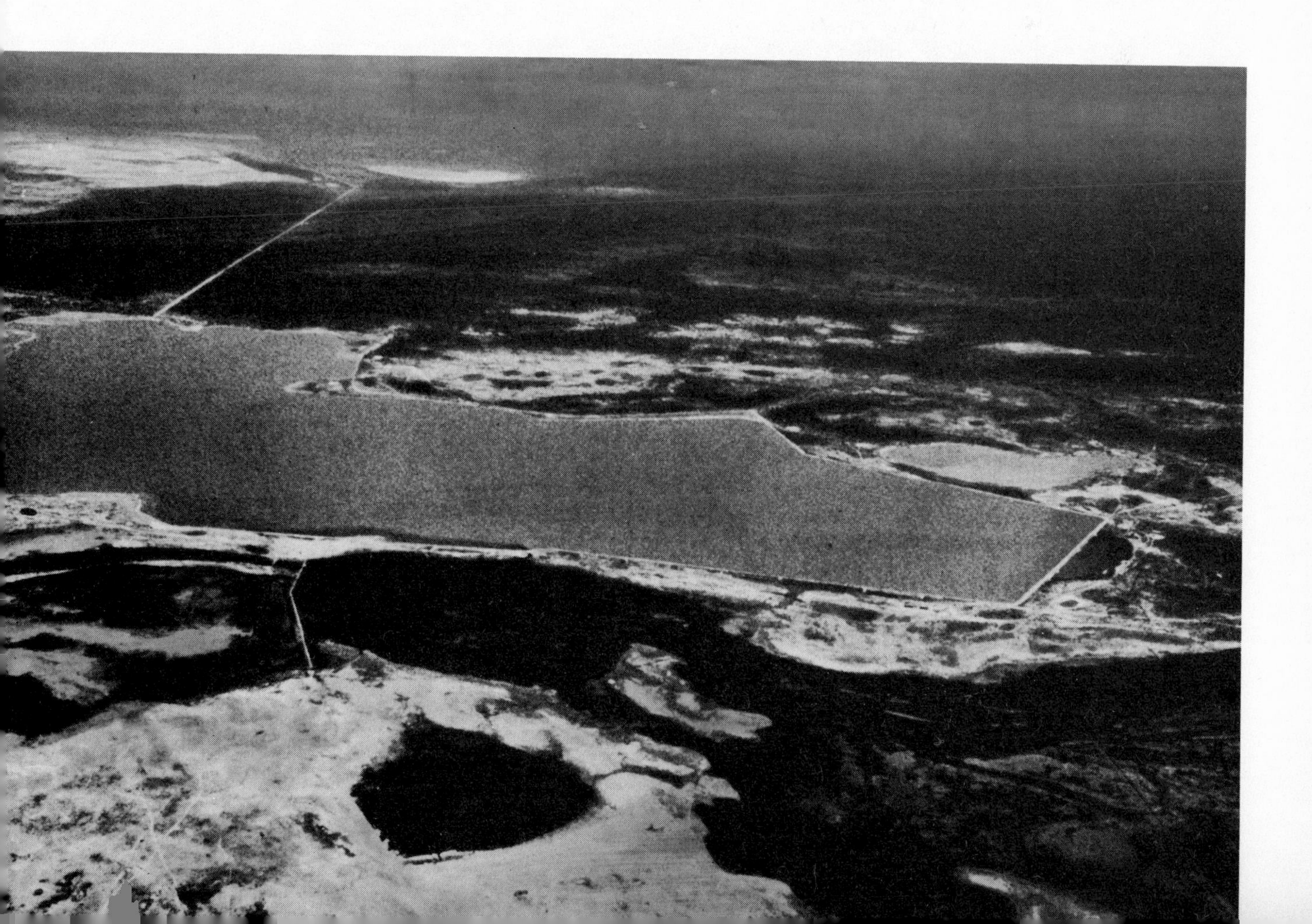

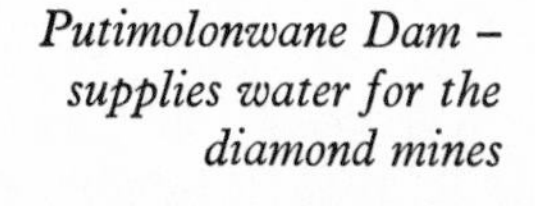

Putimolonwane Dam – supplies water for the diamond mines

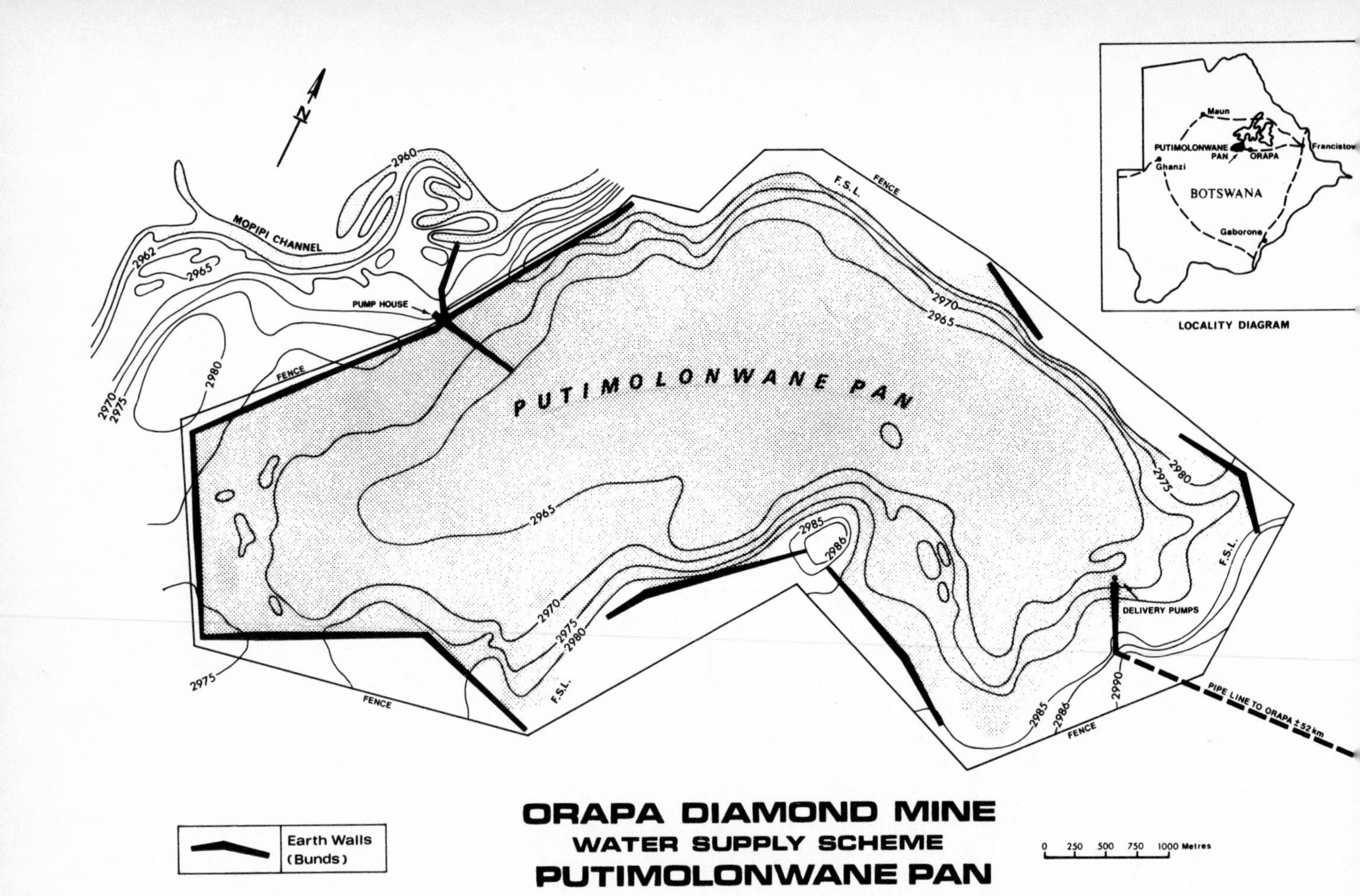

river and swamp behaviour, gave the estimated discharge at Shakawe, at the head of the Swamp, at 243,5 cumecs.

The Swamp, covered mainly with papyrus, is roughly triangular in shape some 161 km long with a base of about 225 km broad at the bottom.

The outflow from the Swamp into the Botletle River is negligible compared with inflow and is unreliable from year to year.

The loss by evaporation, transpiration and from other causes has been estimated at 237 cumecs.

At present there are no major dams in Botswana. However, in order to cope with the water requirements for their expanding diamond mining interests at Orapa, the Anglo American Corporation arranged for the provision of a storage reservoir at Mopipi near the Botletle, making use of the storage capacity of the Putimolonwane Pan by the construction of very long earth walls or "bunds". The design storage is such as to ensure three years' supply for the mining requirements which is achieved by pumping from the pan to Orapa – a distance of some 51 km. The area and capacity of the lake are very large for the height of the walls around the pan which brings the project within the definition of large dams as laid down by the International Commission on Large Dams (ICOLD).

DIMENSIONS

Area-Capacity Area at F.S.L. of storage	2 430 ha
Capacity F.S.L.	$100 \times 10^6 m^3$
Length of wall	11 000 m
Maximum Height	5 m
Volume of Embankment	$365\ 000\ m^3$
Cost	R650 000
Installed Power	500 kW

To supply not less than 13 500 m^3/day to Orapa Diamond Mine with maximum security.

CONSTRUCTION

No dam sites were available on the nearest river, the Botletle, so a decision was made to provide off-channel storage in an adjoining pan. To provide for a 3 year drought between replenishments, a total depth of 6 metres was required – 4,5 metres to provide for evaporation and 1,5 metres over an area of pan to provide the water supply, requiring the natural depth of pan to be increased accordingly. Replenished by 4 pumps with a total capacity of 7 cumecs pumping against a maximum head of 6 metres.

CONSULTING ENGINEERS AND CONTRACTORS

Designed by: B. G. A. Lund & Partner

Constructed by: Grinaker Construction, Botswana (Pty) Ltd. Elandsfontein.

CHAPTER 12 **Lesotho**

LETSENG-LE-TERAI

HISTORY

There are as yet no major dams in Lesotho. For years a major project called the Oxbow Scheme was discussed but this project has been shelved.

There is great potential for dam building in this "Little Switzerland" in Southern Africa.

In 1968 the Cementation Company of South Africa constructed a novel rock and earthfill dam at Letseng-le-Terai for the Rio Tinto Company. The design by Gibb Hawkins and Partners in collaboration with Sir Alexander Gibb and Partners was based on the research work carried out by Dr. H. Olivier in conjunction with the British Hydromechanics Research Association. The design parameters were outlined in his "Paper 7012 – Through and Overflow Rockfill Dams" delivered at the Institution of Civil Engineers, London in April 1967. The design parameters which permit a rockfill dam to be overtopped are: the quantity of water to be accommodated per unit length of crest, the size of rock placed on the downstream face to accommodate the sheet of water, the slope of the downstream rock face and a dimensionless constant referred to as the "packing factor".

DIMENSIONS

The dam is about 7 metres high with a crest length of 170 metres, of which the centre section of 40 metres acts as spillway. The elevation of the crest of the dam is at 3 100 m altitude and the dam was located in an inaccessible part of the country that was the prime reason for the choice of such a simple design, apart from the fact that the cost turned out to be very low compared with more orthodox designs.

The rockfill spillway has been overtopped

Letseng-le-Terai rockfill overtop dam – view looking downstream

Letseng-le-Terai Dam showing crest arrangements for spillway

Letseng-le-Terai Dam – looking upstream, showing spillway in centre of dam and rock protection for spilling section

Bafokeng Dam showing rockfill protection

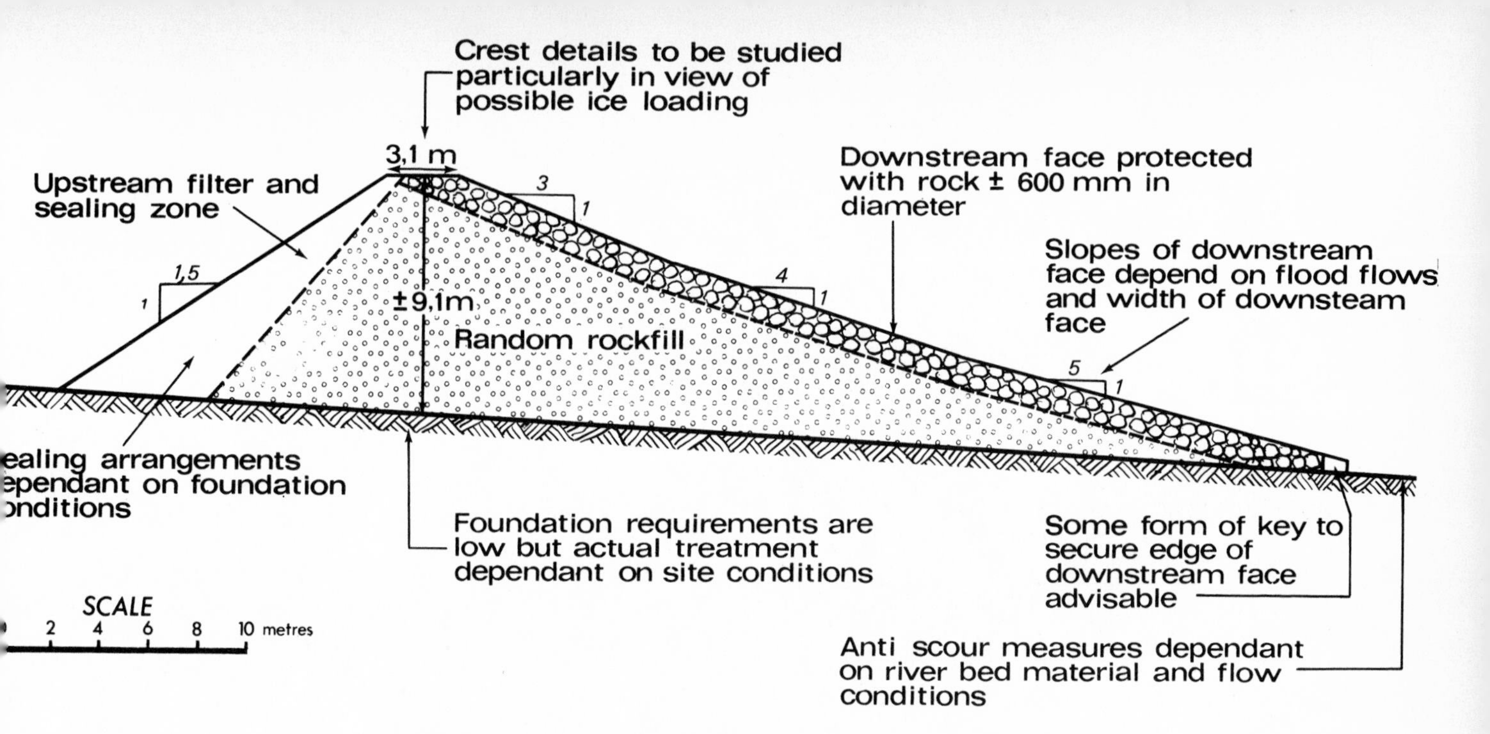

TYPICAL CROSS SECTION

LETSENG-LE-TERAI DAM

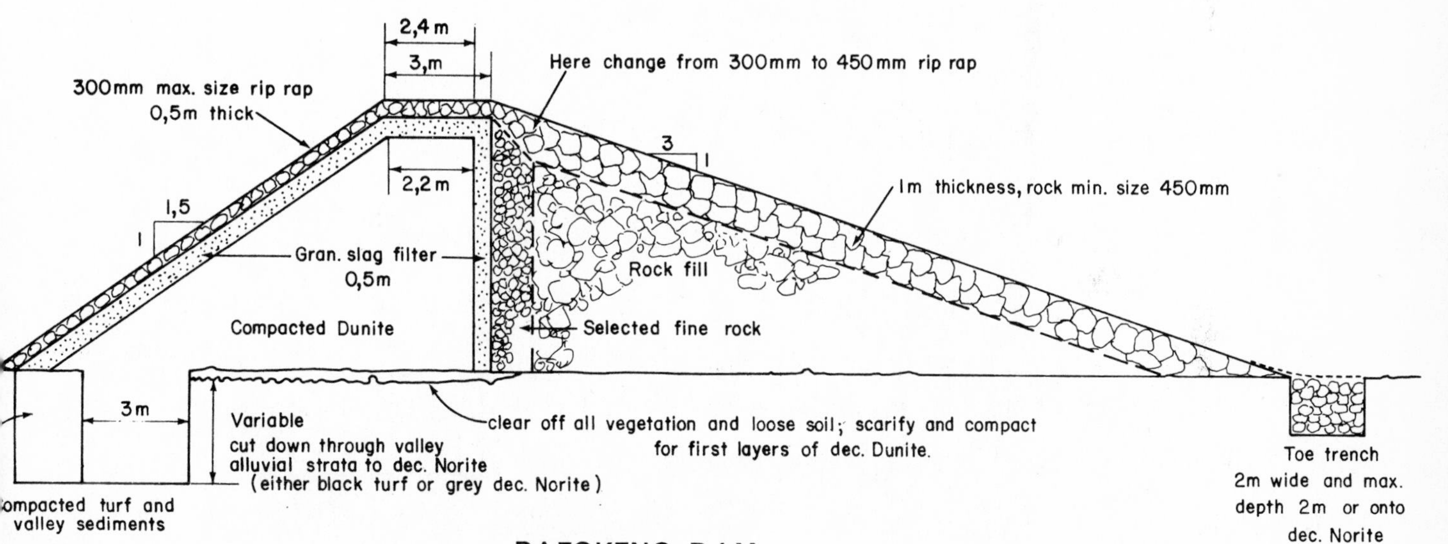

BAFOKENG DAM

many times since then with no deleterious effects. The design flood for the relatively small catchment upstream of the dam but subject to very intensive precipitation was 226 m³/s, or equivalent to 5,7 m³/s per metre of spillway crest length.

The same design parameters have been used since with success, in the construction of the Lesapi dam in Rhodesia and the 40 metre-high upstream diversion coffer dam at Cabora Bassa on the Zambezi that was overtopped twice, in 1972 and 1973. At the peak of the first flood the flow over the crest was some 4 metres deep.

Since then the principles have been applied successfully in the construction of slimes dams in the Republic of South Africa.

A typical case is the design prepared for the Union Corporation Limited by Professors Jennings and Midgley of the University of the Witwatersrand to replace a previous slimes dam that failed at the Bafokeng mine in Rustenburg.

The rock size used for the downstream "working" face at a slope of 1 on 3 was ½ metre cubical fragments.

The dam was completed in the first half of 1975 and has since been overtopped regularly and efficiently. The design has resulted in significant cost savings for the owners.

CONSULTING ENGINEERS AND CONTRACTOR
Cementation Company of South Africa.

CHAPTER 13

Transkei

UMTATA HYDRO-ELECTRIC SCHEME

Transkei became an independent Republic on 26th October 1976. With an estimated population of 3,1 million it has a total surface area of approximately 40 000 square kilometres, which is 1,3 times the size of Lesotho and more than twice the size of Swaziland.

The mean annual rainfall varies from 1 400 mm to 700 mm with an average value over the whole country of approximately 900 mm. The combined mean annual runoff of the three major internal river systems, the Umzimvubu, Umtata and Bashee Rivers, is 4 760 million cubic metres. It is estimated that of the total annual runoff 2 750 million cubic metres per annum, or a continuous flow of nearly 90 m³/s, can be regulated for hydro-electric power, irrigation and other water supply developments.

There are no existing major storage developments on any of the rivers. However, in collaboration with the Transkei Government the South African Government has authorised extensive surveys and exploratory work to establish the viability of developing the hydro-electric potential of the Umzimvubu catchment basin and other rivers for the bulk export of power to the Republic.

The rock and earthfill dam on the Umtata River is scheduled for completion during the latter half of 1977. It is a multi-purpose concept designed to meet the expanding needs of Umtata, the capital of Transkei, as regards

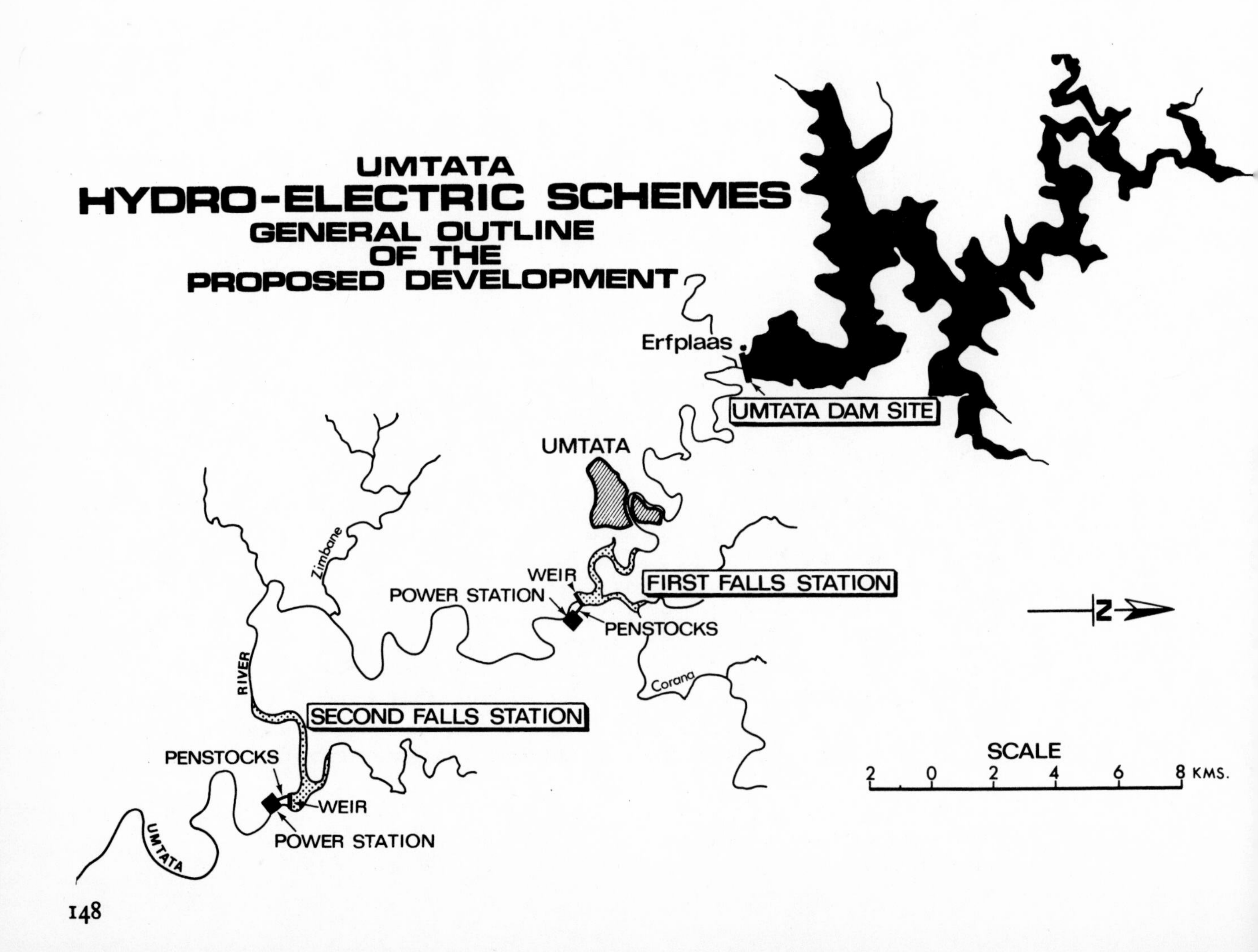

EMBANKMENT CROSS-SECTION

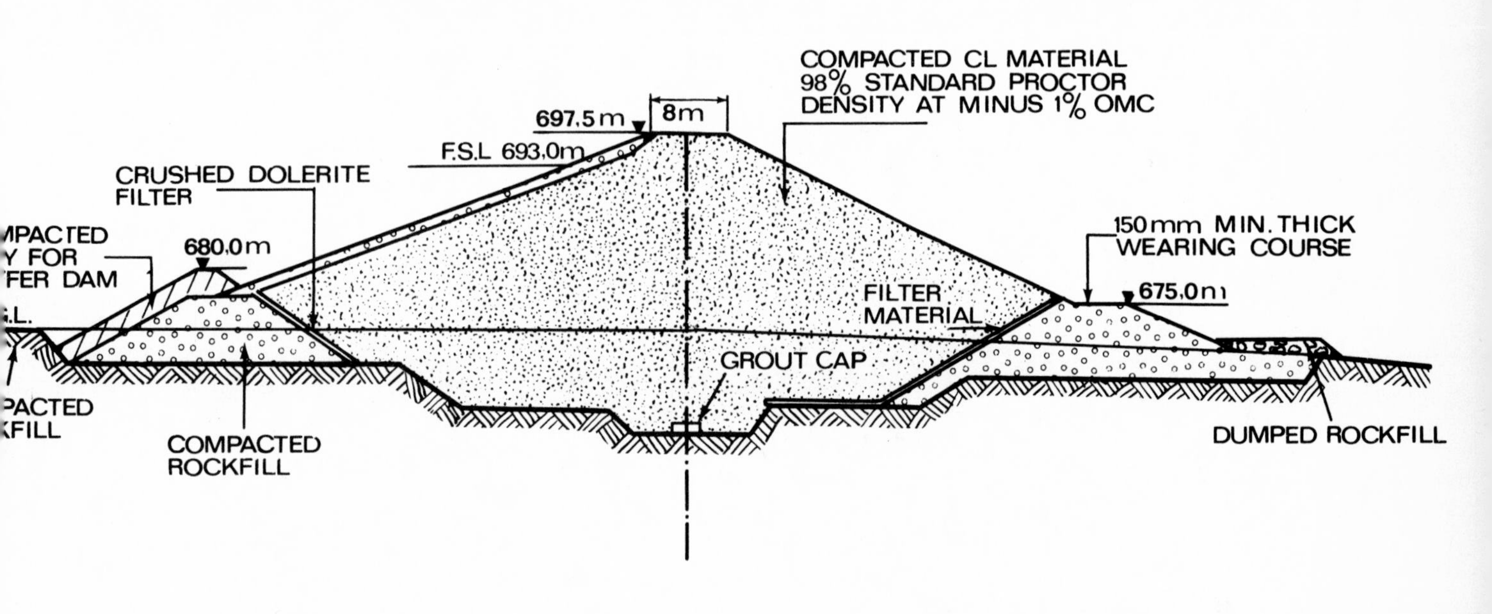

COMPACTED CL MATERIAL
98% STANDARD PROCTOR
DENSITY AT MINUS 1% OMC
697.5 m
8m
F.S.L 693.0m
CRUSHED DOLERITE
FILTER
680.0 m
150mm MIN. THICK
WEARING COURSE
675.0m
FILTER
MATERIAL
GROUT CAP
COMPACTED
ROCKFILL
DUMPED ROCKFILL

GENERAL LAYOUT

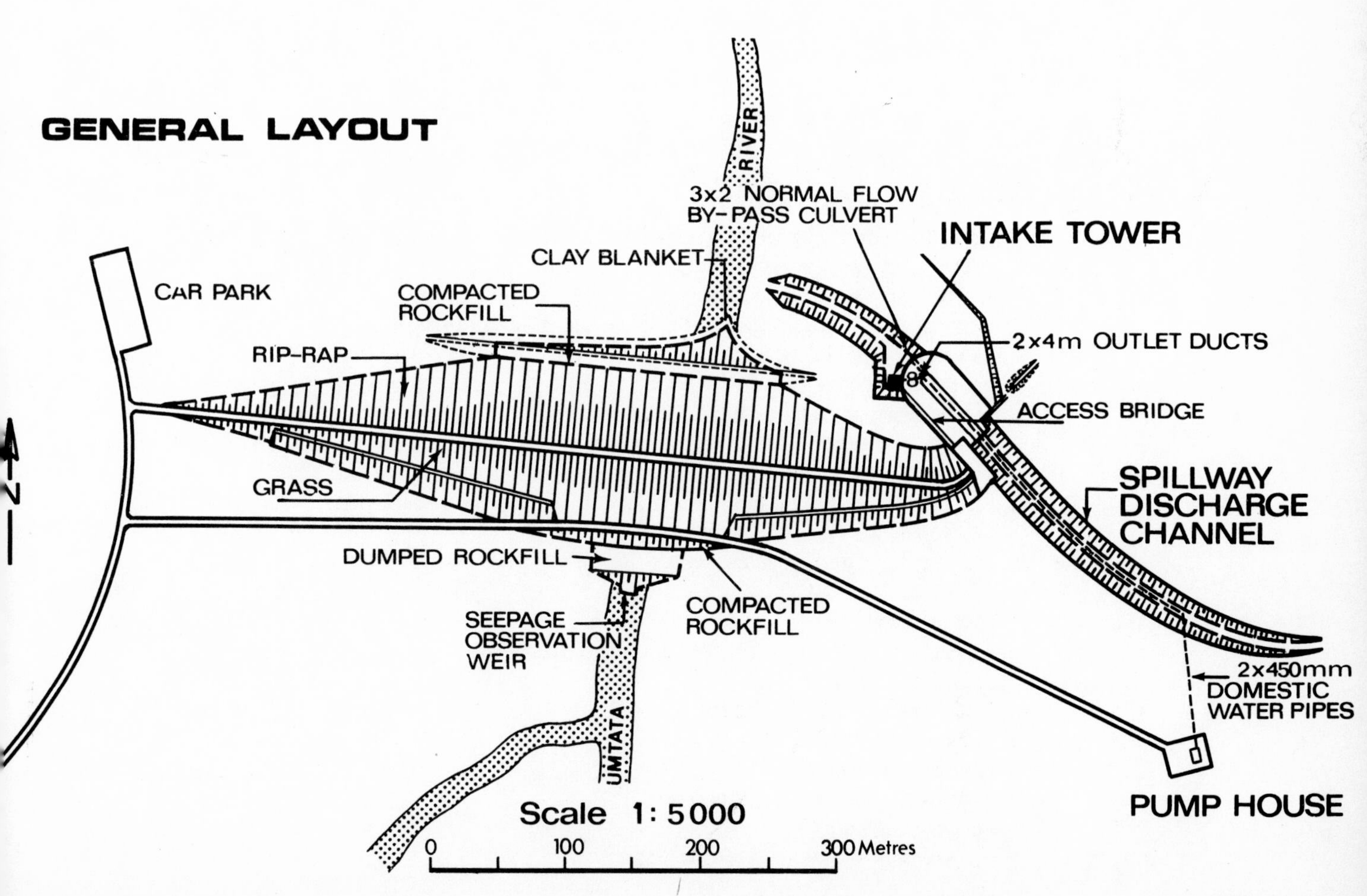

RIVER
3x2 NORMAL FLOW
BY-PASS CULVERT
INTAKE TOWER
CLAY BLANKET
CAR PARK
COMPACTED
ROCKFILL
RIP-RAP
2x4m OUTLET DUCTS
ACCESS BRIDGE
GRASS
N
SPILLWAY
DISCHARGE
CHANNEL
DUMPED ROCKFILL
SEEPAGE
OBSERVATION
WEIR
COMPACTED
ROCKFILL
2x450mm
DOMESTIC
WATER PIPES
UMTATA
Scale 1: 5000
0
100
200
300 Metres
PUMP HOUSE

potable water and electricity supplies.

Key feature in the concept is the construction of the rock and earthfill dam on the Umtata River some 17 km upstream of Umtata. The volume of rock and earthfill is some 800 000 m^3. The gross storage capacity of the reservoir is 260 $m^3 \times 10^6$, or 118 per cent of the mean annual runoff of the river. The maximum height of the dam is 35,5 metres with a crest length of 600 metres. The mean surface area of the lake is 2 500 ha.

Water for urban consumption is supplied from the dam through a pipeline to the Municipal purification plant located on the right bank of the river close to the town.

The two hydro-electric power stations will be located at First and Second Falls on the Umtata River 3 km and 31 km respectively downstream of Umtata. The total head available at these two stations is about 80 metres and the final installed capacity is about 34 MW.

As regards water allocation between the years 1979 and 2003 the urban supply is expected to rise from 5,0 $m^3 \times 10^6$ to 35,6 $m^3 \times 10^6$ while the allocation to hydro generation would fall from 165,0 $m^3 \times 10^6$ to 134,4 $m^3 \times 10^6$.

Apart from direct water power and energy benefits other indirect benefits envisaged are:-

1 Considerable attenuation of flood discharges due to the storage capacity of the reservoir. Thus the 10-year flood peak storm of 24-hour duration will be reduced from 1 065 m^3/s to 363 m^3/s;
2 The reservoir, located so close to the capital of Transkei, will provide numerous opportunities for recreation and relaxation;
3 Development of a commercial fishing industry in the reservoir;
4 The hydro-electric plants will deliver peak energy at a competitive price with imported power or by diesel generation and reduce oil fuel requirements.

The client for the Umtata water and hydro-electric project is the Secretary for Agriculture and Forestry, Transkeian Government.

The concept was developed by Henry Olivier and Associates, Johannesburg.

The dam contract was awarded to Grinaker Construction, Africa (Pty) Ltd., assisted by their consulting engineers, O'Connell, Manthé and Partners.

Tenders for the hydro-electric installations, on a design and construct basis were called for in December 1976.

LUBISI DAM

HISTORY

The Qamata Irrigation Scheme was built in 1968 for the Department of Bantu Administration and Development, and comprises a storage dam, the Lubisi Dam, on the Indwe River in a poort near Southeyville and roughly 53 km east of Queenstown, as well as a diversion weir, the Lanti Weir about 6 km downstream of the dam, and a canal system. The owner is now the Republic of Transkei. The scheme supplies water for the irrigation of 3 420 ha of good ground on the left bank in the vicinity of Qamata and St. Marks, and 860 ha of good ground on the right bank between the Indwe and the White Kei Rivers. A Bantu irrigation settlement on intensive lines is being established by the Department of Bantu Administration and Development.

DIMENSIONS

Type	Double curvature arch
Height above lowest foundation	52 m
Length of crest	268 m
Volume content of dam	77 000 m^3
Gross capacity of reservoir	156 567 000 m^3
Maximum discharge capacity of spillway	1 250 m^3/s
Type of spillway	Uncontrolled

CONSTRUCTION

The storage dam is an arched concrete structure 49 m high and 213 m long with an overspill section 110 m long, and a net capacity of 157,55 million m^3. The diversion weir is a gravity section concrete wall with canal outlet and scour gate on the left flank.

CONSULTING ENGINEERS AND CONTRACTORS

Engineering by	Department of Water Affairs (Republic of South Africa)
Constructed by	Department of Water Affairs (Republic of South Africa)

TSOMO DAM

HISTORY

Tsomo Dam is situated on the Tsomo River some 20 kilometres off the section of National Road N18 between Engcobo and Cofimvaba on the road to the town of Cala. Completed in 1974, it is the first large dam to be constructed on the Tsomo River, which rises in the Indwe and Elliot districts and discharges into the Great Kei River on the south-western border of Transkei.

Tsomo Dam, owned by the Republic of Transkei, stores water primarily for irrigation. Compensation water released from the dam drives a small turbine which supplies electrical power for security lighting, mechanical plant and flow recording equipment.

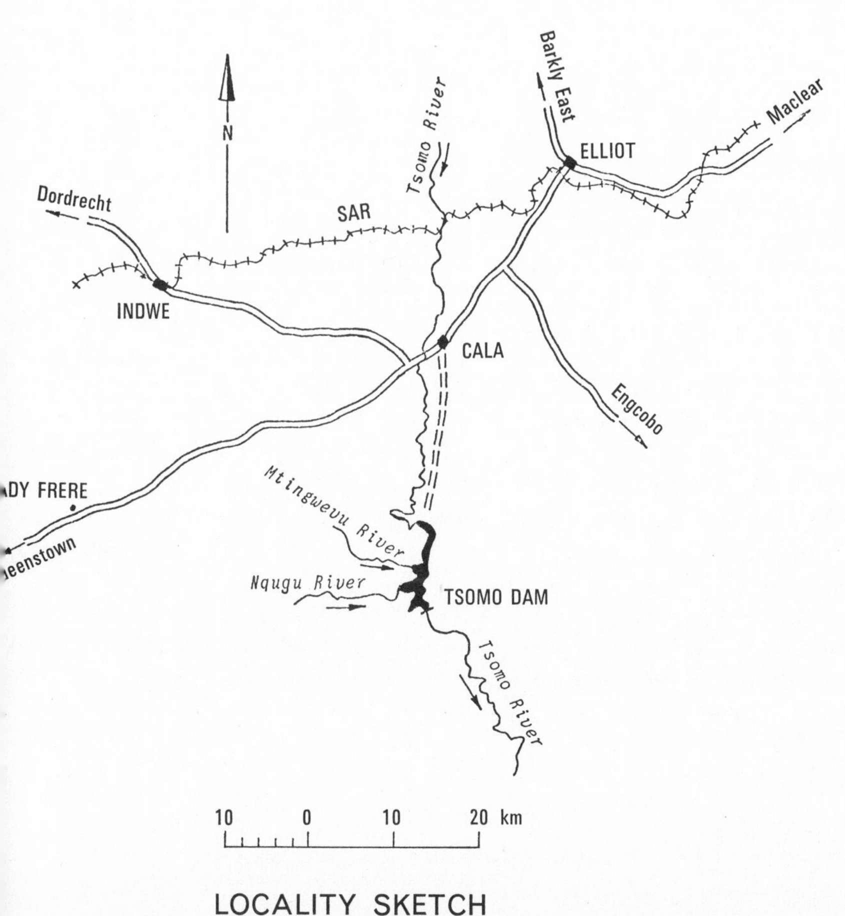

LOCALITY SKETCH

DIMENSIONS

Type	Gravity
Height above lowest foundation	44 m
Length of crest	305 m
Volume content of dam	134 000 m³
Gross capacity of reservoir	181 250 000 m³
Maximum discharge capacity of spillway	1 880 m³/s
Type of spillway	Uncontrolled

CONSTRUCTION

The dam is a concrete mass-gravity structure with a total crest length of 305 metres and centre overspill of 80 metres. The maximum water depth is 32 metres. The spillway has been designed to pass a flood of 1 880 m³/s which, when cognisance is taken of the flood retention characteristics of the dam, is expected to occur statistically once every 100 years.

Tsomo dam will supply water to 5 700 hectares of irrigable ground. The outlet works incorporate two float-controlled butterfly valves which supply water to twin 1 500 mm diameter pipes, which in turn feed a tunnel through the mountain which forms a watershed between the Tsomo and Qamanco Rivers. At the outlet of the tunnel the water is released by float-controlled cylinder valves into the main canal and from the main canal it is piped to the irrigation beds where the water is applied mainly by sprinklers under gravity pressure. The outlet and distribution system is automated for delivery at an individual irrigation plot.

CONSULTING ENGINEERS AND CONTRACTORS

Engineering by	O'Connell, Manthé, Ross & Partners
Constructed by	Concor Construction

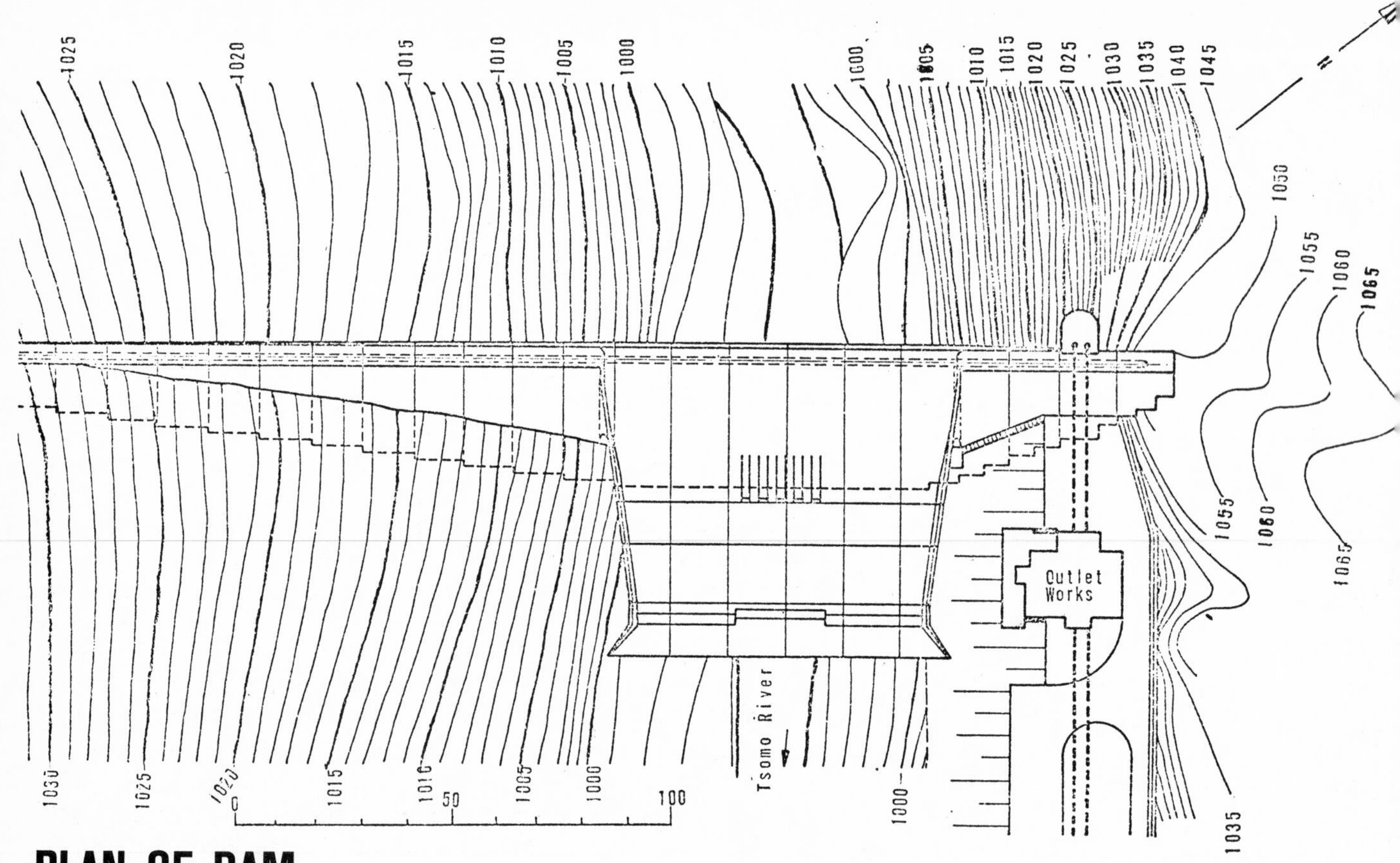

PLAN OF DAM

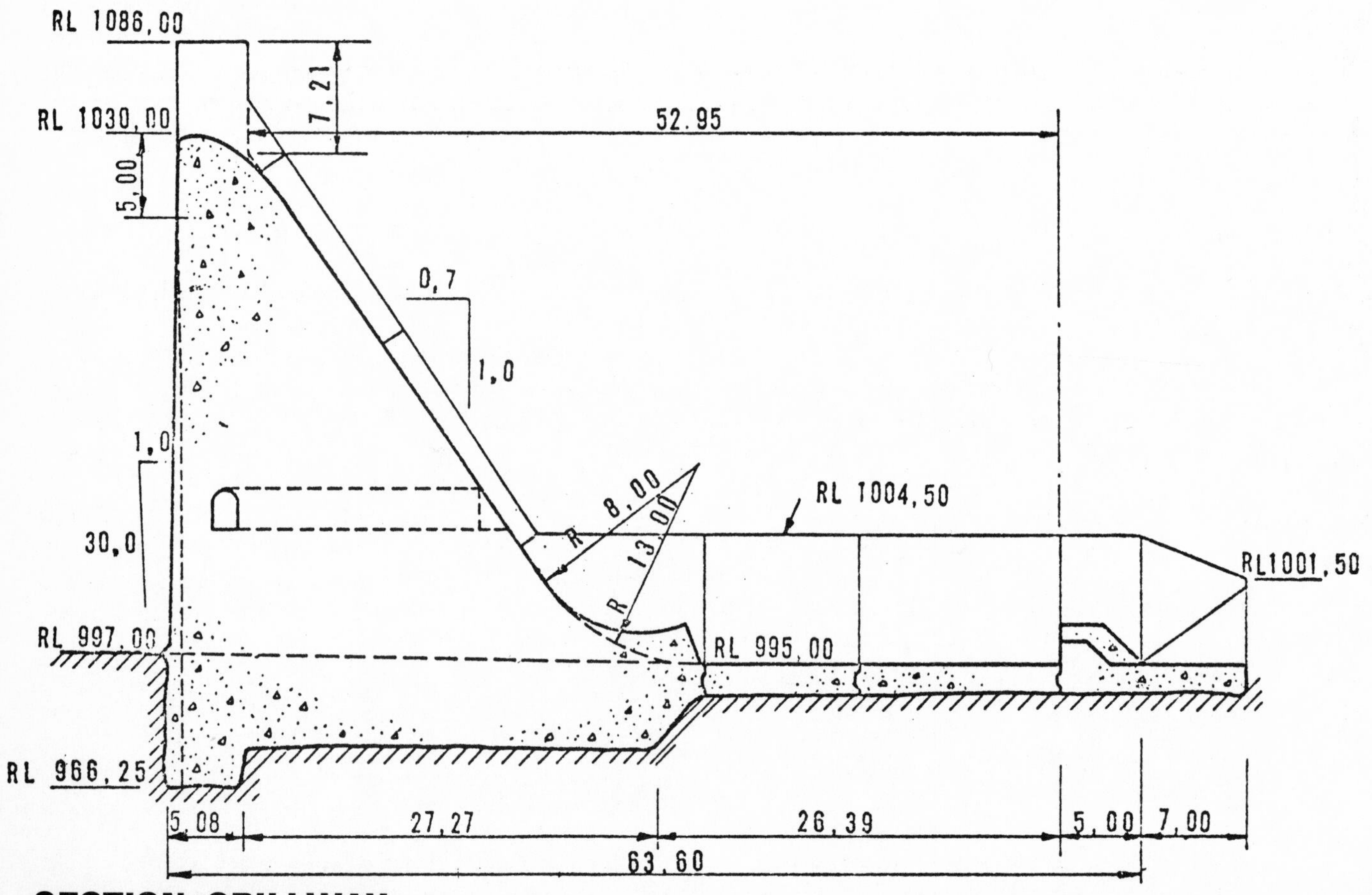

SECTION: SPILLWAY

CHAPTER 14 # South Africa

REPUBLIC OF SOUTH AFRICA

The Republic of South Africa covers an area of approximately 1 221 180 km² which is five-and-one-quarter times the size of Great Britain, and more than the combined area of Germany, France, Italy and Portugal. The Limpopo River forms its northern boundary about 22° south of the Equator.

The Republic is divided into four provinces covering the following areas:

*Cape Province**	676 053 km²
Transvaal	283 863 km²
Orange Free State	129 240 km²
Natal	87 024 km²

South Africa consists of an interior plateau surrounded on three sides, except on the eastern boundary of the Transvaal, by a coastal belt of varying characteristics. The escarpment which separates these two tracts of country is the Drakensberg range stretching from the northern Transvaal through Natal and the Transkei, the Stormberg range of the north-eastern Cape, the Sneeuberg and various other ranges along the west coast. The highest parts of the country are generally on or near the edge of the escarpment; the highest point being in the Drakensberg nearly 3 350 m above sea level. The interior plateau, which represents about 40% of the total area of the Republic, is included on a 1 219 m contour.

It may be said that South Africa lies in the drought belt of the globe – consequently it may be regarded as a dry country. The west is drier than the east and the lower the rainfall the more unreliable it is. Although over the geological ages there have been great changes, the climate of historical times, that is for as long as reliable data have existed, has in general remained much the same. The question therefore is not so much how the drought problems can be solved but rather how best use can be made of the various water resources available.

The average annual runoff from the Republic (i.e. the average annual quantity of water that reaches the rivers) is estimated at 52 m³ × 10⁹ per annum. Because of unavoidable losses by spillage and evaporation from storage and the fact that not all runoff can be diverted for use before reaching the sea, it has been accepted that only 40% of the runoff, or an average of 57 280 × 10³m³/d can be regarded as the assured proportion that can be made available for use through the provision of storage.

About 20 per cent of the total runoff in the Republic is in the Tugela and Umzimvubu River basins.

The best long-term utilisation of these water resources is under active investigation.

Several measures can be applied to enhance appreciably the assured yield of storage works. It may be possible to regulate a somewhat greater proportion of the runoff by means of storage. Thus it should be possible to push up the utilisable surface waters of the Republic to about 50% of the mean annual runoff i.e. to 26 m³ × 10⁹ per annum or 71 370 × 10³m³/day.

To arrive at the total usable natural water resources of the Republic the potential ground water yield must be added. Unfortunately however it is not possible at this juncture to make reliable estimates of the safe yield of underground water resources. It is reasonably certain that present consumption could be doubled if sufficient were known about the occurrences and replenishment of groundwater.

For the sake of conservatism the "Commission of Enquiry into Water Matters" (R.P. 34/1970) took into account only the then current estimated total abstraction of about 3 090 × 10³m³/day from groundwater resources. Given thorough investigation, research and planning, with the necessary financial means the Commission concluded that 27,4 m³ × 10⁹ could be made available per annum. (75 010 × 10³m³/day).

The South African Water Research Commission is proceeding with a nationwide survey of underground water resources with the objective of being able to estimate in the near

* Transkei, previously a part of the Cape Province, became an independent Republic on 26th October 1976. It has an area of approximately 45 000 km².

future, within reasonable margins of error, the extent of the country's available underground water resources, having regard to average rates of annual replenishment.

As regards research in water purification for re-use, the South African Council for Scientific and Industrial Research (C.S.I.R.) has done outstanding pioneer work. It initiated the first application of electrodialysis for the Orange Free State Goldfields and achieved a major breakthrough with the sewage purification techniques adopted for Windhoek, the capital of South-West Africa. Dr. Stander of C.S.I.R. has been president for the second and third terms of the International Association of Water Pollution Research.

The rapid growth rate of the Republic in the past and forecast for the future has made the authorities very conscious of the need for water conservation and "best-use" techniques. This has led over the last two decades to large intercatchment water resources projects such as the Orange River and the Tugela-Vaal schemes.

The following are a few economic indicators of the country's growing industrial strength:

Year	*Installed capacity for generation of electricity (MW)*	*Production of Iron and Cement (1000 metric tonnes)*		*South African Railways Total Traffic Ton-km × 10^6*
		Pig Iron	*Cement*	
1945	1 293 856	556	1051	17 181
1955	3 294 721	1260	2337	30 384
1965	6 565 250	3271	3882	44 306
1975	11 241 500	5180	7176	65 646

In Agriculture, Forestry and Fishing the Gross Production is expected to rise from R1 875 × 10^6 in 1973 to R2 646 × 10^6 by 1979 at an average growth rate of 5,9% per annum.

HENDRIK VERWOERD DAM and ORANGE-FISH TUNNEL

HISTORY

The Hendrik Verwoerd Dam is located on the Orange River, the longest river in Southern Africa, in a narrow gorge some 5 kilometres upstream of Norvalspont in the north-eastern Cape.

It is the largest storage reservoir in the Republic and the key structure in the multi-purpose, intercatchment project conceived by the Department of Water Affairs in collaboration with the Electricity Supply Commission in the White Paper WPX-62 tabled in Parliament during the 1962–3 session. Modifications to the concept were authorised in subsequent White Papers WP.AA–64 and WP. LL–68.

It is not possible to appreciate the function of the Hendrik Verwoerd Dam without reference to a map and to the functions of the other major works, such as the 82 km Orange-Fish Tunnel, the P.K. le Roux dam and its canal system and to the water resources projects which were necessary for the beneficial use of the waters transferred to the eastern province river systems by the Orange-Fish Tunnel from the Hendrik Verwoerd Dam.

The concept of the Orange River Project was the brainchild of Dr. A. D. Lewis who, as Secretary for Water Affairs, launched the idea in 1928.

Scarcity of Water in the Lower Orange River

About 60 per cent of the area of the Republic receives less than 500 millimetres of rain per annum and it is estimated that only 8 per cent of the national rainfall reaches the river. Of this, half is lost through floods and evaporation.

The central part of the country is semi-desert, with very hot summers. The Orange River flows through these areas and, from the areas upstream of the Hendrik Verwoerd Dam, an average of 6 500 million cubic metres of water is carried to the sea annually. For many years it was realised that this valuable water should be used.

The farmers along the banks of the Orange River suffered greatly, since the flow of the river is very inconsistent; in the dry winter months the river can be an insignificant stream which often disappears in the sandy river bed during the early summer.

Thousands of hectares of the most fertile, irrigable land in South Africa are situated on the banks of the river. These areas are wholly dependent upon the river for irrigation and in the past, during the periods of drought, irrigation water had to be released from the Vaal Dam.

The development of the Witwatersrand area and the needs of consumers along the Vaal River have increased to such an extent during the past 20 years that the water from this river had to be reserved for this part of the country only.

A different water supply for the Lower Orange River had to be found. These circumstances greatly influenced the decision to embark on the Orange River Project.

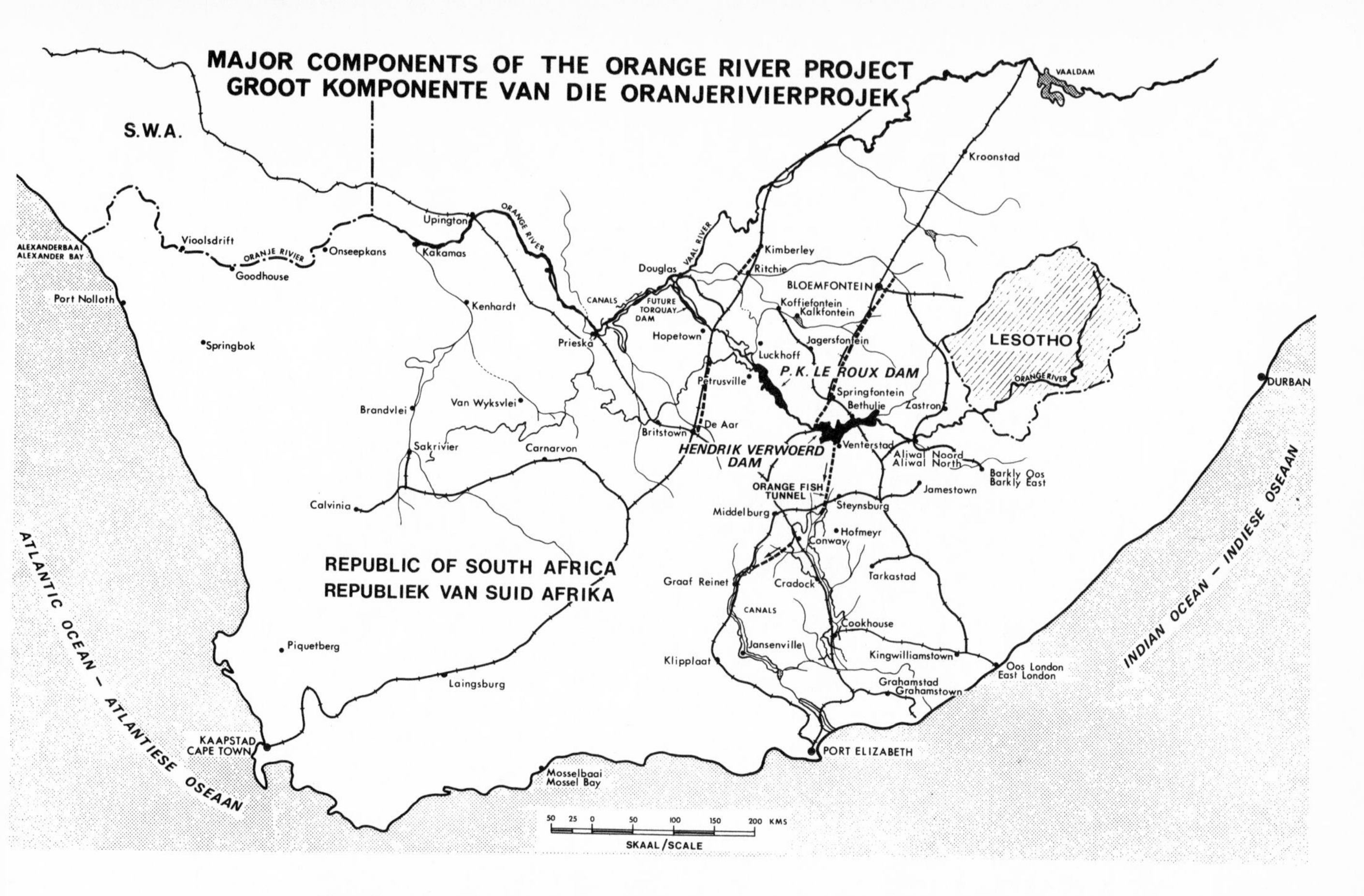

Scarcity of water in the Eastern Cape

In the Eastern Cape the problem was the same: the inconsistent rivers could not fulfil the irrigation water requirements. Early in the 19th Century, this region was developing very rapidly and farmers pressed the Government for assistance in the form of large storage dams to regulate the flow of the Great Fish and Sundays Rivers.

Four large dams were built in the 1920's: Grassridge and Lake Arthur in the Great Fish River Valley, and Van Ryneveld's Pass and Lake Mentz in the Sundays River. However, because of many difficulties these dams did not solve the problem permanently.

By 1942 it became obvious that the only solution to the irrigation problems of the farmers in the Eastern Cape would be the importation of water from the Orange River. However, this would be very costly and for this reason it was impossible to implement the project at that time.

Flood damage

The Orange River is a wild and unpredictable river. Its flood follows no fixed pattern. Heavy rains in the catchment area can cause a raging torrent that crashes over its banks, taking everything with it and annually washes invaluable tons of fertile soil into the sea.

When the floods come, man and animal must flee to safety. Many people have lost their lives. Temporary walls are built feverishly to safeguard the farm lands, but after every flood the scars remain; farmlands stripped of valuable top-soil, full of dongas, and harvests deposited somewhere downstream. The two main dams of the ORP, especially the Hendrik Verwoerd Dam, will reduce these floods to a large degree; but because of the large uncontrolled catchment area below the dams and the nature of rainfall in South Africa, the possibility of heavy floods and damage remains.

Planning

Much of the stored water will first be used for the generation of electricity before it is released for other purposes. During the planning stages much attention was given to the problem of siltation at the Hendrik Verwoerd Dam. The average silt load of the Orange River is not as high as that of the Great Fish River, but nevertheless it is estimated that the capacity of the dam will be decreased by 25 per cent during the first 60 years.

An example of the danger which siltation can bring about – 250 000 tons of silt were carried down the river every hour in the great flood of 1967.

The structure of the dam provides for the wall to be raised twice, and this is how the problem of siltation is to be fought. By considerably increasing the capacity of the dam, the effect of siltation can be counter-balanced.

Orange-Fish Tunnel
This tunnel, 82 km long and 5,3 m in diameter, will carry water from the Hendrik Verwoerd Dam to the Great Fish and Sundays River Valleys. A maximum of 56,5 cumecs of water can flow through this tunnel daily to relieve the need in the river valleys of the Eastern Cape where thousands of hectares of soil had to be deproclaimed in the last few years due to water scarcity.

Design
The design of the Hendrik Verwoerd Dam and the supervision of construction were undertaken by consulting engineers, and the construction by contractors.

The Hendrik Verwoerd Dam is of the combined gravity and arch type, having a central double curvature main arch with an overspill section under the crest road and two gravity abutments, into which the arch section of the dam wall gradually merge. The dam is constructed entirely of concrete, aggregate and sand, which was obtained by crushing dolerite rock quarried at the site.

Preparations
The first major visit to the dam site by engineers of the Department of Water Affairs and the consultants took place during October 1963 and from this time onwards the detailed planning of the amenities and services started. By April 1964 the first of the nearly 3 500 men eventually employed on the project began to arrive on the site; they were housed in temporary accommodation while the permanent accommodation was being built.

The township named Oranjekrag was established to accommodate the construction force and was planned in such a way that after completion of the dam it would become a modern tourist and recreation centre. In order to accomplish this task in time for the award of the main contract in early 1966,

Republic of South Africa
Run-off Diagram (H.O.)

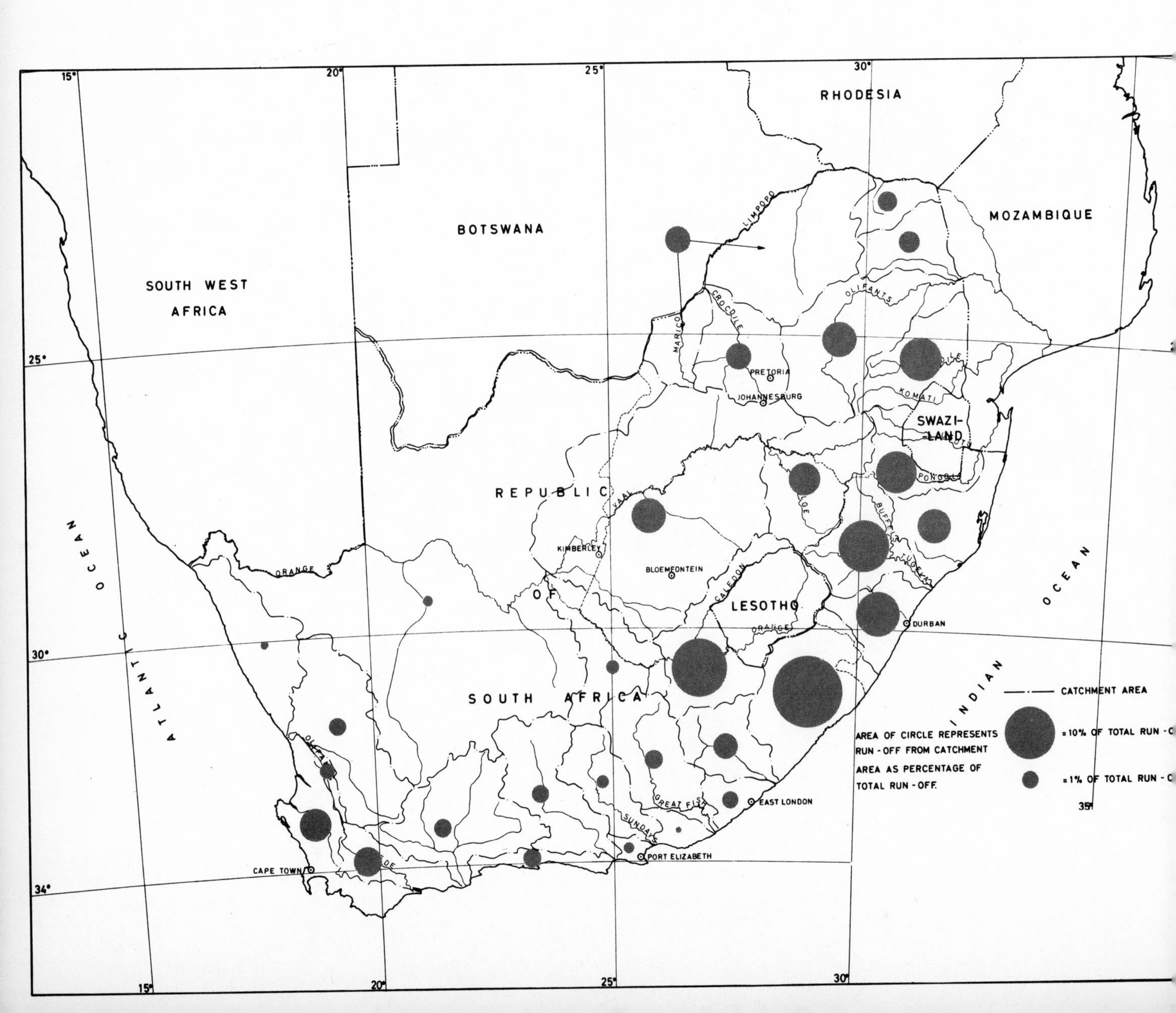

twelve separate contractors were employed on the site during the period 1964 to 1966. The high standard of the accommodation and facilities materially assisted in welding the large number of persons of different nationalities into the strong team required to complete the giant project on schedule.

Contracts awarded

From 1963 onwards twelve preliminary contracts were awarded for site investigation, temporary and permanent housing, road works, river bridges, an aerodrome and the various other aspects of township development at Oranjekrag, including water and electricity supply and reticulation, sewage disposal and amenities for the future residents. This preparation enabled the Contractors to proceed with the main contract for the construction of the dam with the minimum of delay.

Tender documents for the construction of the Hendrik Verwoerd Dam and appurtenant works were issued in August 1965 and by the end of that year, when tenders closed, offers had been received from six groups.

The dam was completed by 3 July 1971. The reservoir filled four days prior to the official opening by the State President on 4 March 1972.

DIMENSIONS

Civil Engineering

River bed level . .	1 191,7 m
Full supply level . .	1 258,8 m
High flood level . .	1 264,9 m
Crest level (roadway) .	1 267,9 m
Maximum height above river bed level . .	77,5 m
Maximum height above foundation level .	90,5 m
Crest length of dam wall	947,9 m
Gross storage capacity	5 960 million cu.m.
Net storage capacity above invert of silt outlets, i.e. R.L.I. . .	203,0 m
	5 943 million cu. m
Area of lake created by dam	37 400 Ha
Volume of excavation	2,1 million cu. m.

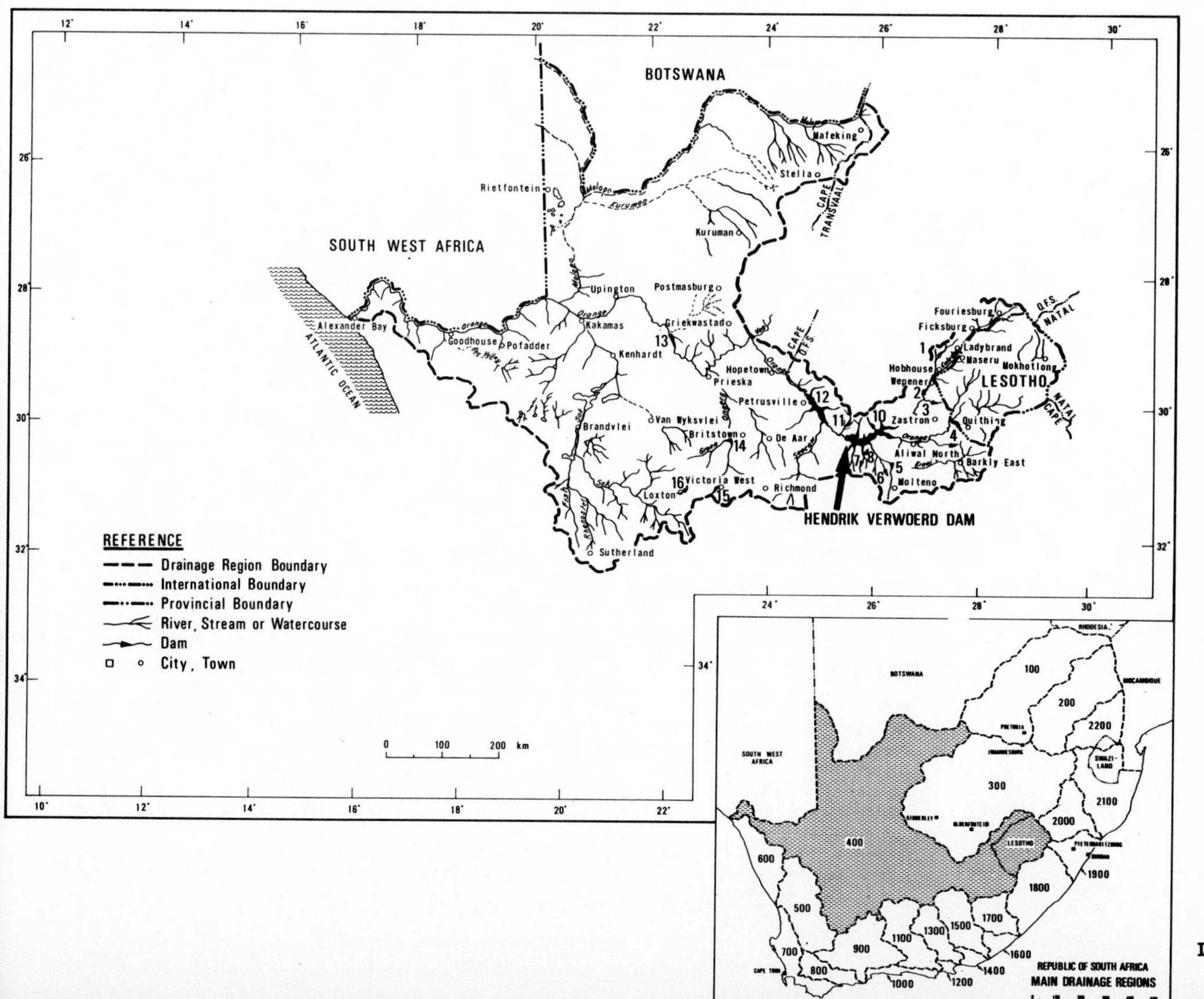

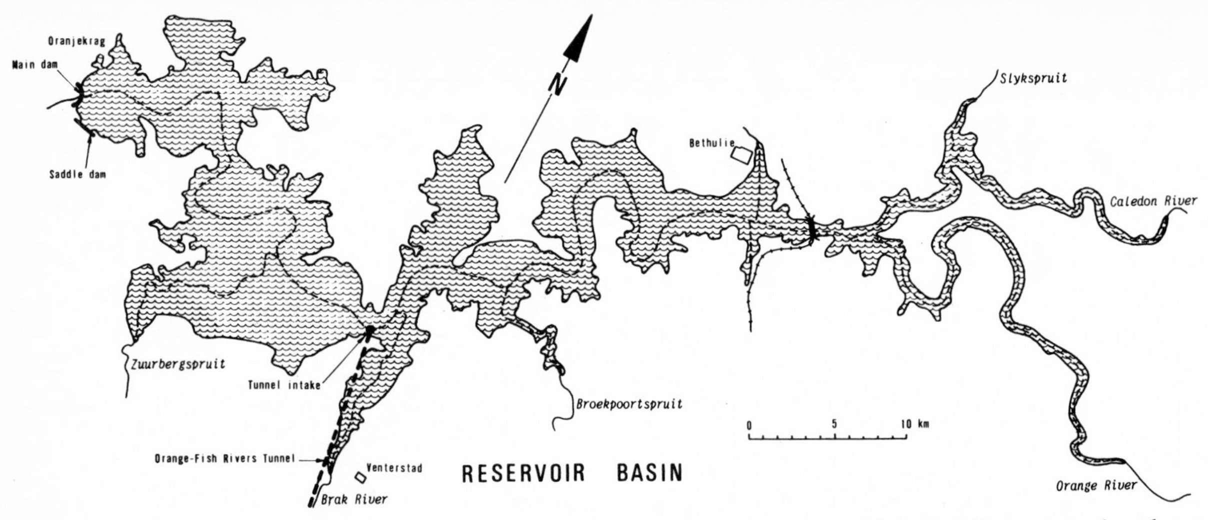

RESERVOIR BASIN

Volume of concrete placed	1,73 million cu. m.
Main Gates provided in dam:	
1. Flood control maintenance gates . .	2 of 12,19 m × 8,53 m. 110 tons
2. Flood control radial gates	6 of 7,62 m × 8,53 m. 86 tons
3. Penstock intake maintenance gates . .	2 of 8,53 m × 5,49 m. 26 tons
4. River outlet maintenance gates . .	6 of 8,53 m × 3,66 m. 21 tons
Main gantries . . .	3 of each with a lifting capacity of 385 tons.

Main gates and valves	*Discharge of each when water is level with overspill crest in cubic metres per second*
4 × 2,29 m. diam. river flow valves	105,50
4 × 2,29 m. diam. penstock bypass valves	105,50
4 × 1,91 m. diam. silt outlet valves .	73
6 Chute Spillways	1 189

Capacity of overflow spillway and outlets at high flood level of 1 264,9 *metres in cubic metres per second*

	Each	*Total*
Overspill crest	7,930	7,930
6-*Chute spillways* . . .	1,385	8,310
8-2,29 *m. diameter valves* .	112	895
4-1,91 *m. diameter valves* .	76,5	305
Total spillway and outlet capacity at high flood level		17,440

Mechanical and Electrical Engineering

Commencement of construction work: December, 1967.

Commencement of commercial service –
First machine: September 1971
Second machine: November 1971

Volume of rock excavation: 175 000 m³

Volume of concrete placed: 50 000 m³

Weight of structural steel: 1 700 metric tons

Terminal output on Generator –

Design: 80 MW

Overload: 90 MW

Power factor: 0,9

Rated speed of machine: 136,4 r.p.m.

Generator voltage: 13 200 volts

Transformer voltage: 13 200/132 000 volts.

Type of hydraulic turbine: Vertical Francis.

Net head above turbine –
Maximum: 62,5 metres
Rated: 48,7 metres
Minimum: 38,0 metres

Water consumption per turbine: up to 200 cumecs.

Inlet diameter to spiral casing: 5,5 metres.

Turbine runner diameter: 5 metres.

Material of turbine runner: Stainless cast steel

Maximum load on thrust bearing: 9 MN

Turbine inlet valve –

Type: lattice door, butterfly

Internal diameter: 5,5 metres

Penstock –

Internal diameter: 7,6 metres

Length: 296 metres

Overhead travelling crane –

Number: One

Lifting capacity: 200/20 metric tons.

CONSULTING ENGINEERS AND CONTRACTORS

Consulting Engineers

International Orange River Consultants (Pty) Company comprising:

France –

Société d'Engineering pour l'Industrie et les Travaux Publics

S.A.R.L. (S.E.I.T.P.)

Coyne et Bellier, Bureau d'Ingénieurs Conseils.

Société Grenobloise d'Etudes et d'Applications

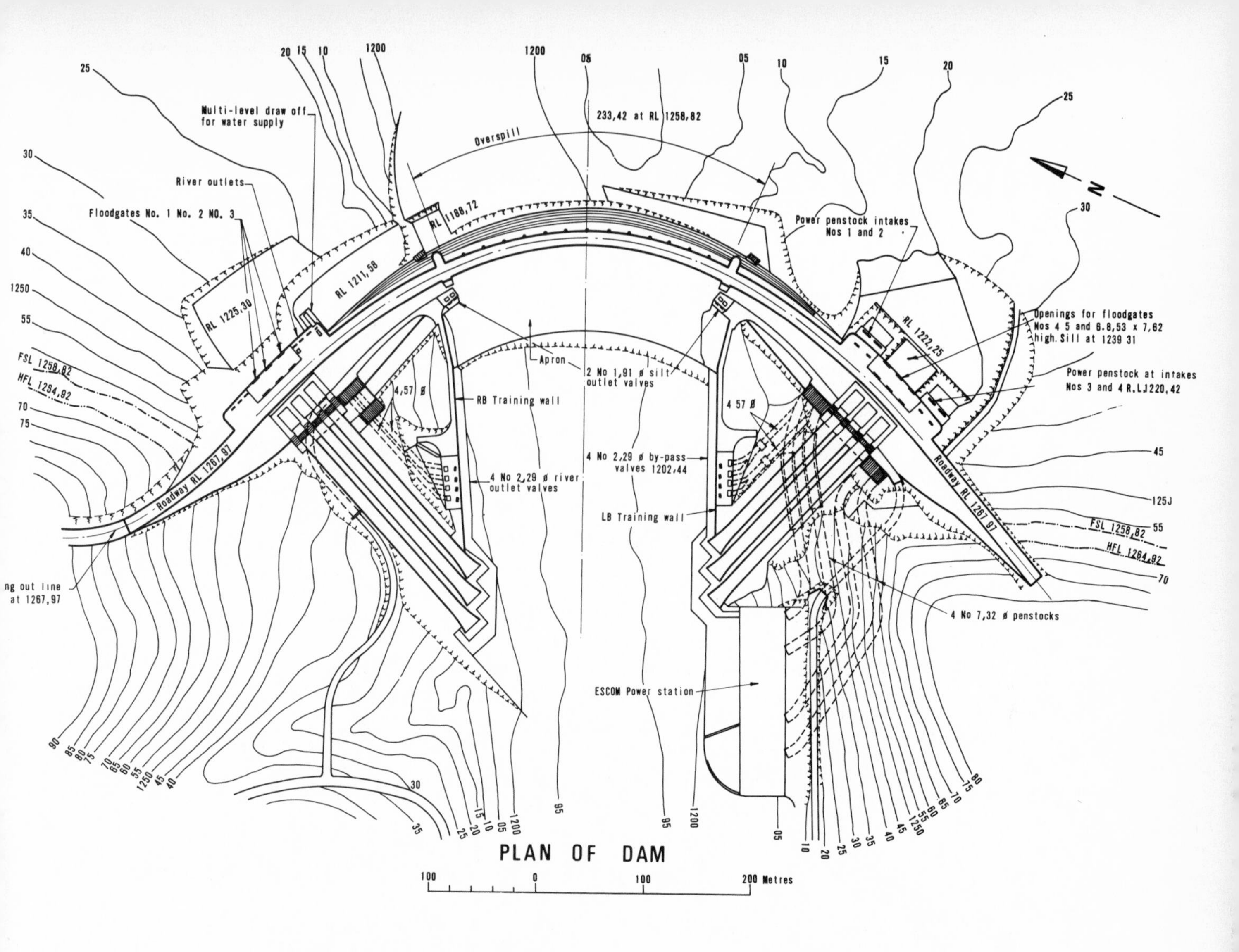

Hydrauliques S.A.R.L. (SOGREAH)
South Africa –
Watermeyer, Legge, Piesold and Uhlmann
Ninham Shand and Partners
Van Niekerk, Kleyn and Edwards
Bird and Robertson
Van Wyk and Louw
Gibb Hawkins and Partners, incorporating:
Sir Alexander Gibb and Partners, United Kingdom
Hawkins, Hawkins and Osborn, South Africa.

Main Contractors

U.C.-Dumez-Borie Dams
Hendrik Verwoerd Dam and Appurtenant Works Saddle Dam
Société B.V.S.
Manufacture and Erection of Gates, Valves, Penstocks and Ancillary Equipment.

Preliminary Works Contractors

Edward L. Bateman – Swimming Pool Filtration and Sewage Treatment Plant.
Engineering Design and Construction Company (Pty) Ltd.
A. G. Burton (S.A.) (Pty) Limited – Construction of Aerodrome.
Cementation (Africa Contracts) (Pty) Limited – Roadwork and Township Site Preparation.
The Cementation Company (Africa) (Pty) Limited – Site Investigation.
Dowson and Dobson – Treated Water Pumping Plant.
Hubert Davies Contracting (Pty) Limited – Electrical Distribution.
James Thompson Manufacturing – Compound and other buildings.
National Trading Company Limited – Distribution Transformers and Substations.
The Roberts Construction Company Limited – Swimming Pool.
Stewarts and Lloyds – Water and Sewage Equipment.
A. Stuart Limited – Road Bridges over River.

Escom employed Gibb Hawkins and Partners as their civil consulting engineers and the main contractors were:-
Civil engineering works: African Batignolles Construction (Pty) Ltd.
Structural steel works: Wright Anderson (South Africa) Ltd.
Hydraulic turbines: Siemens (Pty) Ltd., Fuji Electric Co. Ltd.

Turbine inlet valves: G.E.C. – English Electric of South Africa (Pty) Ltd.
Generator: International Power and Engineering Equipment (Pty) Ltd., representing Tokyo Shibaura Electric Co. Ltd.
Generator Busbars: Hubert Davies Contracting (Pty) Ltd., representing Simel
Generator Transformers: E.T.P. (Pty) Ltd. representing Societa Nazionale Delle Officine Di Savigliano
132 kV Circuit Breakers: Sprecher and Schuh AG.
Low Voltage Switchgear: E. L. Bateman Ltd.
Unit Transformer: First Electric Corporation of South Africa Ltd.
Station Transformers: Reunert and Lenz Ltd. representing ASEA.
Station Batteries: Chloride Electrical Storage Co. S.A. (Pty) Ltd.
Battery Chargers: G. F. T. Products (Pty) Ltd.
Station Cabling: A. E. G. Construction (Pty) Ltd.
Station Lighting: J. F. Smit (Pty) Ltd.
Station Ventilation: C. R. C. Engineering (Pty) Ltd.
Dewatering and Drainage Pumps: Amalgamated Power Engineering S.A. (Pty) Ltd.
Station E. O. T. Crane: Hubert Davies and Co. Ltd. representing Morris Cranes
Draft Tube Gate Crane: Dorman Long (Africa) Ltd.
Station L. P. Pipework: Stewarts and Lloyds of S.A. Ltd.
Lubrication and hydraulic oils: Shell South Africa (Pty) Ltd.

CONSTRUCTION

An inauguration ceremony was held on 18th November 1966 when the Prime Minister, the Hon. B. J. Vorster, signalled the start of construction. As planned, the first large concrete coffer dam was completed on the northern bank in time to provide protection against the first flood season of the construction period. During February 1967 the heaviest recorded flood since 1925 (8 000 cubic metres per second) occurred and the recently completed coffer dam was submerged for two days. Apart from the deposition of silt inside the coffer dam no significant damage was caused but it served as a timely warning of the moods of the Orange River and justified the engineers' proposals for the control of the river during the critical construction period.

As a further means of controlling the river an additional coffer dam was constructed on the south bank and completed in May 1967. By June 1968 the flow of the river was diverted through the temporary openings which had been provided in that portion of the main wall protected by the coffer dam on the northern bank.

By December 1968 the large coffer dam protecting the central riverbed sections, was successfully completed. Apart from the flood of February 1967 only minor floods occurred during the construction period on 10th March 1969 and 30th October 1969 thus enabling the main concrete production phase of the project to proceed without any major delays.

Closure of the temporary openings by a system of grilles, stop logs and steel plates commenced on 31st May 1970 and was completed on 1st September 1970. Concreting of the temporary openings was completed in December 1970.

The first concrete was poured during May 1967, and from this time onwards the monthly volume of concrete poured steadily increased to a peak of 58 000 cubic metres during October 1968. The average volume of concrete poured during the 50-month period of concreting was 36 300 cubic metres per month.

The rock for manufacturing the coarse and fine aggregate for the concrete was obtained from a quarry suitably located at a high elevation on the north bank. The aggregate production plant, which was normally capable of producing 550 tons per hour and was located between the top of the dam and the quarry, supplied aggregate to the concrete batching plant, normally capable of a production rate of 155 cubic metres per hour.

Four cableways with a carrying capacity of $4\frac{1}{2}$ cubic metres each transferred concrete from a loading platform at the batching plant to the main wall. Those portions of the wall which could not be covered by the cableways were served by cranes. Work on the project was carried out by day and night for six days a week, but in spite of the high tempo of operations the record of serious and fatal accidents was low for a project of this magnitude. The certificate of completion dated 4th July 1971 signified that the works had been accepted as essentially complete and that the maintenance period had commenced. The south bank saddle dam was accepted as complete with effect from 31st August 1969.

Valves and gates:

The two concrete gravity flank walls on either side of the main arch house the intake blocks which contain the equipment for controlling releases of water from the reservoir. The largest openings are those for the chute spillways – three on each bank. The six openings are each 8,33 metres wide by 5,44 metres high and together with the central overspill crest will pass flood water of the Orange when these exceed the storage capacity of the reservoir. On the Cape Province side further intakes draw water from the reservoir to feed the Escom hydro-electric power station through steel-lined penstocks. On both banks large pipes lead to the river outlets in the training walls below the dam, which are provided to supply water for users downstream when the power station is not operating.

In the main arch low down there are also four silt draw-offs – two on each side – for evacuating silt from the reservoir during flood periods. The various outlets described above are controlled by a complex arrangement of valves and gates which have been supplied and erected by the French firm Société B.V.S., but most of the equipment provided was actually manufactured by South African firms.

Generating electric power

The Electricity Supply Commission, generally known as Escom, was first approached in 1961 by the Department of Water Affairs with the suggestion that it should use surplus water from the proposed dams of the ORP to generate electric power. Subsequent studies showed that power generation could be substantially increased if the dam height were raised by an extra 18,3 metres, the height to which the wall would have been raised after twenty years. It was decided to raise the wall immediately to this height and Escom agreed to pay a fixed unit tariff to the Department of Water Affairs as a contribution towards the additional costs incurred for the purpose of increasing the hydro-power potential of the dam. The studies also included the optimisation of the hydro-electric potential of the dams, taking into account the fact that this electricity would be fed into Escom's interconnected system. The final design required the provision of four 80 MW machines, the first two of which went into commercial service in 1971. Thus the ultimate generating capacity at Hendrik Verwoerd Power Station will be 320 MW.

Operating the station

As the natural flow of the Orange River fluctuates over a wide range from drought to flood, and the amount of water diverted for irrigation and domestic use will increase with time, less and less water will become available for power generation. This complicated the design of the station as the amount of power generated is governed by both the amount of water in the dam and estimates of the inflow and draw-off in the future. The station has been designed as a peakload station, providing power during peak-demand periods on the interconnected network and not operating at other times; it will only operate on average for about one-third of the time. In fact, it is not possible to operate the station continuously because if there were no inflow into the dam, one of the four sets would empty the full dam in 10 months of continuous operation. Due to the location of the station, which required it to have long penstocks, fast-response frequency control of the turbines cannot be achieved and they are therefore able to cope with slow changes in load only.

The design

In the design stage the suitability of the rock for supporting the station was determined by trial borings. Model tests were carried out to establish the most efficient method of discharging water from the turbines, to determine what the height of the power station wall should be to protect the station during maximum discharge conditions from the dam, and to verify the height and speed of waves travelling down the Orange River when starting four machines simultaneously.

The building is protected from flooding to a level 1,5 metres above the main floor, and the design of the structure allows for the transfer of the forces onto the opposite wall during flood conditions.

Below the main floor the concrete is designed to withstand the large forces produced by the water passing through the turbine, in addition to normal loadings. For instance, the force exerted horizontally by the penstock during operation of the turbine is 42 meganewtons. Special precautions have also been taken to ensure watertightness because the lower floors are permanently below river level.

Construction

Work started on the station in December 1967 with the excavation of the power station site. The lowest point of the excavation is 15 metres below the river bed, and a concrete coffer dam was built before construction started to give protection against a possible flood of up to 5 700 cubic metres per second. This coffer dam was 11 metres high and 120 metres long and demolished after the walls of the power station were sufficiently high to protect the excavations.

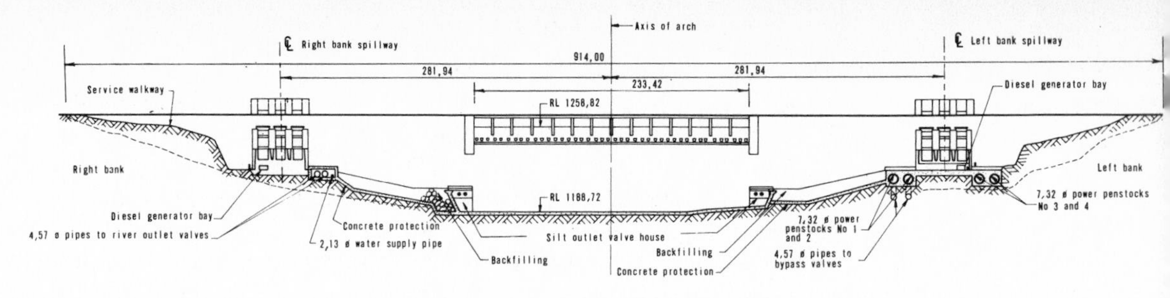
Axis of arch
℄ Right bank spillway
℄ Left bank spillway
914,00
281,94
281,94
233,42
Service walkway
Diesel generator bay
RL 1258,82
Right bank
Left bank
RL 1188,72
7,32 ø power penstocks No 3 and 4
Diesel generator bay
4,57 ø pipes to river outlet valves
Concrete protection
2,13 ø water supply pipe
Silt outlet valve house
Backfilling
7,32 ø power penstocks No 1 and 2
Backfilling
4,57 ø pipes to bypass valves
Concrete protection

DOWNSTREAM ELEVATION

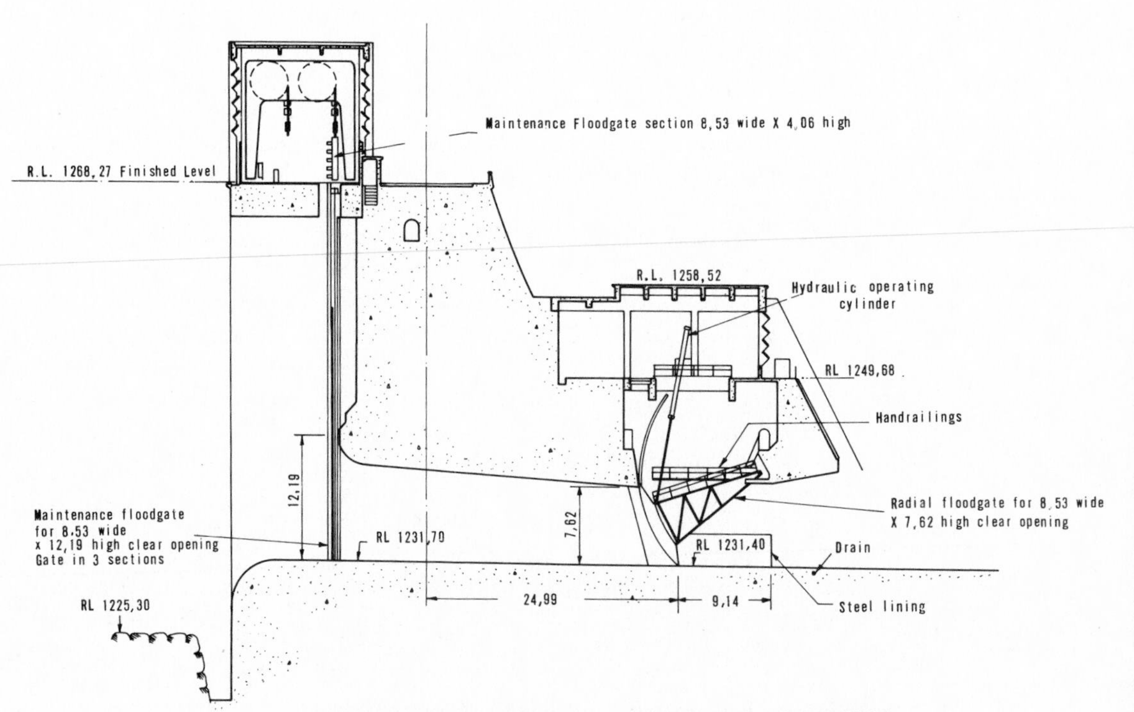
Maintenance Floodgate section 8,53 wide X 4,06 high
R.L. 1268,27 Finished Level
R.L. 1258,52
Hydraulic operating cylinder
RL 1249,68
Handrailings
12,19
7,62
Maintenance floodgate for 8,53 wide x 12,19 high clear opening Gate in 3 sections
RL 1231,70
RL 1231,40
Radial floodgate for 8,53 wide X 7,62 high clear opening
Drain
RL 1225,30
24,99
9,14
Steel lining

SECTION THROUGH FLOODGATE OPENING

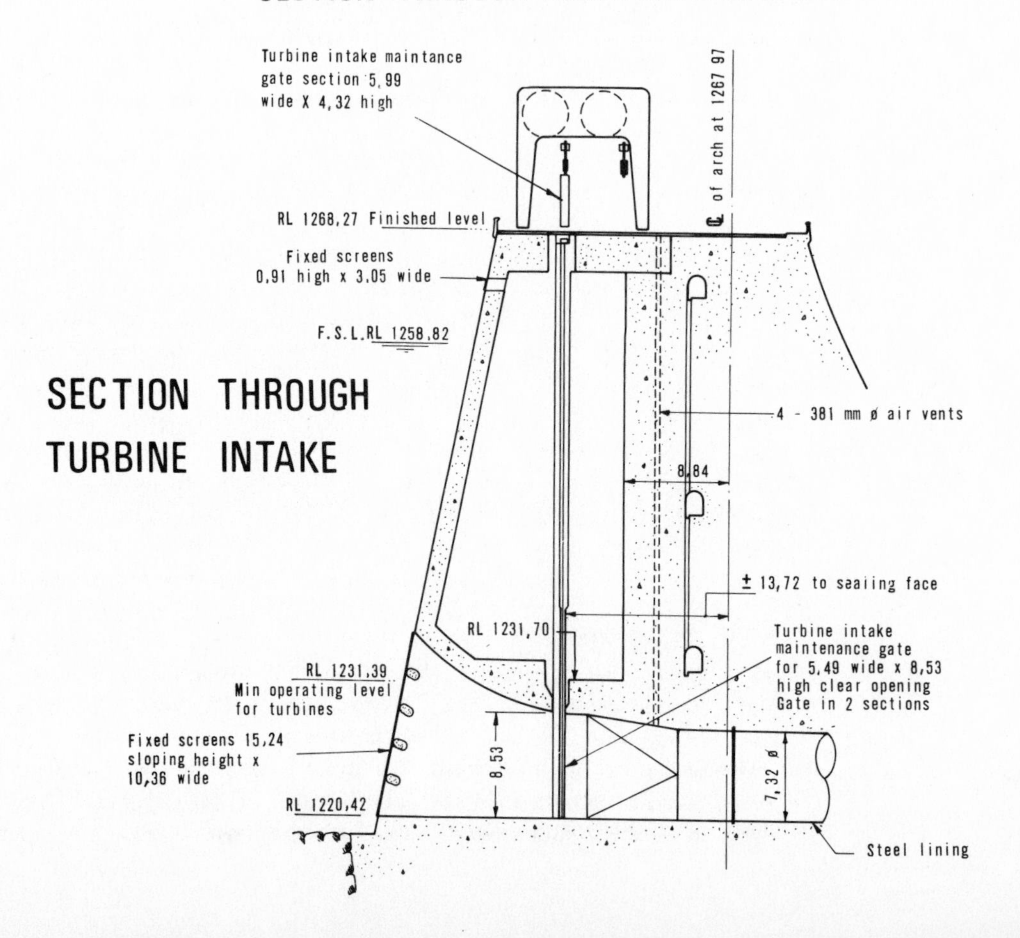
Turbine intake maintance gate section 5,99 wide X 4,32 high
℄ of arch at 1267 97
RL 1268,27 Finished level
Fixed screens 0,91 high x 3,05 wide
F.S.L. RL 1258,82
4 - 381 mm ø air vents
8,84
± 13,72 to sealing face
RL 1231,70
RL 1231,39 Min operating level for turbines
Turbine intake maintenance gate for 5,49 wide x 8,53 high clear opening Gate in 2 sections
Fixed screens 15,24 sloping height x 10,36 wide
8,53
7,32 ø
RL 1220,42
Steel lining

SECTION THROUGH TURBINE INTAKE

Turbines

The Francis, Kaplan and Deriaz hydraulic turbine types were investigated, and the Francis-type was selected. These turbines have lower part-load efficiencies, but this is not of importance at Hendrik Verwoerd because they will be operated mainly at full load either during the daily system peak or when employed for standby and reserve duties.

The turbines are of the verticalshaft type, rotating at 136,4 revolutions per minute and consuming up to 200 cubic metres of water per second.

The turbine runner is a solid integral casting of stainless steel, 5 metres in diameter, and having a mass of 52 metric tons. The spiral casing conveying the water to the turbine runner has an inlet diameter of 5,5 metres and an overall width of 15,5 metres. The spiral casings, each consisting of 150 metric tons of plate steel, were manufactured in South Africa. After the water energy is converted into mechanical energy in the turbine runner, the water is discharged into the tailrace pool and then into the river via the draft tube. The draft tube steel liners which were also manufactured in South Africa, are 19 metres high and have a discharge area of 12 metres wide and 6 metres high.

Generators

Each turbine drives a directly coupled generator via a shaft of 0,97 metre in diameter and 8,2 metres in length, having a mass of 55 metric tons. A thrust bearing situated below the generator carries the weight of the rotating components as well as the hydraulic thrust which, combined, result in a load of 9 meganewtons.

The generators are of salient pole construction having 44 poles on the rotor and a synchronous speed of 136,4 r.p.m. The output rating of the generators is 100 000 kVA at a terminal voltage of 13 200 volts. Power losses in the generators appear as heat within the machines. To conduct this heat away, the generators are cooled by a closed-circuit ventilation system whereby air circulating within the machines conducts the heat from the various machine components to air-to-water heat exchangers mounted in the generator foundation block.

Excitation power for the generator field windings is provided by the exciter, which is a directly-driven alternating-current generator mounted above the main generator rotor. The alternating current output of the exciter is rectified in a bank of silicon-controlled rectifiers before being fed to the generator field windings.

Each generator is directly connected to a 100 000 kVA step-up transformer which raises the voltage to 132 000 volts. The power is then transmitted by overhead line to the Ruigtevallei Distribution Station, situated approximately 1,5 km from the power station. Thereafter the power is fed into Escom's high voltage transmission network at Hydra Distribution Station near De Aar, which is one of the distribution stations fed by the transmission lines linking the Western Cape with the Transvaal.

THE ORANGE-FISH TUNNEL

HISTORY

This is the longest continuous water tunnel in the world. Its purpose is the inter-basin transfer of water from the Orange to the Fish River valleys. By any standards this is a major project. The tunnel has a lined diameter of 5,35 m and is 82½ km long.

The tunnel was officially opened by the Hon. B. J. Vorster, the Prime Minister on August 22, 1975, and is serving areas in the Fish River valley which could become one of the most agriculturally productive areas in the Republic.

CONSTRUCTION

Water enters the tunnel via an inlet control tower located within the basin of the Hendrik Verwoerd Dam and fitted with intakes at varying levels so that water may be drawn from the dam at different depths. The invert or bottom of the tunnel at its inlet is 30,5 m below the full supply level of the dam and can draw off 88 per cent of the full capacity of the dam. It has a constant floor slope of 1 in 2 000, which means that the invert at the outlet is 41 m lower than at the inlet, or 71, 5m (235 ft) below the full supply level of the dam.

Besides permitting the drawing off of water at varying depths, the inlet control tower will also allow the closing of the tunnel in the event of an emergency or when repairs have to be made. Daily control is from the downstream or outlet end where an intricate system of control and bypass valves has been installed. The maximum discharge of the tunnel is 57 m^3/s, (cumecs). The gradient is 1:2 000 (dynamic).

The rock through which the tunnel was driven consists of silt-stones, mudstone and sandstone of the Karoo system and between 10 and 15 per cent of intruding dolerite. Apart possibly from the dolerite, the stone is not

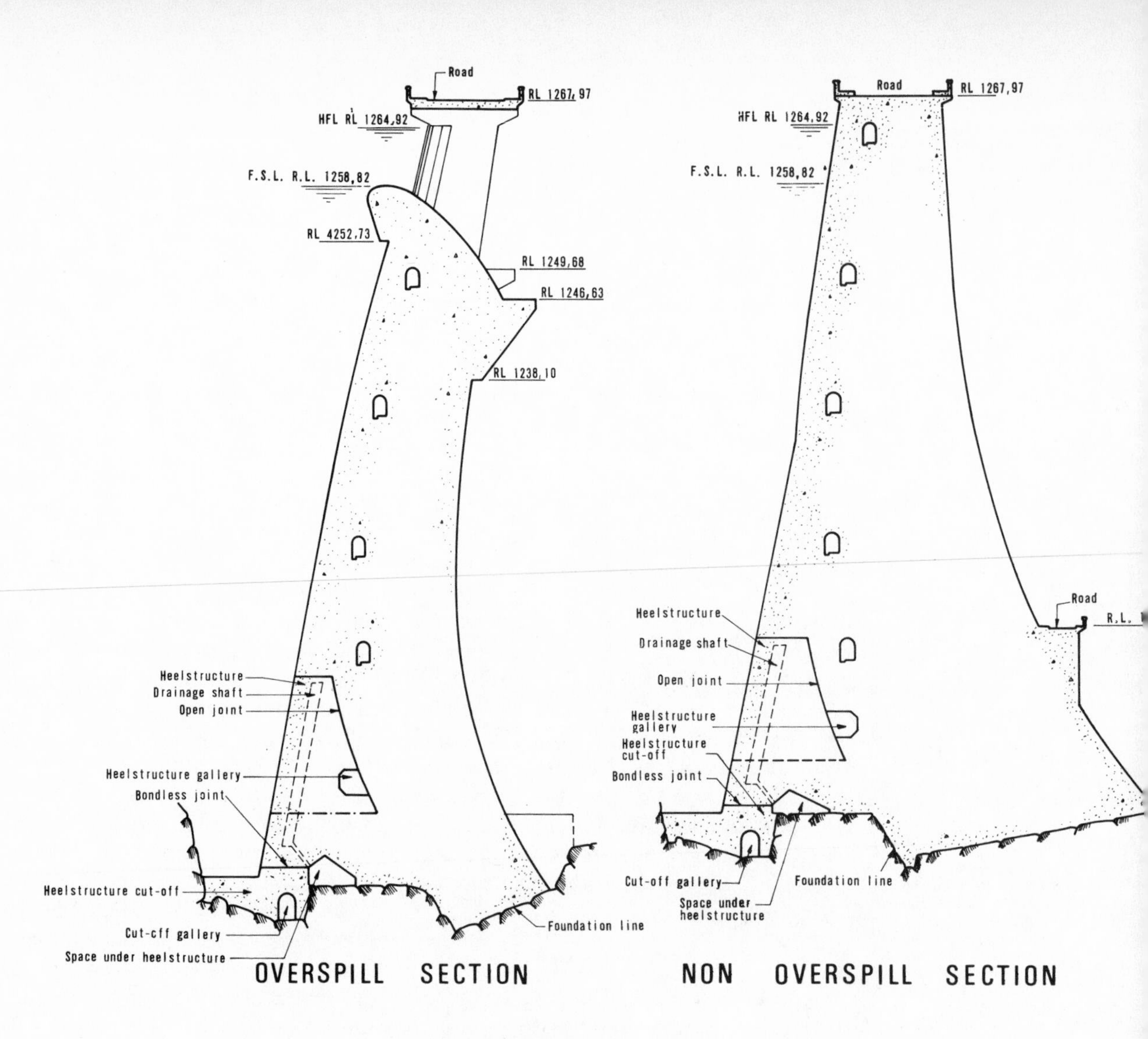

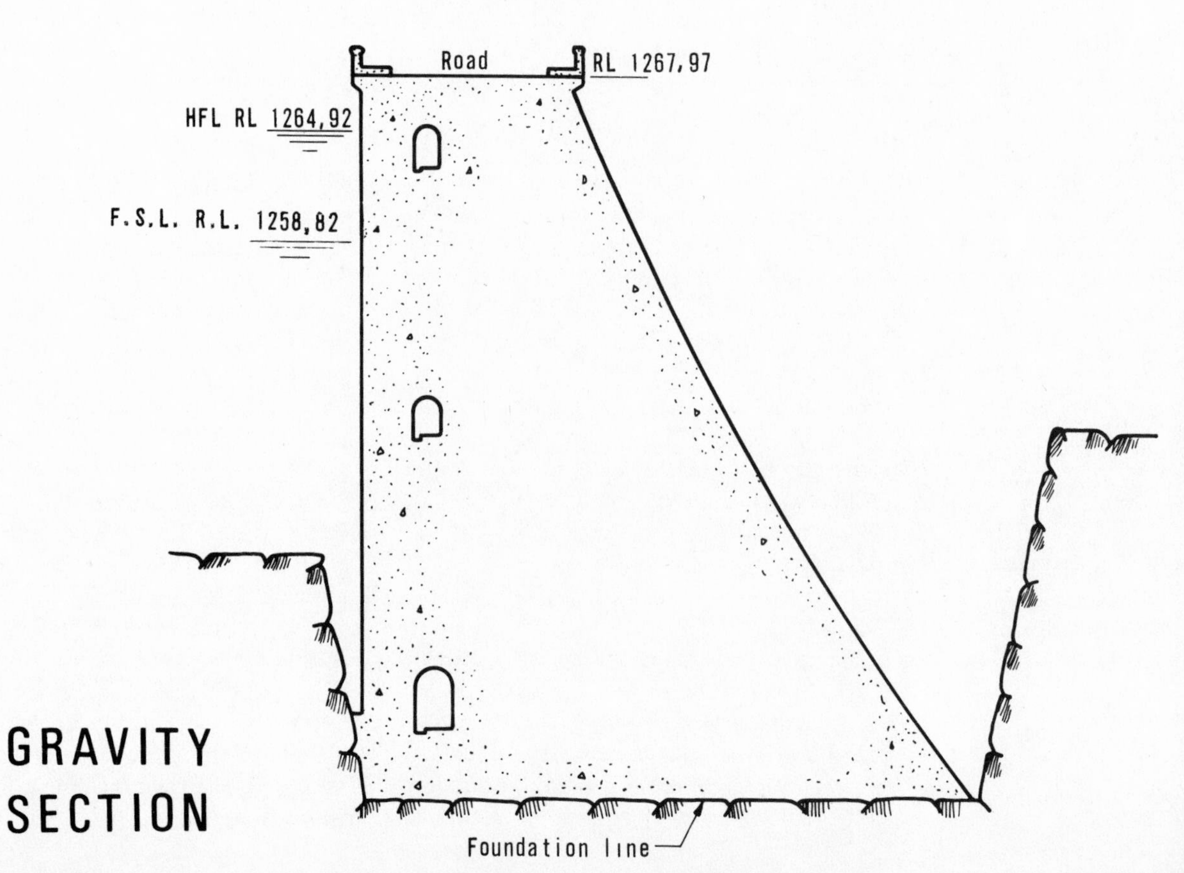

GRAVITY SECTION

HENDRIK VERWOERD DAM

sufficiently durable to withstand the weathering and the scouring effects of the flowing water and it was necessary to provide the tunnel with a concrete lining with a minimum thickness of 25 cm.

The outlet control works of the tunnel are situated some 7 163 m from where the actual tunnel emerges near the Teebus and Koffiebus koppies. This is necessary because the cover on the tunnel in this last reach is not sufficient to balance the internal pressure of the water in the tunnel. It emerges into a concrete-lined canal 2 896 m long which, in turn, discharges into the Teebus Spruit, a tributary of the Great Fish River.

In view of its great length it was impractical to drive or excavate the tunnel from two ends only. To have done so – even with modern equipment – would have taken something like 30 years.

Work was therefore carried out from the bottom of seven vertical shafts which, together with the inlet and the outlet faces, provided 16 faces from which to work. With this number of working faces it took 49 months to complete the 82½ km of excavation. The seven shafts vary in vertical depth from 75 m to 378 m. The two shallow ones near the intake are inclined but this is simply because the depth made such a preference possible.

In order to ensure its timeous and successful completion the tunnel project was divided into three separate contracts as follows:-

Inlet section awarded in January 1968 for the construction of 27 km of tunnel and the inlet control tower.

Plateau section awarded in October 1968 for the construction of 31 km of tunnel.

Outlet section awarded in July 1968 for the construction of 24 km of tunnel, the outlet control works and the outlet canal.

As at the Hendrik Verwoerd Dam, the Department of Water Affairs moved to the site some four years before the main tenders were awarded and during this period organised the contracts for the preliminary works. This involved a tarred road down the length of the tunnel route and three complete townships known as Oviston, Midshaft and Teebus.

A "full face" attack by conventional drilling and blasting was adopted by the contractors for the tunnel excavation. The "burn cut" was generally used and the lengths of round, comprising some 65 to 85 drilled holes, varied according to the rock conditions generally from 2,5 m to 3,5 m.

Verwoerd Dam – no floods discharging

Verwoerd Dam – main and side sluice discharging showing ESCOM power station on right (Department of Information)

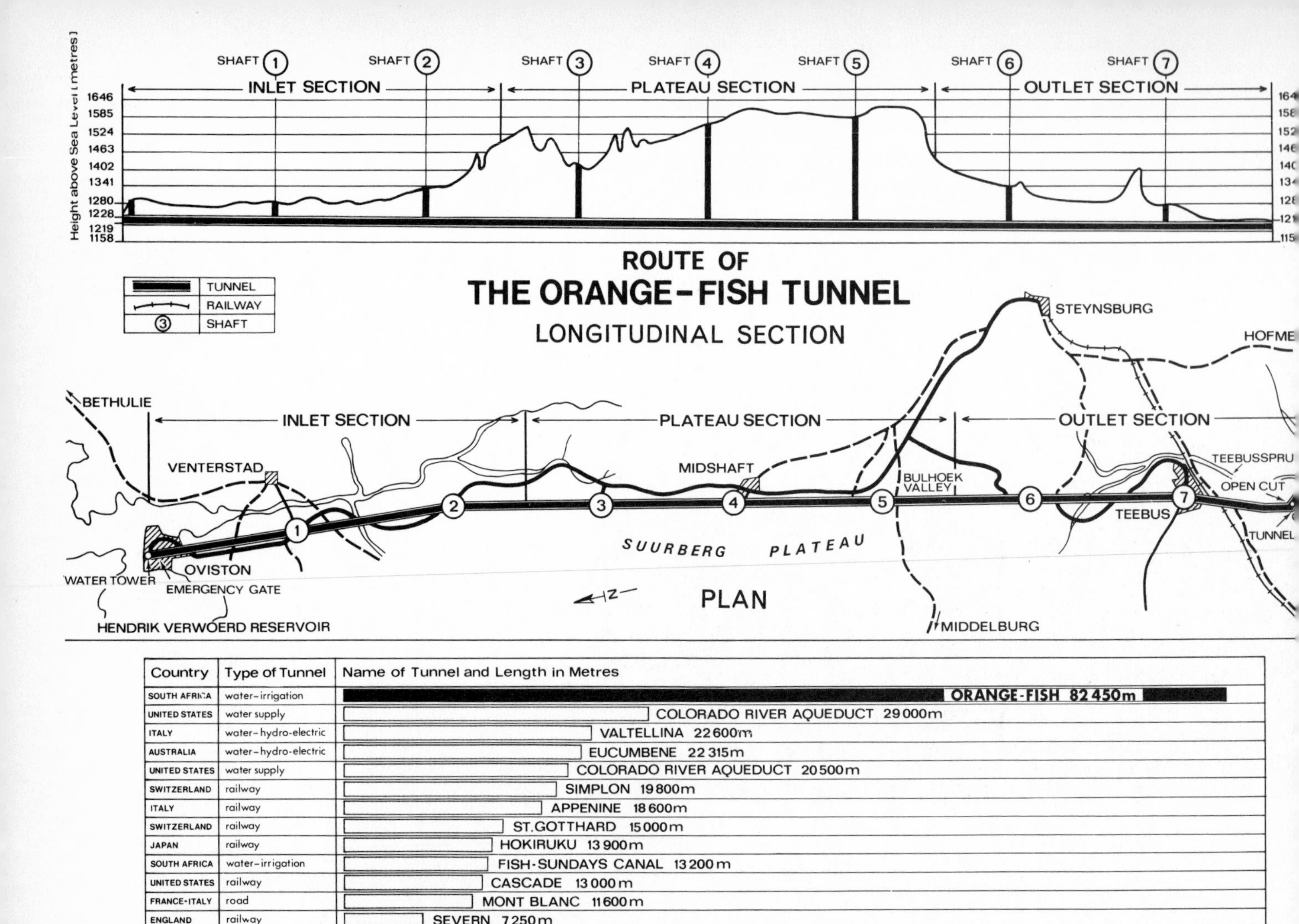

Country	Type of Tunnel	Name of Tunnel and Length in Metres
SOUTH AFRICA	water-irrigation	ORANGE-FISH 82 450m
UNITED STATES	water supply	COLORADO RIVER AQUEDUCT 29 000m
ITALY	water-hydro-electric	VALTELLINA 22 600m
AUSTRALIA	water-hydro-electric	EUCUMBENE 22 315m
UNITED STATES	water supply	COLORADO RIVER AQUEDUCT 20 500m
SWITZERLAND	railway	SIMPLON 19 800m
ITALY	railway	APPENINE 18 600m
SWITZERLAND	railway	ST.GOTTHARD 15 000m
JAPAN	railway	HOKIRUKU 13 900m
SOUTH AFRICA	water-irrigation	FISH-SUNDAYS CANAL 13 200m
UNITED STATES	railway	CASCADE 13 000m
FRANCE-ITALY	road	MONT BLANC 11 600m
ENGLAND	railway	SEVERN 7 250m
SOUTH AFRICA	railway	HILTON 6 000m
ENGLAND	road	MERSEY 3 200m
		0km 20km 40km 60km 80km

COMPARISON OF WORLD'S LONG TUNNELS

Rail-mounted drill "jumbos" of portal design, each equipped with seven hydraulically operated arms, were used for the Central and Outlet sections. Similarly equipped gantries on the Inlet contract travelled on a mucking track of 1,1 m gauge. All of the contractors used electrically-powered Conway 100–2 muckers, with a bucket capacity of about one cubic metre, to load the spoil into muck cars.

Generally, two muck trains, each train comprising ten 15-ton cars, were required to remove a 3 m round. Cars on the Inlet section were changed by hydraulically operated "cherry pickers", and on the Central and Outlet sections, respectively, by 90 m long Californian switches and 100 m long sliding floors.

On the Inlet section the cars were side-tipped into a large storage bin located below the tunnel floor at the bottom of each inclined shaft, and the rock spoil was then hauled out of the tunnel by conveyer belt into elevated steel bins on the surface.

On the Central section the bottoms of the cars, hinged at one end, were opened automatically at the shaft bottom, allowing th muck to fall into a substantial storage bin be low the tunnel floor. The spoil was then fe through a measuring flask into 6 ton skip wound to the surface and discharged auto matically into muck hoppers housed in th headgear.

On the Outlet section the cars were ove turned in a hydraulically operated tippler int storage bins of 15-tons capacity. The spo was then chuted into rock skips of equa volume, wound to the surface and automat cally emptied into spoil bins in the headgea On all three sections the excavated rock wa discharged from the spoil bins in the head gear into dump trucks and carted to the spo heaps.

The tunnel lies wholly within the geologic Karoo Basin, and the rocks belong to th general Karoo System which is the mo dominant and wide-spread geological syste in South Africa. The topography reflects geological structure of horizontally bedde sedimentary rocks, chief of which are gree

Orange Fish Tunnel inlet tower at Oviston

shale, purple red mudstone and grey sandstone, intruded by sheets and dykes of dolerite, one of the hardest rocks in existence. The rock through which the tunnel was driven consists of 10 per cent dolerite, 15 per cent sandstone, 15 per cent green shale, and 60 per cent mudstone.

The geological conditions in the tunnel provided many challenging problems, and a number of important safeguards had to be adopted. The mudstones and muddy siltstones tended to be friable (readily crumbled, brittle), and the tunnel crown, or roof, required reinforcement throughout most of the length of the tunnel. The tunnel roof was supported with steel arches, rockbolts, steel mesh or sprayed concrete (shotcrete) used singly or in combination.

The whole length of the tunnel is concrete-lined. The smooth lining provides a better hydraulic effect than would a rough, rocky lining. The concrete lining primarily serves the purpose of increasing the carrying capacity of the tunnel. In addition, it protects the rock from weathering and consequent deterioration, and provides physical support to the rock in certain areas. Furthermore, it ensures that possible loss of water by seepage into the underground formations, which generally are very tight, will be minimised.

The internal diameter of the concrete lined tunnel is 5,33 metres. Nearly 2½ million cubic metres of rock were removed from the tunnel excavation. Maximum weekly advances of up to 138,3 m were achieved.

After the whole tunnel had been excavated, the concrete lining operation was carried out around the clock, six days a week. A minimum lining thickness of 250 millimetres was applied throughout, and a total of 842 000 cubic metres of concrete was placed, making it the biggest operation of its kind ever carried out in South Africa. Telescopic steel forms and pneumatic concrete placing equipment were used throughout the operation; there was no hand placing of concrete. It was a specified requirement of the work that the lining should have a smooth finish in order to obtain the maximum flow of water through the tunnel.

On 26 August, 1969, during a normal tunneling operation, the contractors working at a depth of 110 metres (360 feet) in a section south of Shaft 2 fired a charge that opened up an unexpectedly strong flood of water which could not be controlled by the tunnel pumps. Fortunately, there was no loss of life in the initial inrush of water which rose rapidly and flooded the tunnel for a length of 1 752 metres, and up to 18 metres from the ground surface. The flood attained a flow of 1,25 m^3/s, and Shaft No. 2 had to be temporarily abandoned.

Indications by the Geological Survey were that methane gas might be encountered underground, and safeguards similar to those observed in fiery mines were taken by the contractors. In spite of these precautions, however, a methane gas fire broke out on 23 October, 1971, during blasting of the face between Shafts 4 and 5. The necessary precautions for containing a methane fire were immediately implemented by sealing it off, and although the fire continued to burn fiercely, the ventilation system was adequate to deal with the products of combustion. After various methods of extinguishing the fire had been carefully considered, it was decided that "oxygen starvation" was the best tactic to adopt. This works by the simple expedient of depriving the fire of oxygen in which to burn.

A concrete bulkhead was constructed 19 metres from the face to extinguish the fire by oxygen starvation. The fire area having been effectively sealed off, and the fire extinguished, the space between the bulkhead and the tunnel face was filled with cement grout, and the joints in the surrounding area of rock were sealed by injecting cement through an aureole of drill holes drilled from outside the concrete plug.

After tunnelling through the area, steel arch ribs were erected for a distance of 16 m on either side of the seat of the fire, and concrete was placed between the arch ribs and rock, and in the invert (tunnel floor). Further holes were then drilled radially from the tunnel to intersect any gas-bearing fissures which were sealed by the injection of cement grout. Subsequently this area was monitored by methane detectors.

CONSULTING ENGINEERS AND CONTRACTORS

The Consulting Engineers appointed by the Department of Water Affairs were The Orange-Fish Tunnel Consultants, the membership being Sir William Halcrow & Partners, London, and Keeve Steyn & Partners of Johannesburg.

The main contractors were:-

Inlet Section:

Batignolles-Cogefar-African Batignolles, comprising –

(1) Société de Construction des Batignolles of France
(2) Construzione Generali Farsura-Cogefar of Italy
(3) African Batignolles Constructions (Pty) Ltd. South Africa

Plateau Section:

Orange River Contractors (Orco) comprising –

(1) LTA of Johannesburg, South Africa
(2) Boart & Hard Metals of Johannesburg
(3) Boyles Bros. Drilling Company of the United States
(4) Compagnie de Constructions Internationales of France

Outlet Section:

J.C.I. DI PENTA (J.D.P.) comprising –

(1) Johannesburg Consolidated Investment Company (JCI)
(2) Impresa Ing. Di Penta of Italy

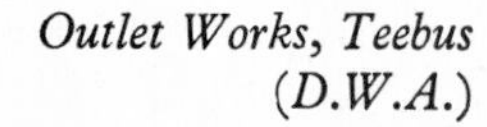

Outlet Works, Teebus (D.W.A.)

P. K. LE ROUX DAM and HYDRO-ELECTRIC SCHEME

HISTORY

The construction of the P.K. le Roux Dam was originally scheduled to be started during 1967 – soon after the start of the Hendrik Verwoerd dam project – and much of the construction of the preliminary works was, in fact, carried out at the same time as those for the Hendrik Verwoerd Dam. A start on the main civil works for the dam was, however, delayed for a number of reasons, and in October 1967 the Government decided to postpone its construction for a further indefinite period as part of its programme to curb inflation.

In 1970 tenders were again called for, but they were considerably higher than expected and the government therefore decided early in 1971 that the dam should be built by the Department of Water Affairs. An immediate start was made and the project was scheduled for completion in June 1977.

Height above lowest foundation	107 m
Length along the crest	765 m
Volume of concrete in wall	1 100 000 m³
Provision for flood discharge	8 500 m³/s
Storage capacity	3 185 m³ × 10⁶
Power generated	220 MW
Level of river bed	1 095,0 m
Left bank canal invert level	1 147,10 m
Right bank canal invert level	1 146,67 m

CONSTRUCTION

The dam will be similar in design to the Hendrik Verwoerd Dam, except that it will have four sluice-gates on the left bank and none on the right. The site is in fact more efficient than that of the Hendrik Verwoerd Dam in that it is narrower and more suited to an arch structure. This is shown by the fact that while the dam wall will be 30 m higher than the Hendrik Verwoerd Dam it will require only about two-thirds the volume of concrete.

An underground hydro-electric power station with an installed capacity of 220 MW is being built. This work, as at Hendrik Verwoerd, is being managed by Escom.

The water level in the basin can be con-

Excavation of river August, 1973, showing diversion of river (D.W.A.)

Showing progress after twelve months (August, 1974) taken from the left bank (D.W.A.)

trolled by releasing water from the Hendrik Verwoerd Dam. Because of its more favourable surface evaporation characteristics, the water level in the basin will, as far as possible, be maintained within a metre or two of the overspill crest. Releases to the river to provide for requirements below the dam will be made through the hydro-electric turbines during times when peak power demands have to be met.

To gain the maximum advantage from the hydro-electric power stations at both the Hendrik Verwoerd and P.K. le Roux Dams, releases into the river for irrigation and other purposes should be confined to periods of peak power demand. Thus the river will be subjected to daily fluctuations of from 10 to 800 m^3/s. This large variation in the river flow will create problems which will require special solutions and the Department of Water Affairs is at present studying this problem.

Van der Kloof Canals

The P.K. le Roux Dam will, in addition to its other functions of storing and releasing water to the lower Orange River, divert water into a system of canals on both the left and the right bank of the river, downstream of the dam site.

Looking upstream August 1974 (D.W.A.)

Vanderkloof Irrigation Canal on the south bank of the Orange River (D.W.A.)

Dam ready for final closure of temporary openings (D.W.A.)

In the initial proposals provision was made to command irrigable ground in the Beervlei, Great Brak River, Sak River and Witsands areas. Subsequent investigations have shown, however, that because of the high potential salinity of the soils in these areas and the negative cost benefit ratios, these proposals cannot be economically justified, especially as there are more than enough suitable soils in the triangle between the Orange and the Vaal Rivers that can be more economically irrigated.

The first phase of the project will now make provision for the irrigation of the following areas:

On the southern bank of the Orange River, between the P.K. le Roux Dam and Hopetown .	13 600 ha
On the north bank of the river from the dam down to the border between the Cape and the Orange Free State	8 800 ha
TOTAL:	22 400 ha

Future development of this canal and irrigation scheme will depend on circumstances and the growth in demand for products which can be produced in this area. There is more than sufficient ground for the water available. The most significant feature of the scheme is that the first 14 km on the right bank will have a capacity of 57 m^3/s, and provision has been made to increase its depth so that its eventual capacity will be 114 m^3/s. This is by far the largest irrigation canal in Southern Africa. As will be appreciated, there will be no clear ending to this scheme for many years to come.

WEMMERSHOEK DAM

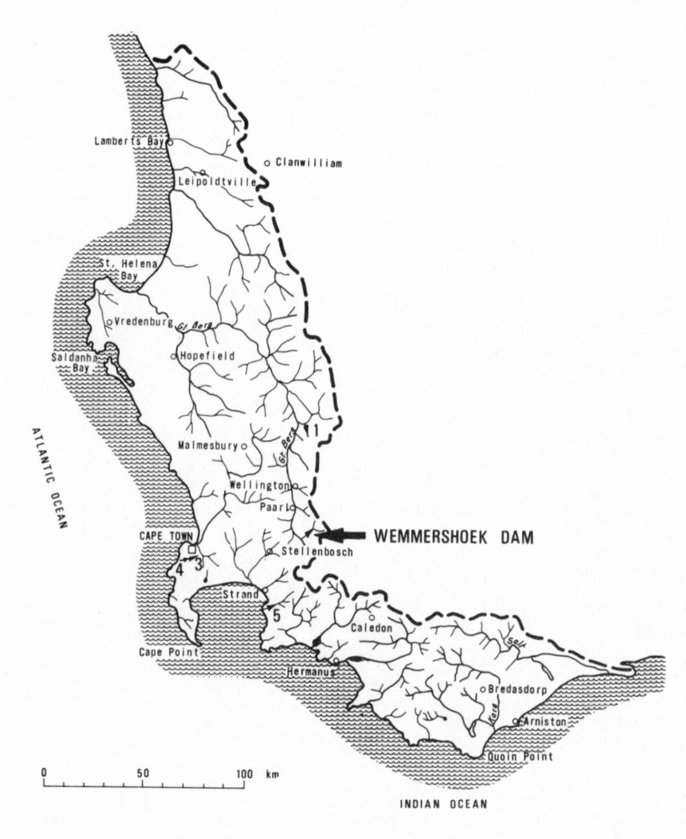

Drainage region and location

HISTORY

Wemmershoek Dam is part of a scheme comprising the dam, a treatment plant, 50 km prestressed concrete pipeline, a service reservoir and a number of subsidiary pipelines for the city of Cape Town and local authorities embracing an area of about 440 square kilometres. The owner is the Municipality of Cape Town.

The Wemmershoek scheme makes available a supply of more than 136 000 m^3d for urban purposes and also quantities (0,10 m^3/s) of compensation water for irrigation. This comparatively large supply is made available from a relatively small catchment area of some 125 square kilometres. This catchment is a mountainous one, surrounded by a ring of spectacular peaks rising in some cases to nearly 1 800 m. The area has abundant winter rainfall and in some places the annual rainfall exceeds 3 500 mm. It was completed in 1957.

DIMENSIONS

Type	Earth
Height above lowest foundation	53 m
Length of crest	488 m
Volume content of dam	2 886 million m^3
Gross capacity of reservoir	58 643 million m^3
Maximum discharge capacity of spillway	1 065 m^3/s
Type of spillway	Controlled

CONSULTING ENGINEERS AND CONTRACTORS

Engineering by	Board of Engineers, Dr. S. S. Morris, J. L. Savage
Constructed by	Geo. Wimpey & Co. (London).

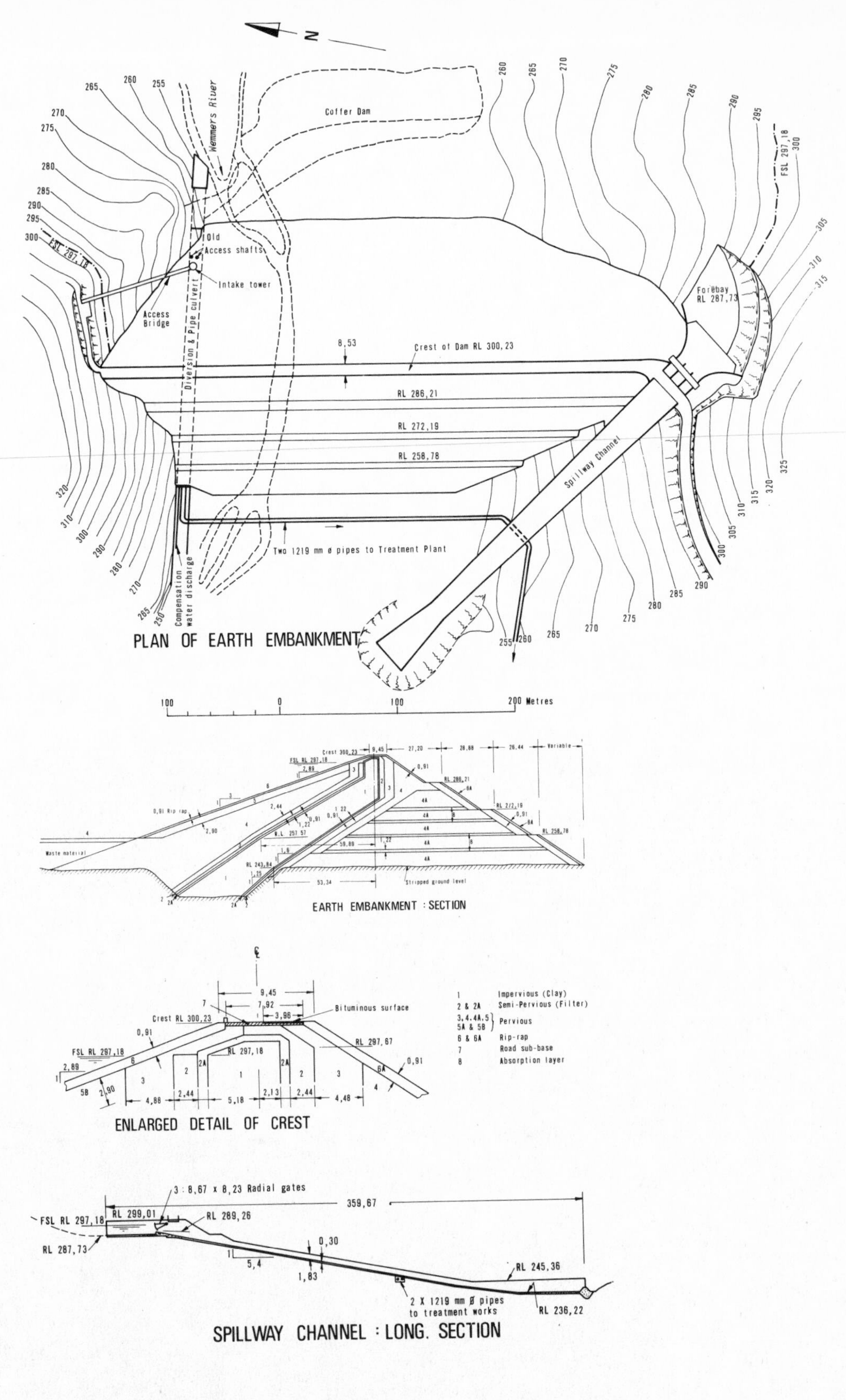
N
Coffer Dam
Wemmers River
Old
Access shafts
Intake tower
Access
Bridge
Diversion & Pipe culvert
FSL 297,18
Forebay
RL 287,73
8,53
Crest of Dam RL 300,23
RL 286,21
RL 272,19
RL 258,78
Spillway Channel
Two 1219 mm ø pipes to Treatment Plant
Compensation
water discharge
PLAN OF EARTH EMBANKMENT
100
0
100
200 Metres
Stripped ground level
EARTH EMBANKMENT : SECTION
9,45
7,92
3,96
Bituminous surface
Crest RL 300,23
0,91
FSL RL 297,18
2,89
RL 297,18
RL 297,67
4,88
2,44
5,18
2,13
2,44
4,48
ENLARGED DETAIL OF CREST
1 Impervious (Clay)
2 & 2A Semi-Pervious (Filter)
3, 4, 4A, 5, 5A & 5B Pervious
6 & 6A Rip-rap
7 Road sub-base
8 Absorption layer
3 : 8,67 x 8,23 Radial gates
359,67
FSL RL 297,18
RL 299,01
RL 289,26
RL 287,73
0,30
5,4
1,83
RL 245,36
2 X 1219 mm ø pipes
to treatment works
RL 236,22
SPILLWAY CHANNEL : LONG. SECTION

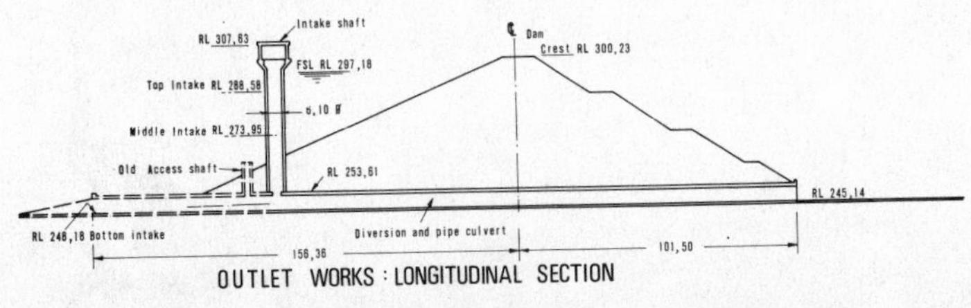
Intake shaft
RL 307,83
Dam
Crest RL 300,23
FSL RL 297,18
Top Intake RL 288,58
6,10 ø
Middle Intake RL 273,95
Old Access shaft
RL 253,61
RL 245,14
RL 246,18 Bottom intake
Diversion and pipe culvert
156,36
101,50
OUTLET WORKS : LONGITUDINAL SECTION

VAAL DAM

HISTORY

Vaal Dam is one of two main storage units – the other being Bloemhof dam – on the Vaal River and is situated about 72 km upstream of the town of Vereeniging. Water is released from the dam in accordance with the requirements of the Rand Water Board and of the various municipalities, water supply schemes and other users along the Vaal River. Before the construction of the Hendrik Verwoerd Dam, Vaal Dam water was let down the Vaal River to augment the flow in the Orange River during drought periods. Likewise, before the construction of Bloemhof Dam, flow from Vaal Dam was released for the Vaal-Hartz Irrigation Scheme. It is owned by the Department of Water Affairs which also carried out the engineering and construction.

The Rand Water Board, whose barrage is situated 80 km downstream of Vaal Dam, supplies water for domestic and industrial use to an area of more than 16 000 km² with a population of over three million persons.

DIMENSIONS

Type . . .	Earth and Gravity
Height above lowest foundation . .	52 m
Length of crest .	2 570 m
Volume content of dam . .	1,041 m³ × 10⁶ (concrete 0,246 m³ × 10⁶ earth 0,795 m³ × 10⁶)
Net capacity of reservoir . .	2 330 m³ × 10⁶
Maximum discharge capacity of spillways . . .	5 097 m³/s
Type of spillway	Controlled

Raised 6 m in 1956.

CONSTRUCTION

Vaal Dam was built in 1938; by 1951 the demand on the dam had grown to such an extent that it was decided to raise the dam by 6 m. This was achieved by increasing the height of the concrete overspill section by 3 m, and the

Vaaldam Reservoir

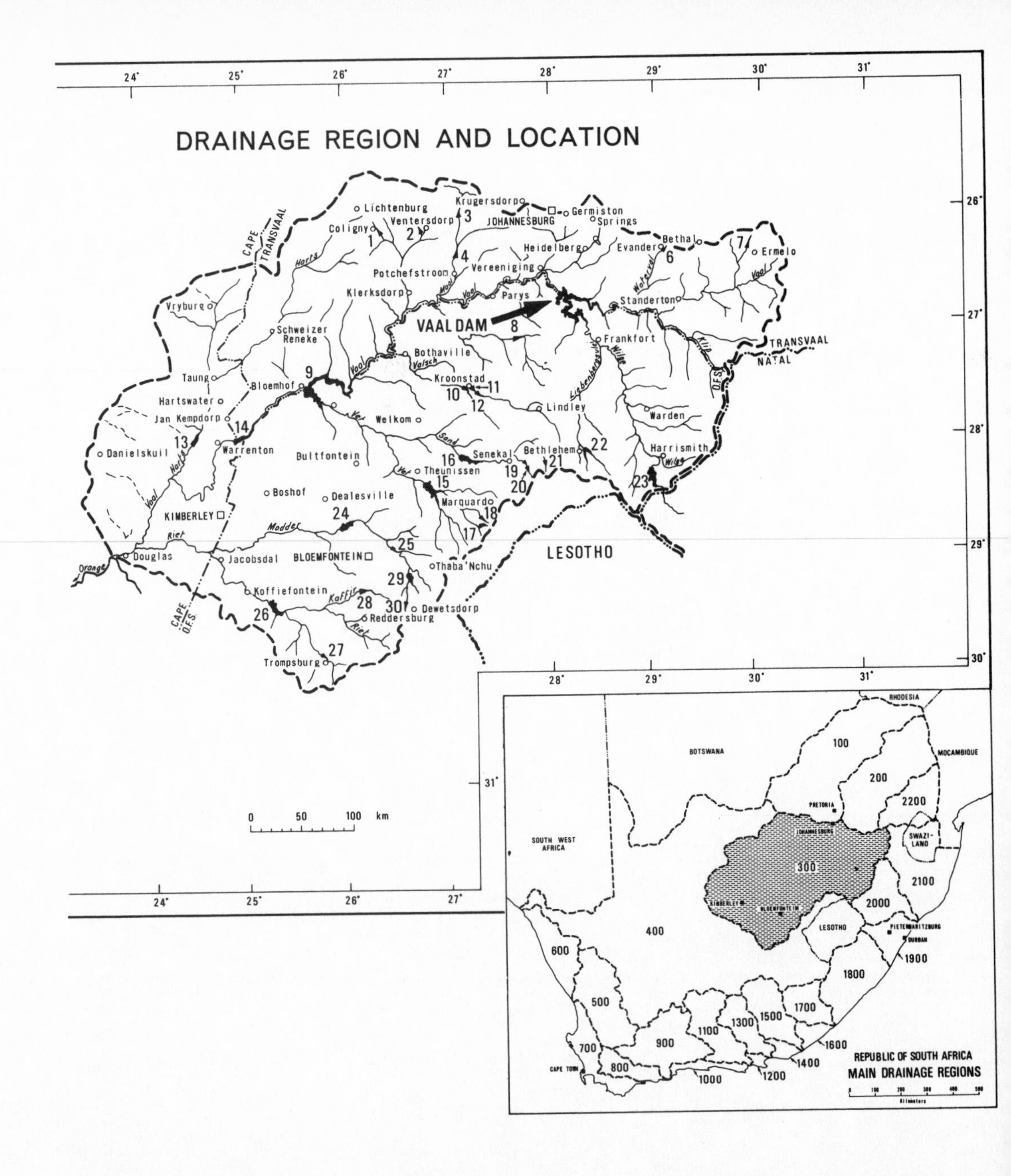

installation of crest gates 3 m high, on the raised overspill section. The earthern saddle dam, which was constructed to the right of the concrete dam in 1938, was also raised and is now linked to the concrete dam to form a continuous wall 2 570 m long. During the raising works an additional earth embankment was constructed across a saddle to the north-east of the main earth dam. This dam is 732 m long, 9,5 m high at maximum depth and contains 116 000 m³ of earth. The two earth embankments are only 427 m away from each other.

The outlet block is an integral part of the main concrete wall and provision is made for 24 river outlets in tiers of four, each 3 m apart. The two lowest tiers are fitted with jet dispersion valves, sleeve type, which discharge directly into the Vaal River. The upper four tiers are blanked off to be used when silt accumulation renders the lower outlets ineffective. Three special outlets are also provided for the Rand Water Board and these are connected to a pipeline which runs down the right bank of the Vaal River to the Suikerbos pumpstation near Vereeniging.

With the building of the Bloemhof Dam, 478 km downstream of Vaal Dam, the demand on Vaal Dam for irrigation water has been reduced, thus enabling more water to be made available for use in the "Vaal River Triangle". Future augmentation to Vaal Dam will be made from the Upper Tugela River, from where water is to be diverted into the Vaal catchment and stored in the Sterkfontein Dam, now under construction near Harrismith, and released down the Wilge River to Vaal Dam as and when required.

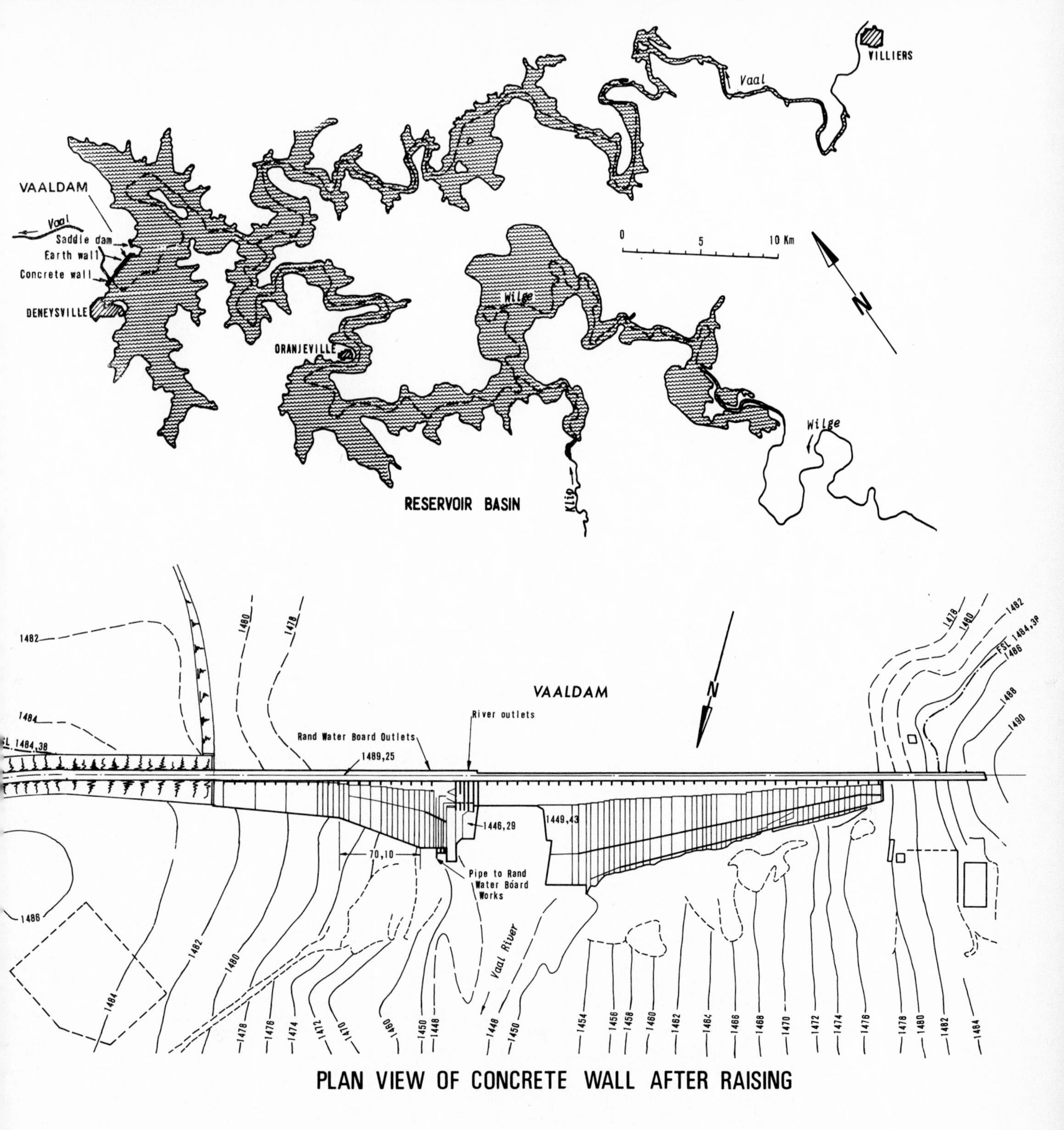

PLAN VIEW OF CONCRETE WALL AFTER RAISING

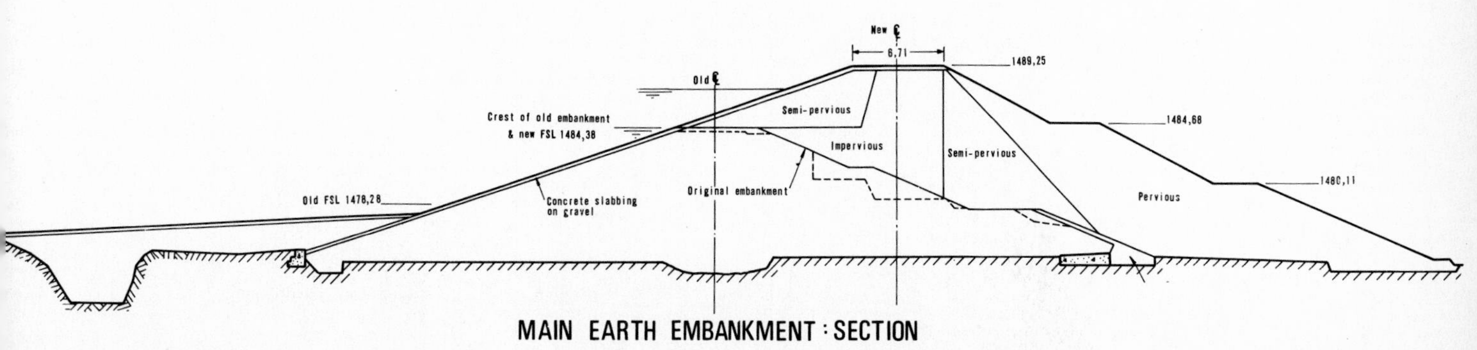

MAIN EARTH EMBANKMENT : SECTION

Jet disposition valves discharging into the Vaal River (D.W.A.)

Vaal Dam (G. Douglas)

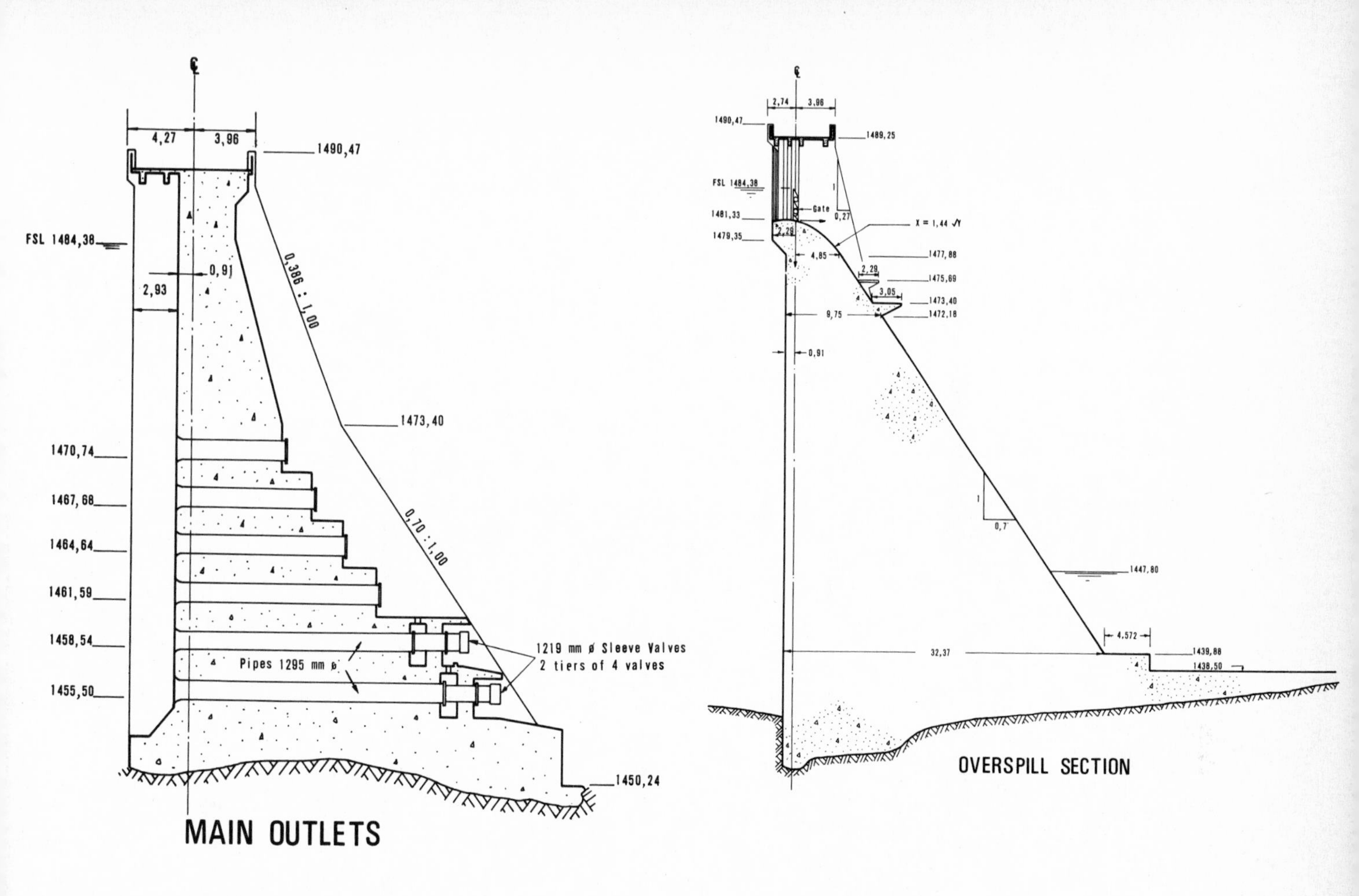

STEENBRAS DAM

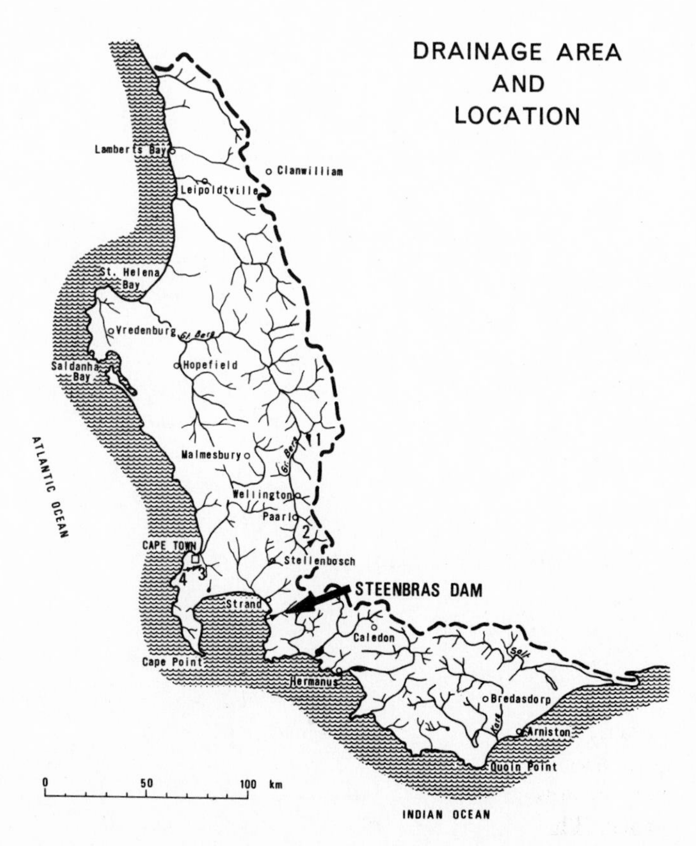

HISTORY

The Steenbras River Valley is about 60 km east of Cape Town. It is a spectacular narrow valley close to the sea, bounded by steep-sided mountains of the Hottentots Holland range.

The dam, completed in 1921 at the south end of the valley, is fed by two streams, the Steenbras and the Kogel Berg. These have a total catchment area of 60 square km producing an average inflow during the wet season of about 27,3 million m^3. The purpose of the 335 ha reservoir is to augment the water supply of the Cape Peninsula, and it is owned by the Municipality of Cape Town.

DIMENSIONS

Type	Constant Radius
Height above lowest foundation	36 m
Length of crest . . .	412 m
Volume content of dam .	0,051 $m^3 \times 10^6$
Gross capacity of reservoir	32,240 $m^3 \times 10^6$
Maximum discharge capacity of spillway . . .	481 m^3/s
Type of spillway . . .	Uncontrolled

Raised 6 m in 1924
11,5 m in 1927
1,35 m in 1954
6 m in 1957

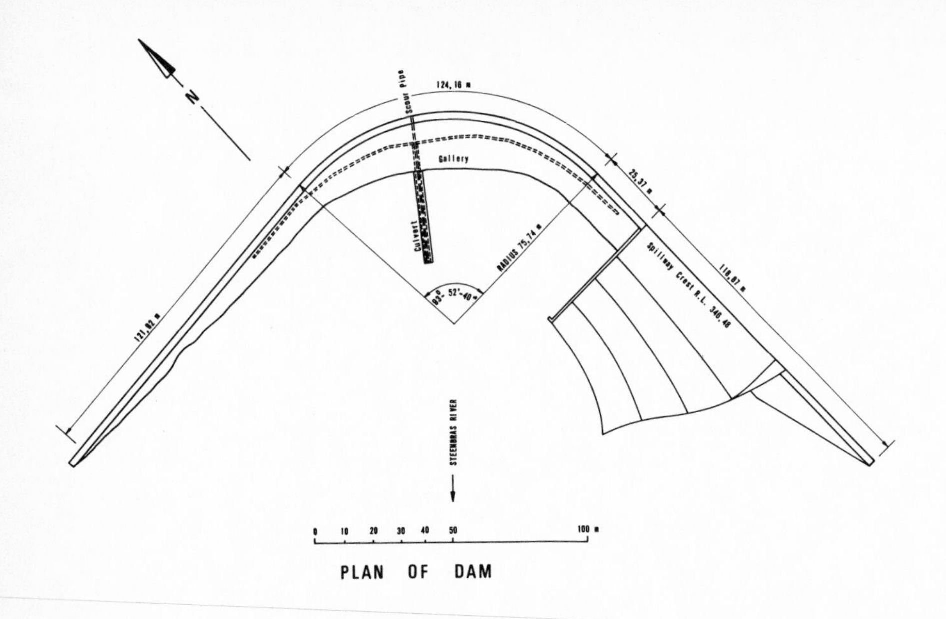

PLAN OF DAM

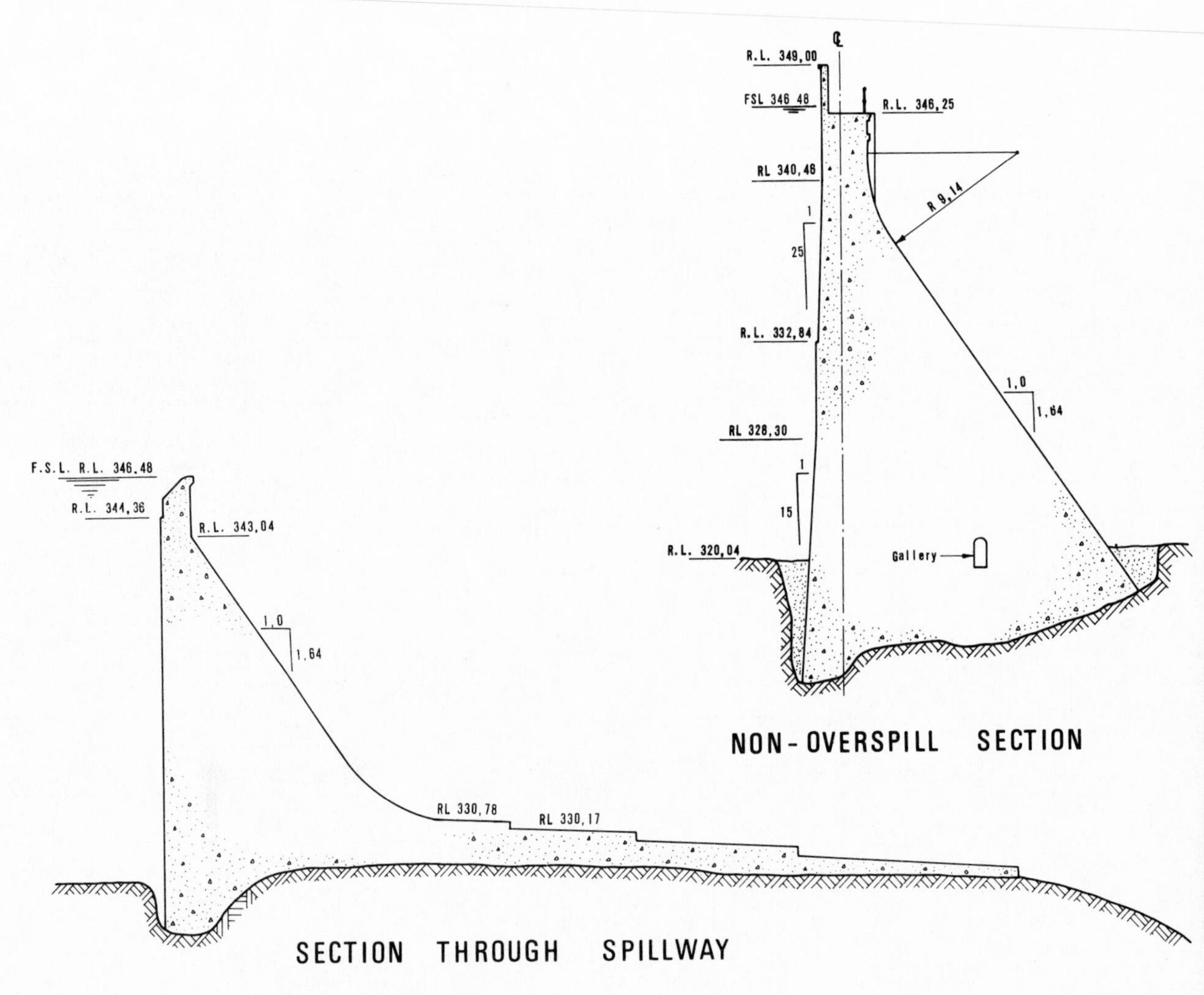

NON-OVERSPILL SECTION

SECTION THROUGH SPILLWAY

CONSTRUCTION

The dam is of concrete gravity design.

Post stressing of dam prior to raising in 1957 was done with high tension steel cables anchored into the bed rock. This application of the Coyne System was unique in South Africa. After this a 3 metre retaining wall can safely be added.

ENGINEERS CONTRACTORS

Engineering by . . F. E. Kanthack
Constructed by . . Cementation Co. Ltd.

ROODE ELSBERG DAM

HISTORY

Roode Elsberg Dam, completed in 1968, is situated on the Sanddrift River about 9 km west of the town of De Doorns, and 30 km from Worcester. The dam is the lower of the two storage dams constructed on the Sanddrift River to provide supplementary water so as to create a more assured supply for 2 535 ha of land under irrigation – mainly export table grapes – in the Hex River Valley and to the regions of De Wet, Over Hex, Aan-de-Doorns, Nooitgedacht and Nonna, comprising some 3 126 ha, and situated to the south of the Hex River Valley. It is owned and was engineered and constructed by the Department of Water Affairs.

DIMENSIONS

Type	Double Curvature
Height above lowest foundation	72 m
Length of crest	274 m
Volume content of dam	0,116 $m^3 \times 10^6$
Gross capacity of reservoir	8,210 $m^3 \times 10^6$
Purpose	Irrigation
Maximum discharge capacity of spillway	708 m^3/s
Type of spillway	Uncontrolled

CONSTRUCTION

The Roode Elsberg Dam is a concrete cupola arch structure with a centrally situated overspill section. A concrete lined tunnel, 1,98 m in diameter and 5,18 km long, conveys water from Roode Elsberg Dam to the Hex River Valley where distribution is by means of an enclosed pipe system with special offtakes. Water is also released to the river for abstraction lower down the Hex River.

Downstream view

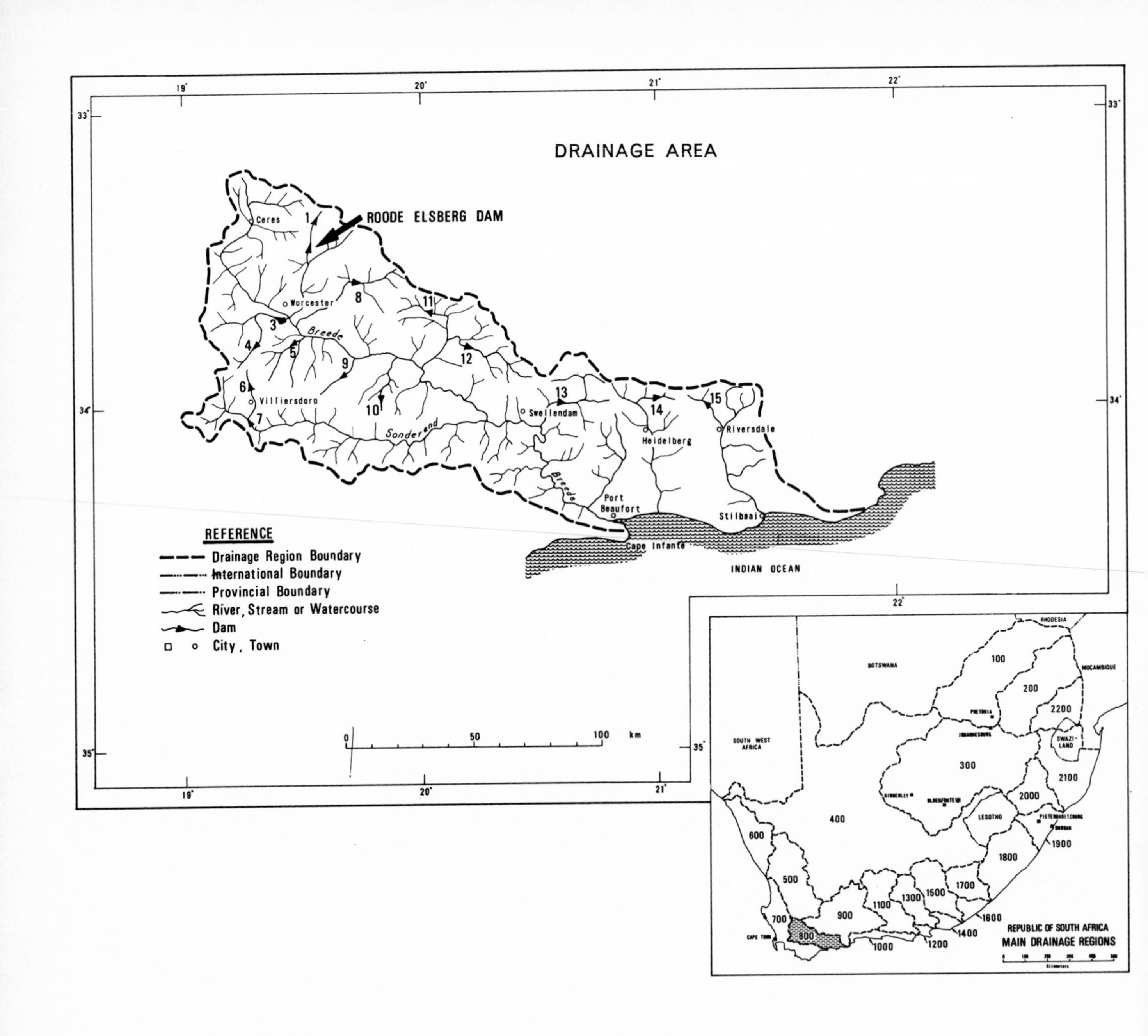

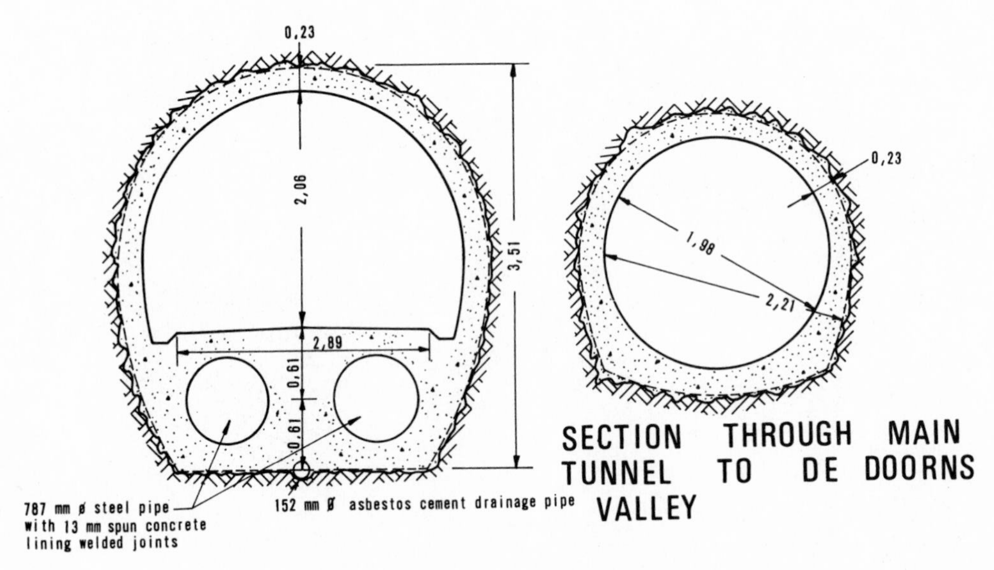

SECTION THROUGH MAIN TUNNEL TO DE DOORNS VALLEY

SECTION THROUGH ACCESS TUNNEL

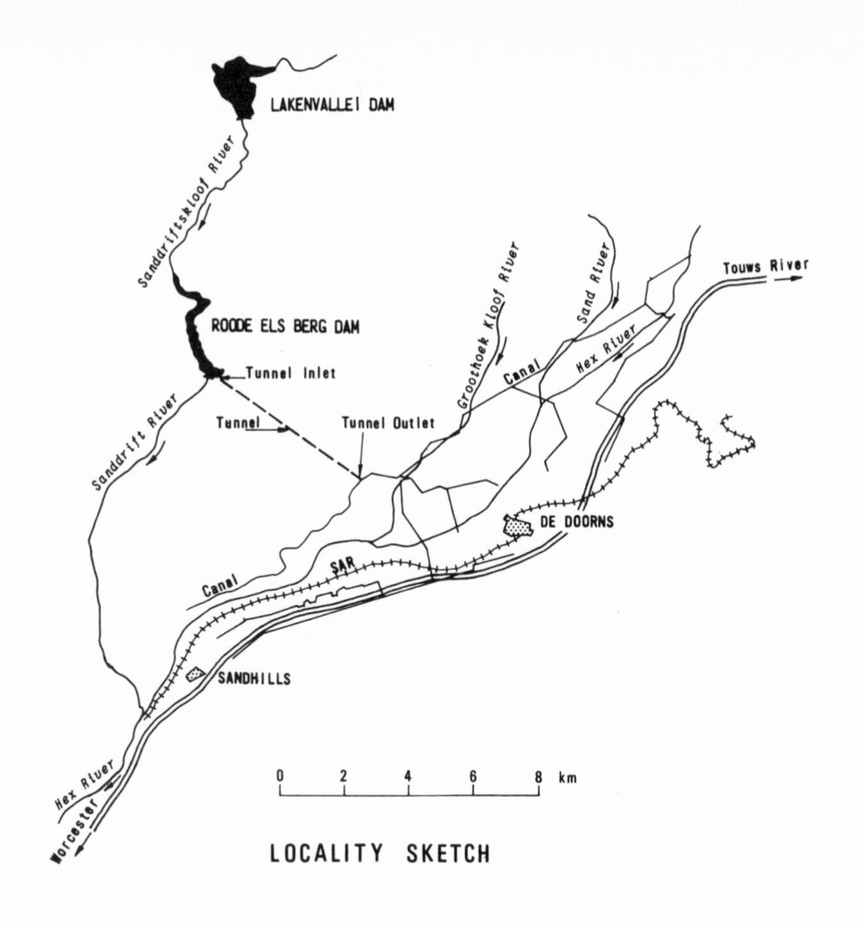

LOCALITY SKETCH

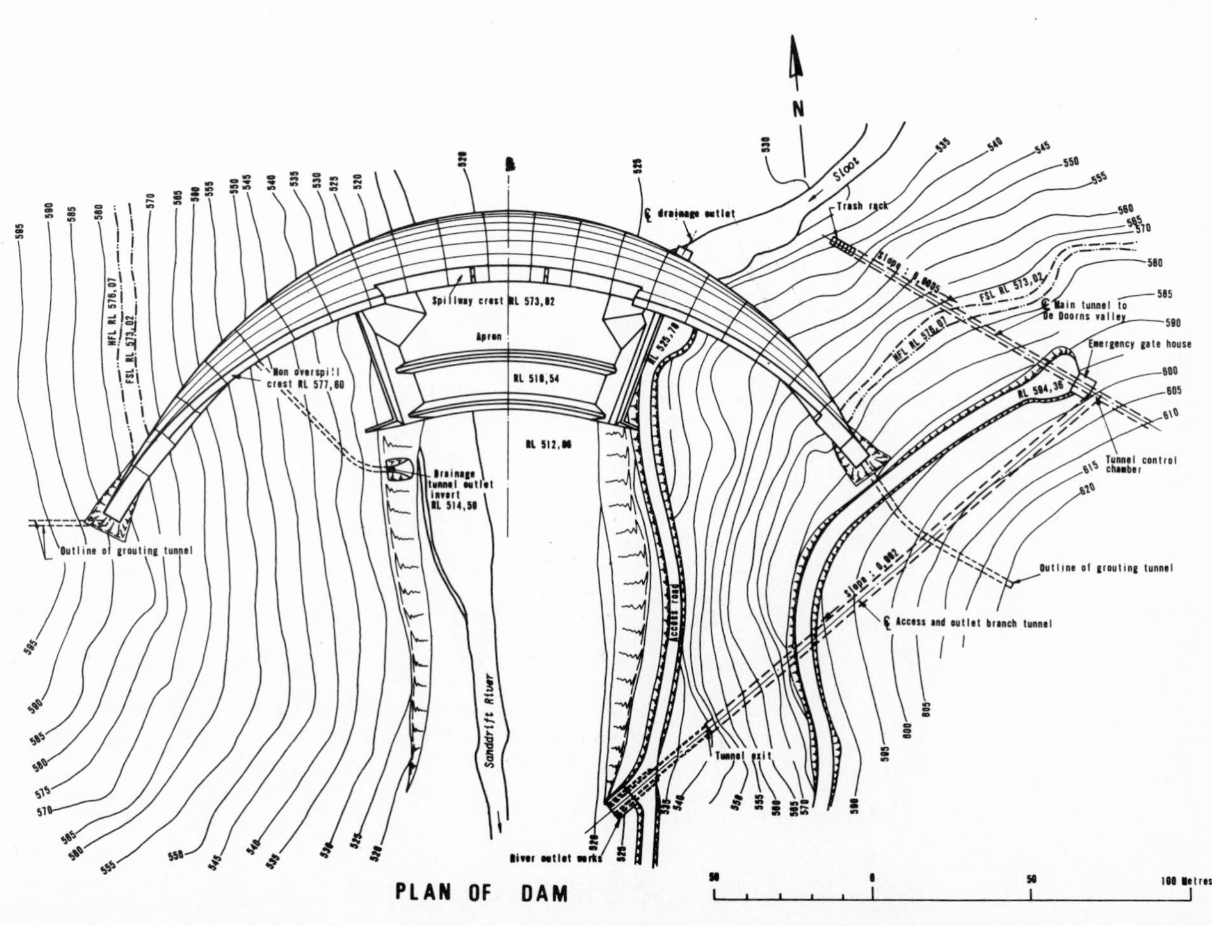

PLAN OF DAM

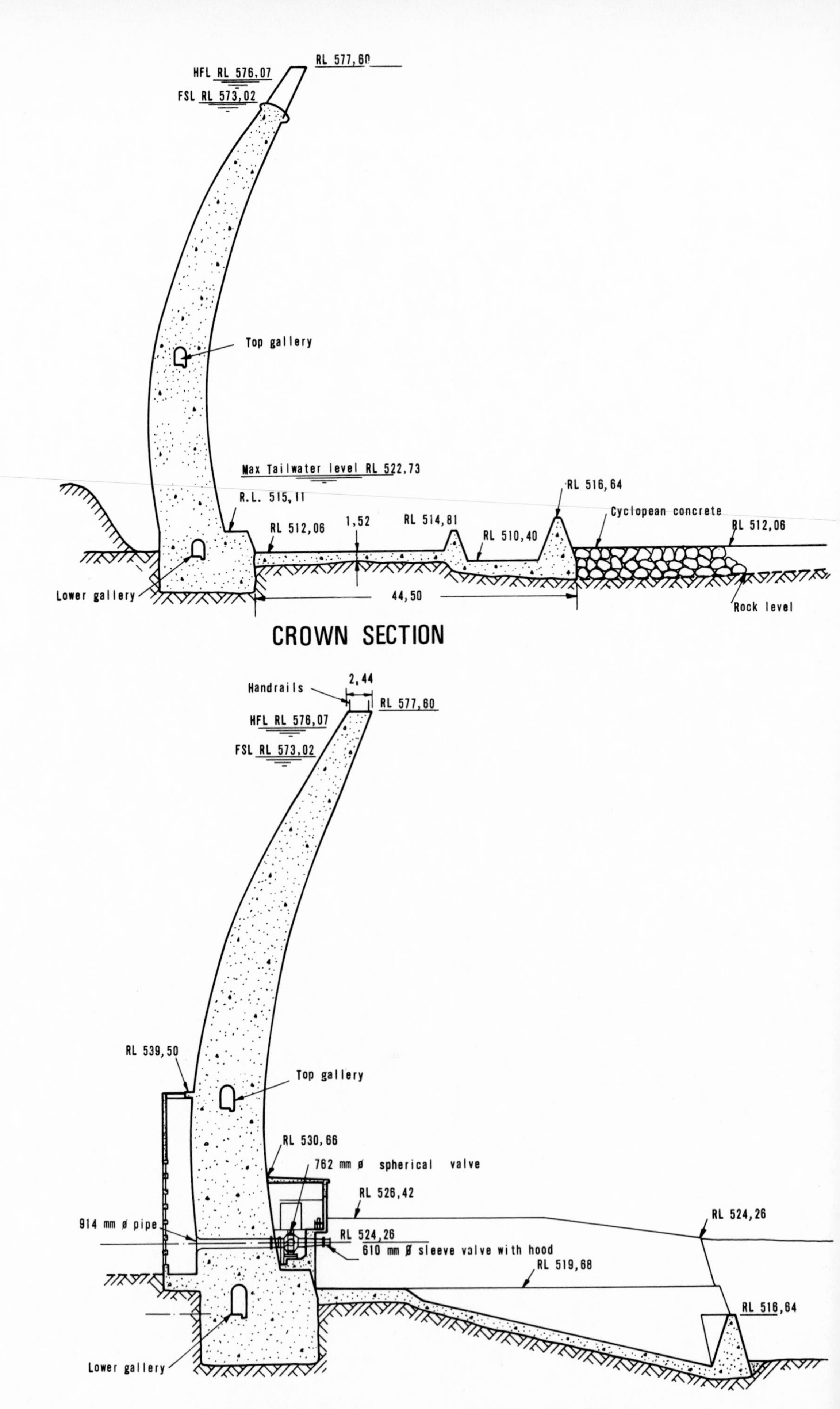
RL 577,60
HFL RL 576,07
FSL RL 573,02
Top gallery
Max Tailwater level RL 522,73
R.L. 515,11
RL 516,64
Cyclopean concrete
RL 512,06
1,52
RL 514,81
RL 510,40
RL 512,06
Lower gallery
44,50
Rock level
CROWN SECTION
Handrails
2,44
RL 577,60
HFL RL 576,07
FSL RL 573,02
RL 539,50
Top gallery
RL 530,66
762 mm ø spherical valve
RL 526,42
RL 524,26
914 mm ø pipe
RL 524,26
610 mm ø sleeve valve with hood
RL 519,68
RL 516,64
Lower gallery
SECTION ALONG ℄ OF DRAINAGE OUTLET

Looking downstream
(S.A. Tourist Corp.)

PAUL SAUER DAM

HISTORY

The Paul Sauer Dam, completed in 1969 and formerly known as Tweerivieren or Kouga Dam, is the first large cupola type dam constructed in the Republic of South Africa. The dam is situated on the Kouga River about 5 km upstream of the Groot-Kouga River's confluence, 27 km west of the town of Patensie, and some 120 km from Port Elizabeth.

The original design, submitted to Parliament in 1957, was for a straight mass gravity concrete structure 73,15 m high, with an overspill section in the centre and a capacity of 10,051 million m³. After further investigation this design was abandoned in favour of the arch dam that would be more suited to the site and less expensive.

In 1961 it was decided to increase the height of the dam to 82 m thereby increasing the capacity to 132,320 million m³. It is owned and was engineered and constructed by the Department of Water Affairs.

DIMENSIONS

Type	Double Curvature
Height above lowest foundation . . .	82 m
Length of crest . .	317 m
Volume content of dam	0,167 m³ × 10⁶
Gross capacity of reservoir	132,320 m³ × 10⁶
Purpose	Irrigation and Water Supply
Maximum discharge capacity of spillway	4 248 m³/s
Type of spillway . .	Controlled and uncontrolled

CONSTRUCTION

The dam is built on Table Mountain limestone and quartz alternated by shale and tillite layers, which required intensive foundation grouting and strengthening along the right flank by means of post-tension cables sunk from 24 to 46 m into the rock to bind the rock layers. Three tunnels, each 145 m long, were constructed in the right flank for drainage and pressure alleviation. A chute spillway is provided on the left flank for flood control when the dam is raised to its ultimate height, i.e. a raising of 15,2 m. At present the chute spillway inlet invert level is only some 3,81 m below the full supply crest of the dam.

Paul Sauer Dam is the main storage unit for irrigators in the Gamtoos Valley, the main crops produced being tobacco, citrus and vegetables. Water is taken into a canal through a tunnel 2,44 m diameter and pipe extensions to three turbo-alternators, each 1 750 kVA (1 200 kVA, present F.S.L.) then into the canal system. Water is also provided for Port Elizabeth via the canal system to the Loerie Dam at the tail-end of the scheme.

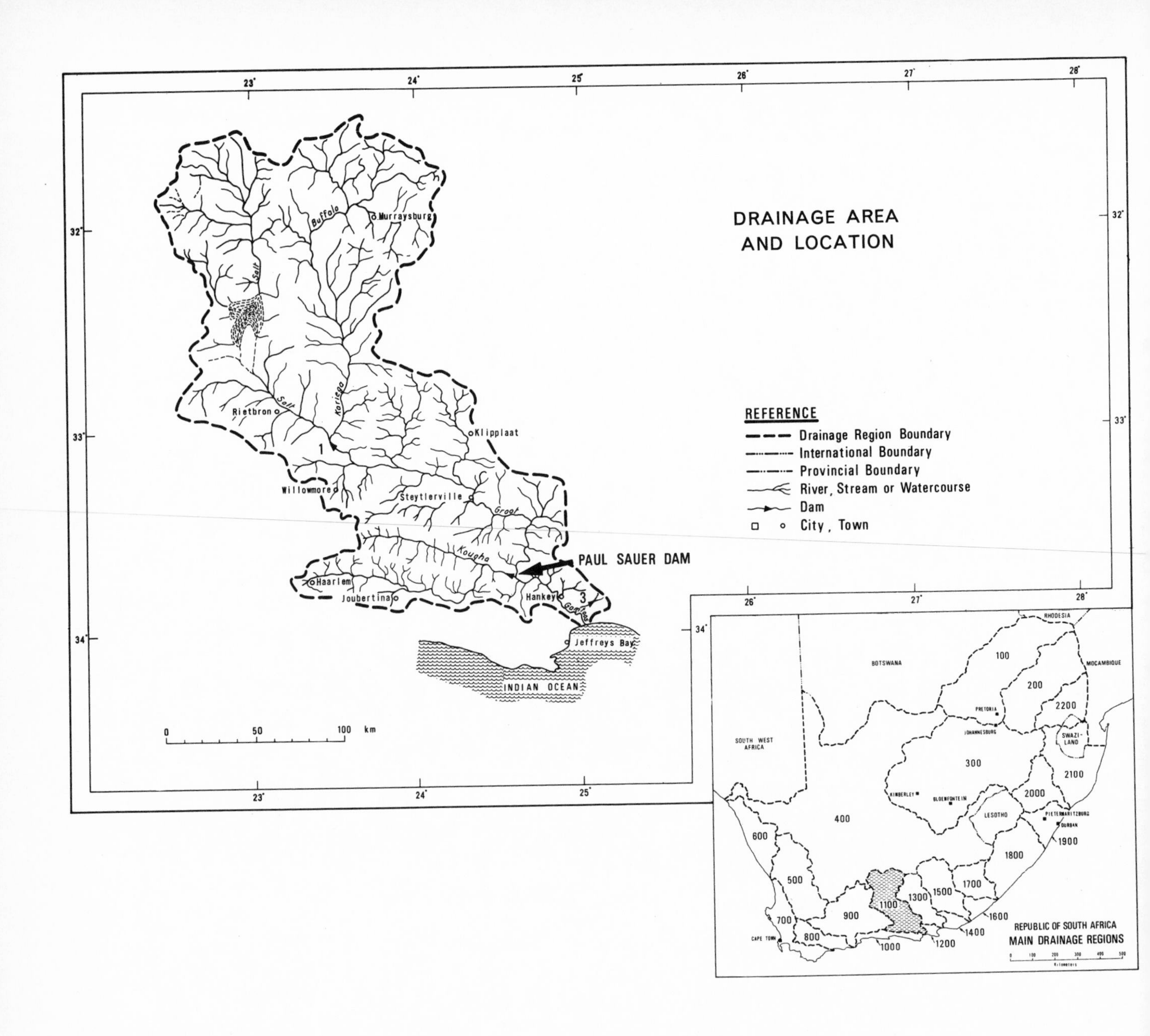

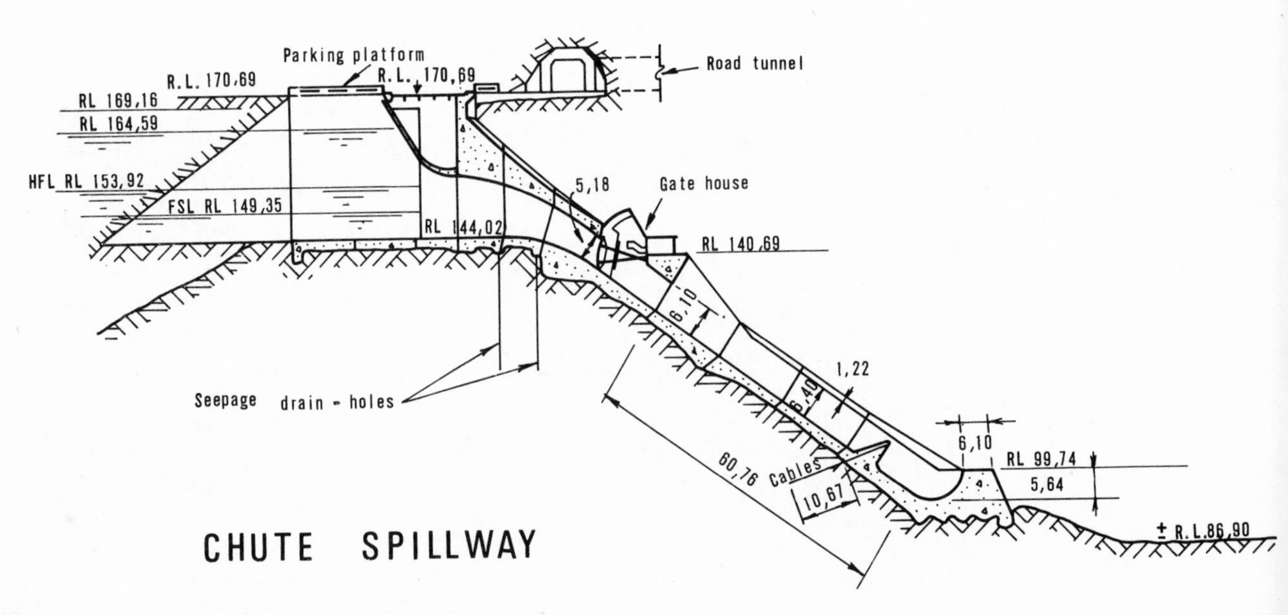

CHUTE SPILLWAY

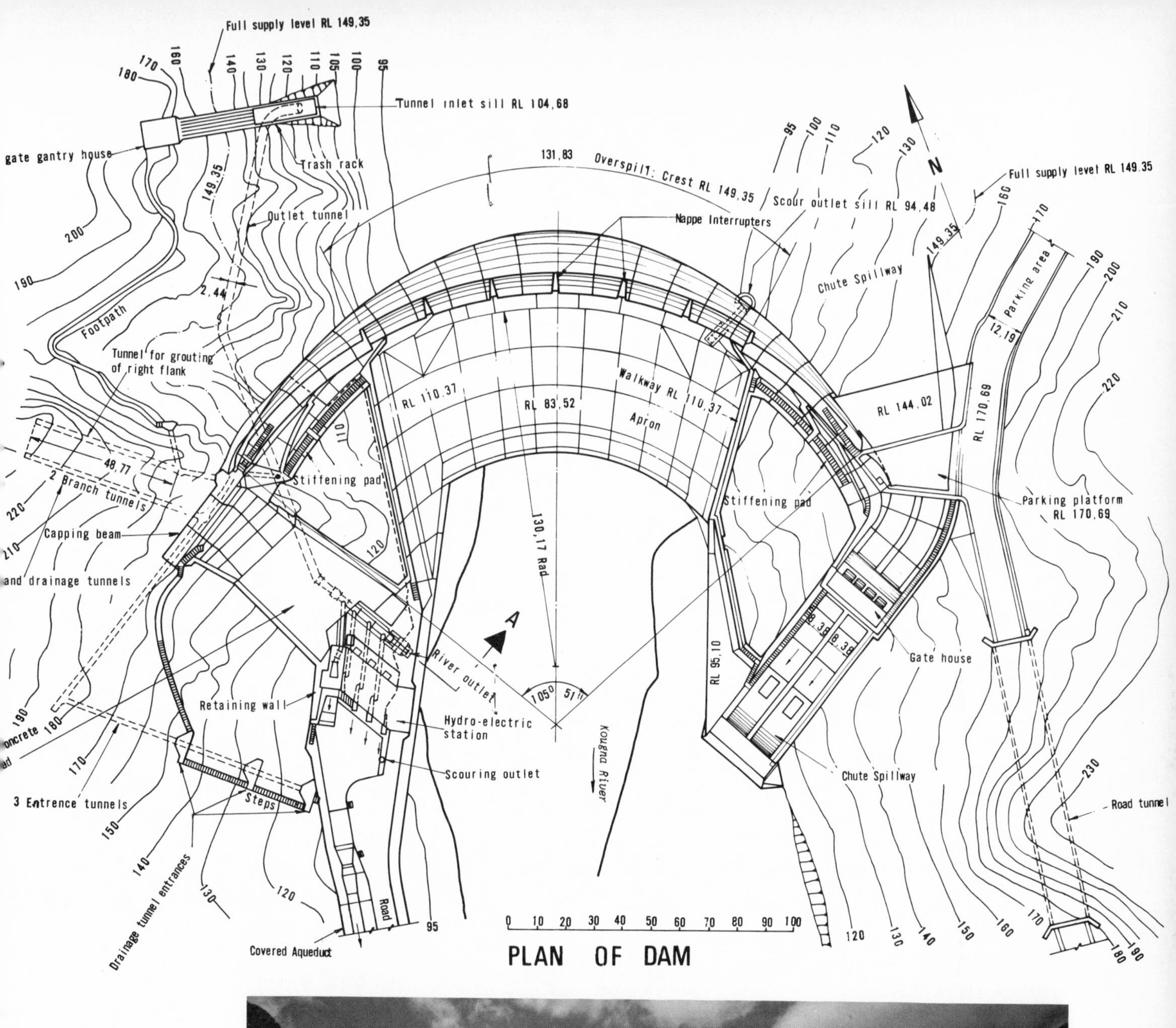
Full supply level RL 149,35
Tunnel inlet sill RL 104,68
gate gantry house
Trash rack
Outlet tunnel
149,35
2,44
Footpath
Tunnel for grouting of right flank
48,77
2 Branch tunnels
Capping beam
and drainage tunnels
Stiffening pad
Retaining wall
3 Entrence tunnels
Steps
Drainage tunnel entrances
Covered Aqueduct
Road
131,83
Overspill: Crest RL 149,35
Scour outlet sill RL 94,48
Nappe Interrupters
N
Full supply level RL 149.35
Chute Spillway
Parking area
12,19
RL 110.37
RL 83,52
Walkway RL 110,37
Apron
RL 144.02
RL 170,69
Parking platform RL 170,69
Stiffening pad
130,17 Rad
A
River outlet
105°
51°
Hydro-electric station
Scouring outlet
Kougha River
RL 95,10
8,38
8,38
Gate house
Chute Spillway
Road tunnel
0 10 20 30 40 50 60 70 80 90 100

PLAN OF DAM

Looking upstream
(*D.W.A.*)

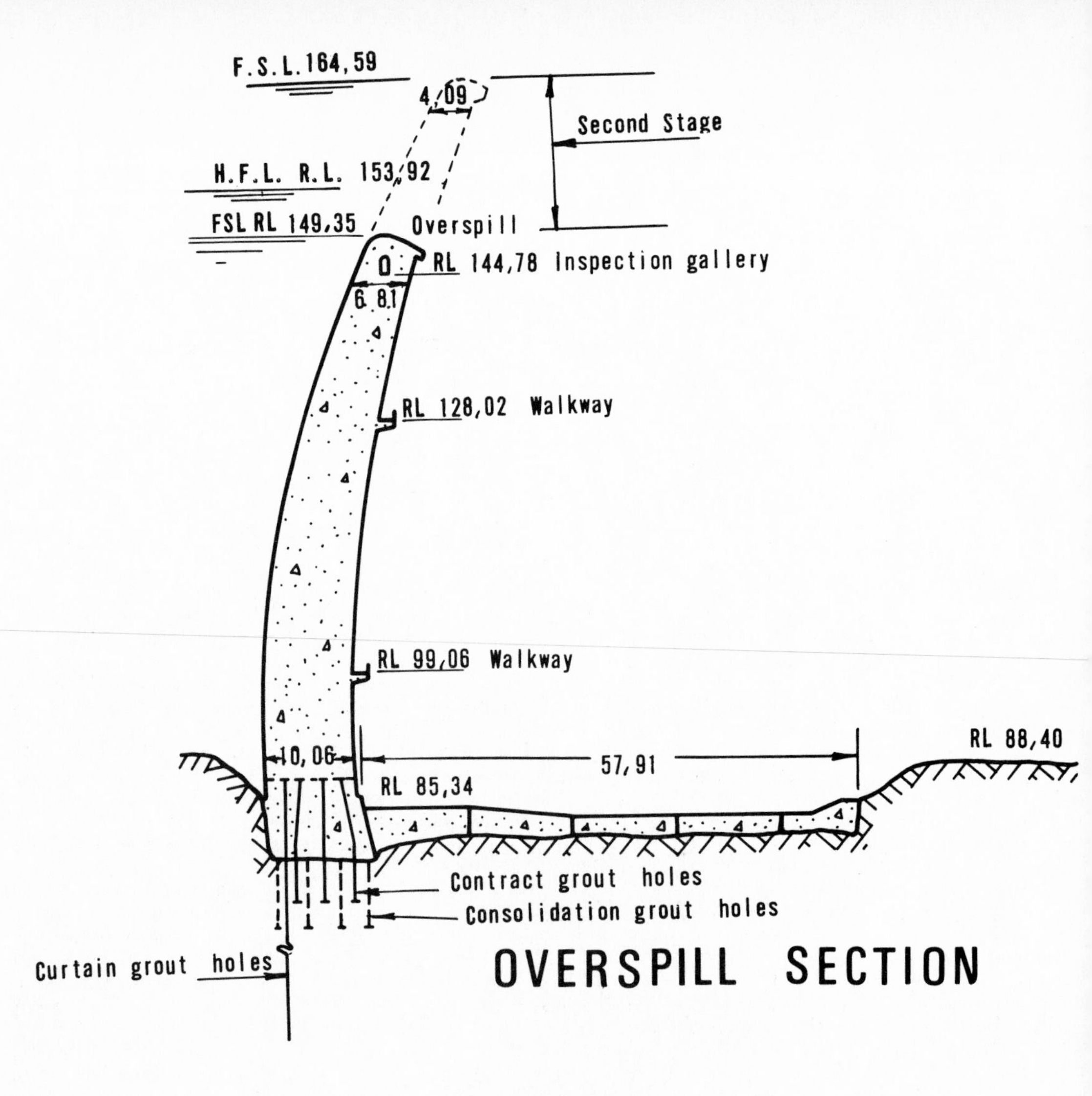
F.S.L. 164,59
4,09
Second Stage
H.F.L. R.L. 153,92
FSL RL 149,35
Overspill
RL 144,78 Inspection gallery
6.81
RL 128,02 Walkway
RL 99,06 Walkway
10,06
57,91
RL 88,40
RL 85,34
Contract grout holes
Consolidation grout holes
Curtain grout holes
OVERSPILL SECTION

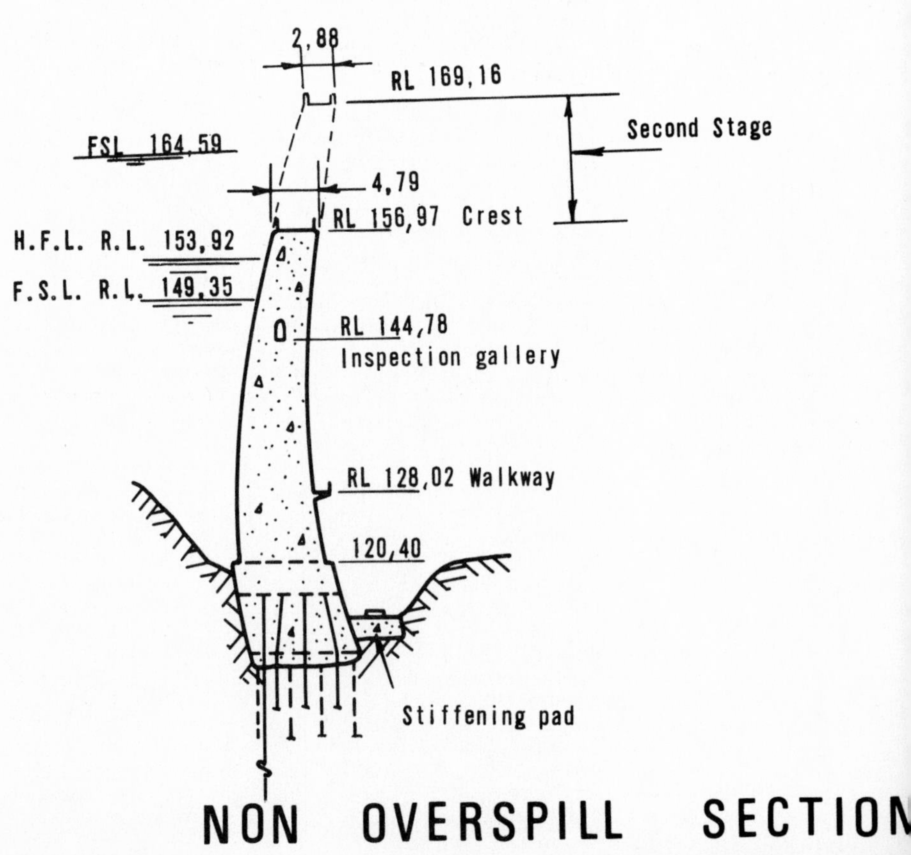
2,88
RL 169,16
Second Stage
FSL 164,59
4,79
RL 156,97 Crest
H.F.L. R.L. 153,92
F.S.L. R.L. 149,35
RL 144,78
Inspection gallery
RL 128,02 Walkway
120,40
Stiffening pad
NON OVERSPILL SECTION

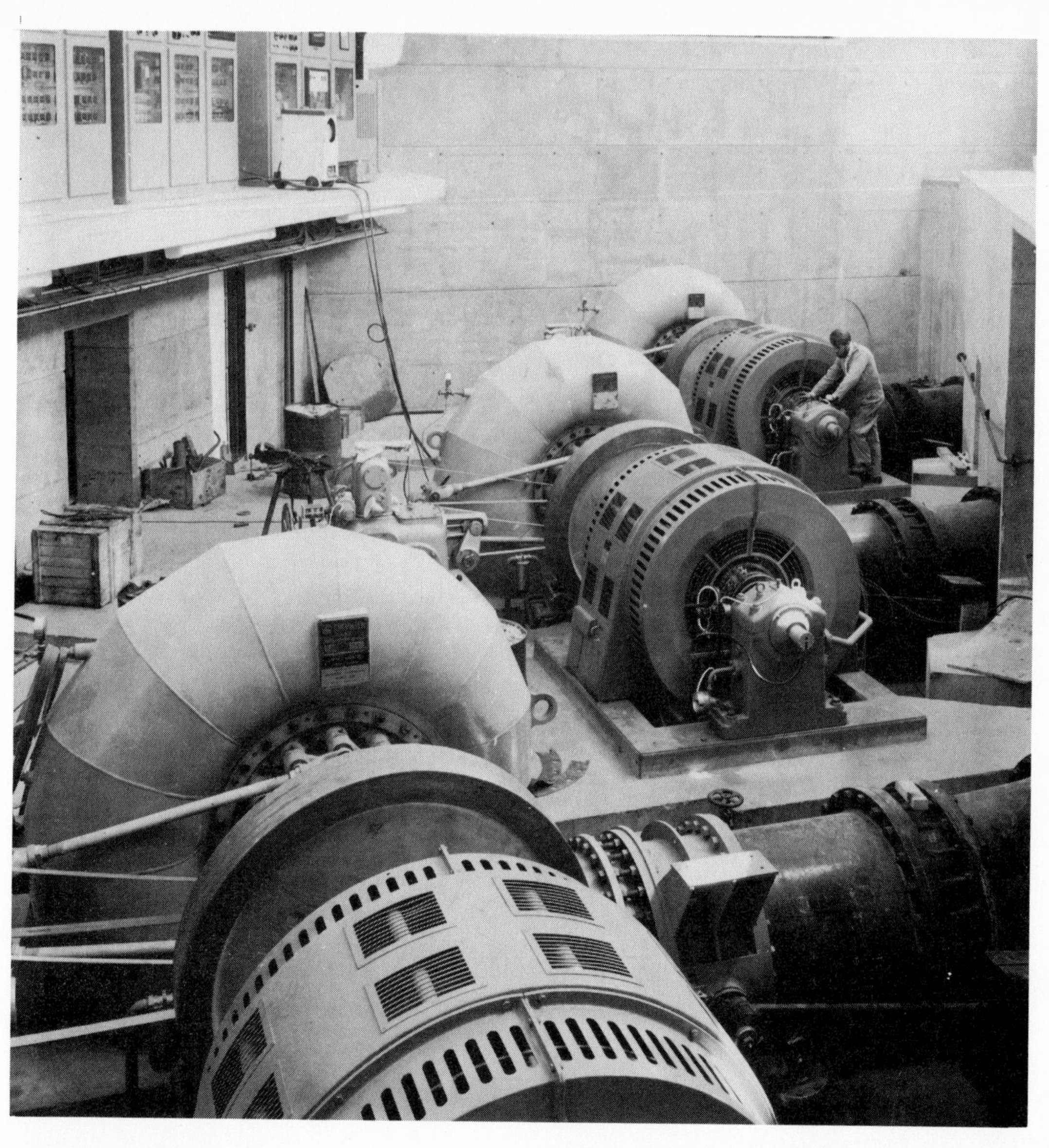

Three turbo-alternators feeding irrigation canals

MIDMAR DAM

Midmar Dam, completed in 1965, is situated on the Mgeni (Umgeni) River 3 km south-west of the town of Howick, and is one of the major storage units serving the metropolitan area from Pietermaritzburg to Durban. It is owned, and was engineered and constructed by the Department of Water Affairs.

The dam consists of a mass gravity concrete overspill section astride the river bed with earth embankments on its flanks. The concrete section has an exceptionally good foundation on a dolerite sill. The earth embankments are, due to the absence of suitable pervious materials, built throughout of red doleritic clayey soil. This has called for the provision of special drainage layers in the embankments. Provision is made in the design of the dam for the installation of crest gates, 4,57 m high to enable the dam to conserve a further 80 931 million m³ when the need arises.

Water from the dam is conveyed by pipeline to the urban and industrial areas along the railway line between Howick and Pinetown, also down the Mgeni River to Durban's Nagle Dam, a distance of 56 km, where it is abstracted and purified for use in Durban and its environs.

DIMENSIONS

Height above lowest foundation . .	32 m
Length of crest .	1 338 m
Volume content of dam . .	$0{,}917 \text{ m}^3 \times 10^6$ (concrete: $0{,}077 \text{ m}^3 \times 10^6$ (earth: $0{,}840 \text{ m}^3 \times 10^6$
Gross capacity of reservoir .	$159{,}254 \text{ m}^3 \times 10^6$
Maximum discharge capacity of spillway . . .	$2\ 550 \text{ m}^3/\text{s}$
Type of spillway	Uncontrolled

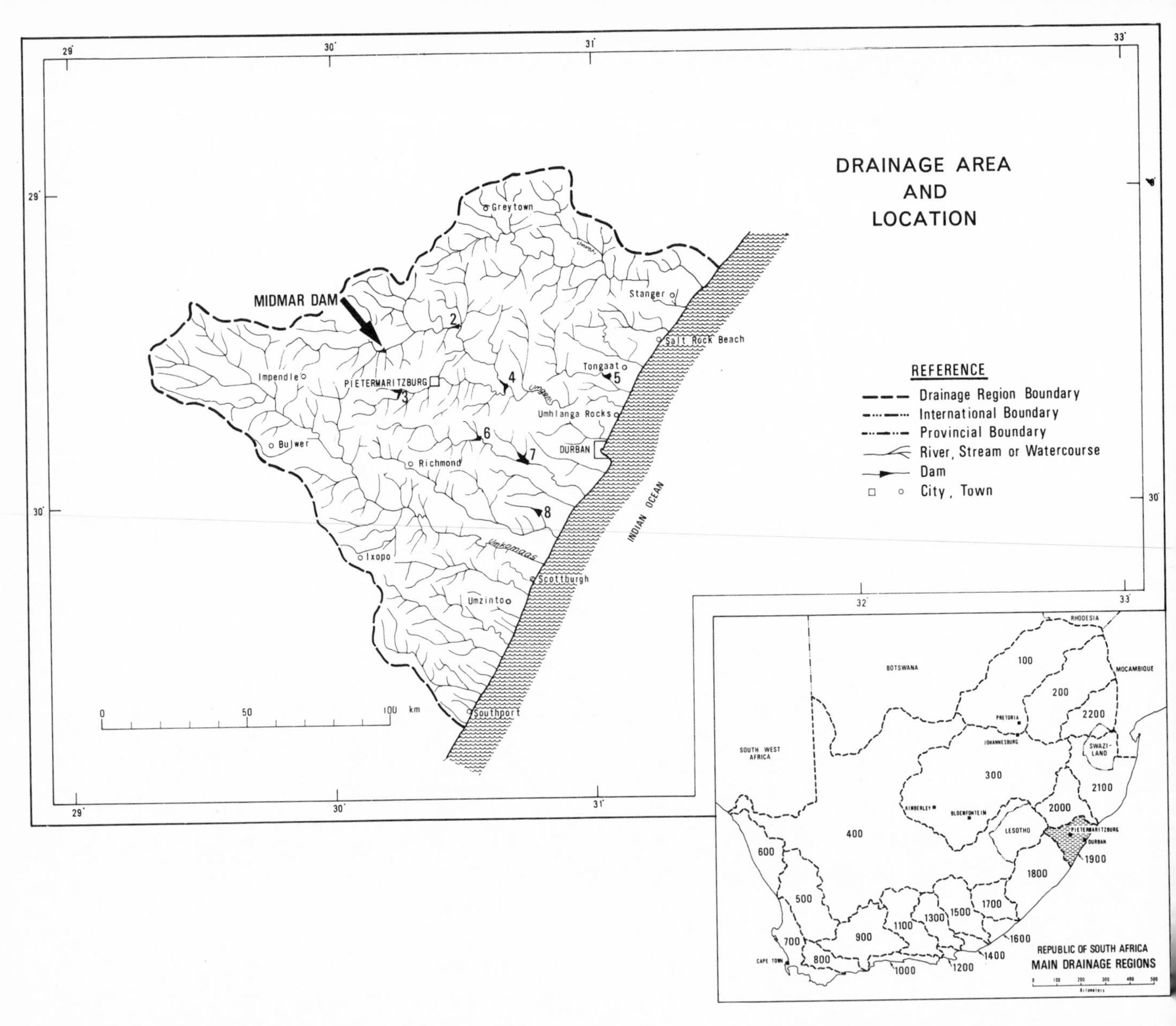
DRAINAGE AREA
AND
LOCATION
MIDMAR DAM
Greytown
Stanger
Salt Rock Beach
Tongaat
Impendle
PIETERMARITZBURG
Umhlanga Rocks
Bulwer
DURBAN
Richmond
INDIAN OCEAN
Ixopo
Umkomaas
Scottburgh
Umzinto
Southport
0
50
100 km
REFERENCE
Drainage Region Boundary
International Boundary
Provincial Boundary
River, Stream or Watercourse
Dam
City, Town
BOTSWANA
RHODESIA
MOCAMBIQUE
SOUTH WEST AFRICA
SWAZI-LAND
LESOTHO
CAPE TOWN
REPUBLIC OF SOUTH AFRICA
MAIN DRAINAGE REGIONS

Midmar Dam

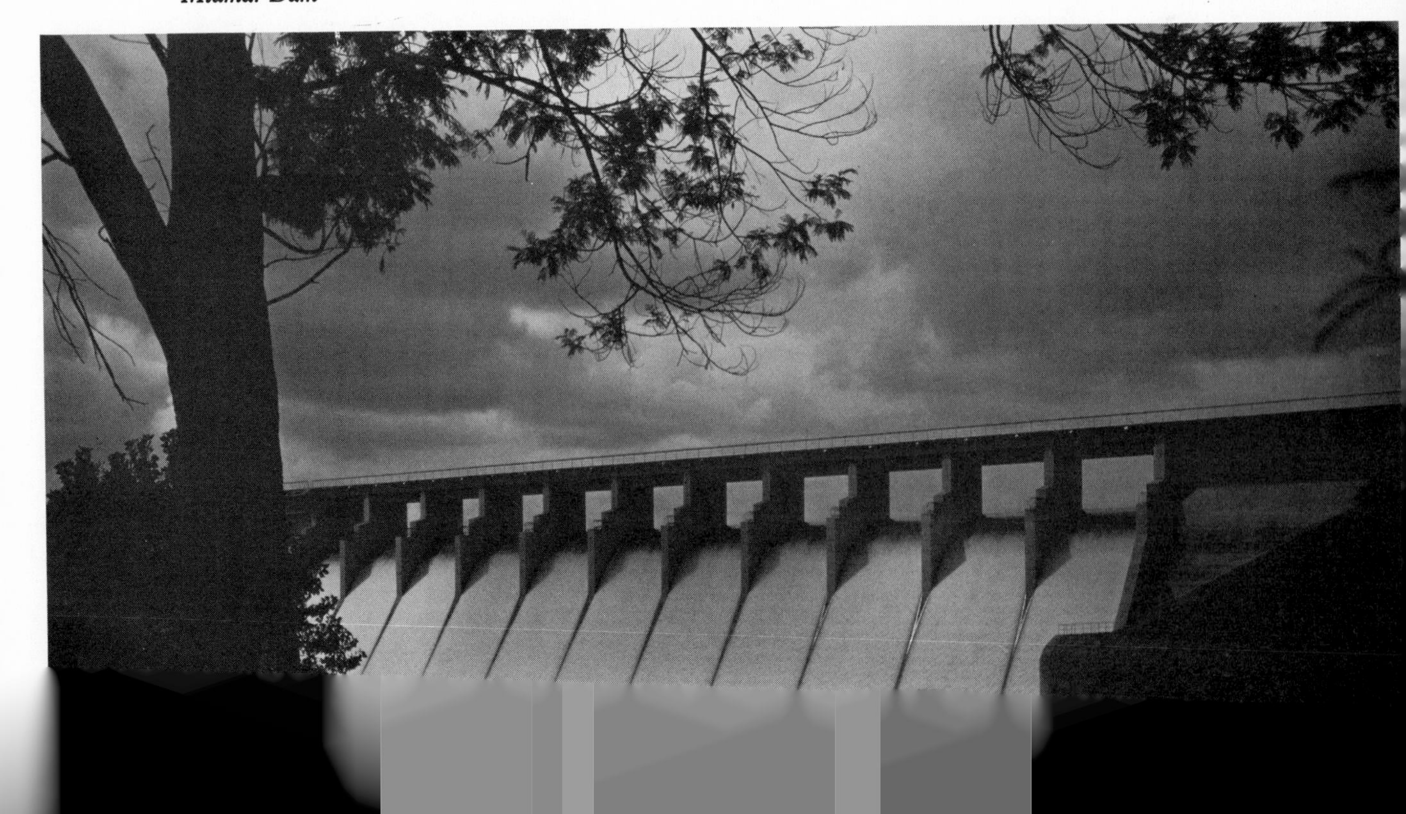

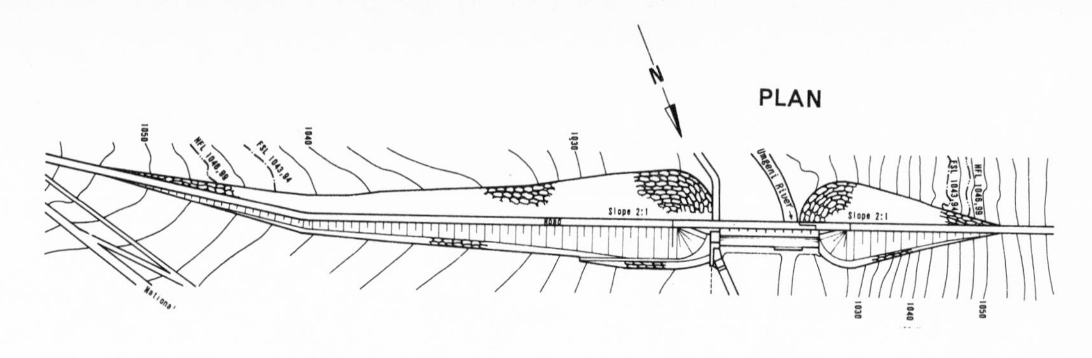
PLAN
N
Slope 2:1
Slope 2:1
Umgeni River
HFL 1046,99
FSL 1043,94
1050
1040
1030
National

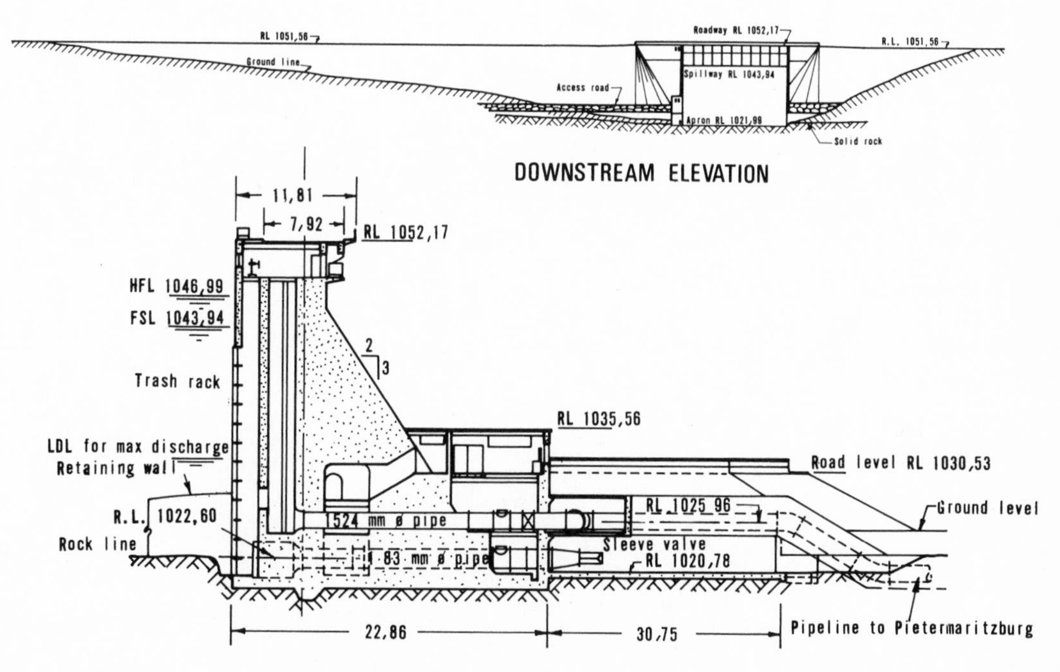
RL 1051,56
Roadway RL 1052,17
R.L. 1051,56
Ground line
Spillway RL 1043,94
Access road
Apron RL 1021,99
Solid rock
11,81
7,92
RL 1052,17
HFL 1046,99
FSL 1043,94
2
3
Trash rack
RL 1035,56
LDL for max discharge
Retaining wall
Road level RL 1030,53
R.L. 1022,60
RL 1025,96
Ground level
1524 mm ø pipe
Rock line
Sleeve valve
83 mm ø pipe
RL 1020,78
22,86
30,75
Pipeline to Pietermaritzburg

DOWNSTREAM ELEVATION

SECTION: OUTLET WORKS

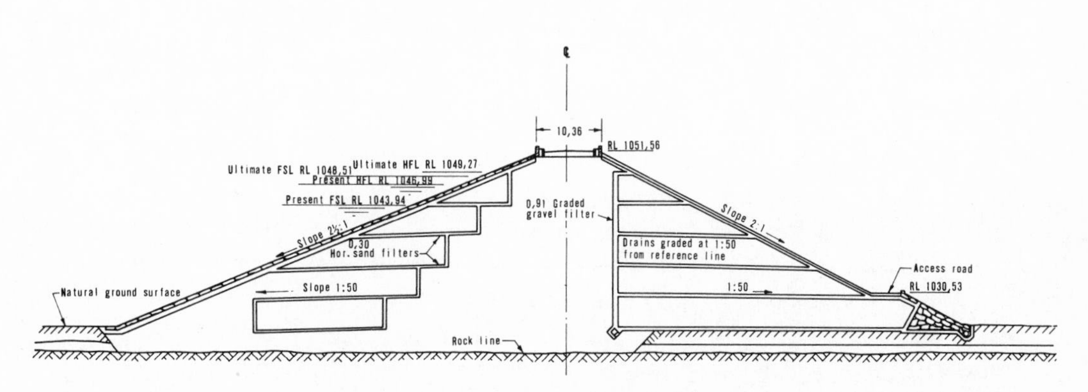
10,36
RL 1051,56
Ultimate FSL RL 1048,51
Ultimate HFL RL 1049,27
Present HFL RL 1046,99
Present FSL RL 1043,94
0,91 Graded gravel filter
Slope 2:1
Slope 2½:1
0,30 Hor. sand filters
Drains graded at 1:50 from reference line
Access road
RL 1030,53
Natural ground surface
Slope 1:50
1:50
Rock line

SECTION: EARTH EMBANKMENT

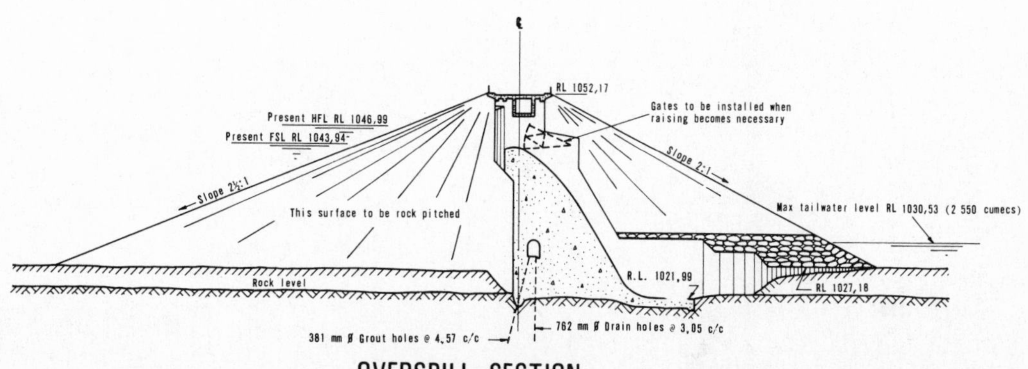
RL 1052,17
Present HFL RL 1046,99
Present FSL RL 1043,94
Gates to be installed when raising becomes necessary
Slope 2:1
Slope 2½:1
This surface to be rock pitched
Max tailwater level RL 1030,53 (2 550 cumecs)
R.L. 1021,99
Rock level
RL 1027,18
762 mm Ø Drain holes @ 3,05 c/c
381 mm Ø Grout holes @ 4,57 c/c

OVERSPILL SECTION

LOSKOP DAM

HISTORY

After the severe depression of the early 1930's the Government decided that public works were urgently required to give employment to many thousands of jobless Europeans. Amongst other projects three major irrigation schemes were to be constructed. They were the Vaaldam-Vaalhartz complex on the Vaal River, the Kalkfontein scheme in the Orange Free State on the Riet River, and the Loskop scheme.

Work on all three was started by the Department of Irrigation (now Water Affairs) during 1934.

A large number of returned soldiers as well as civilians were eventually placed. In this way more than 500 families came to live in an area where only a handful had lived before. Most of the previous owners had their residences and major interests on the Middelburg, Bronkhorstspruit and Carolina highveld and had used these farms for ranching and as squatters' farms to augment their labour supplies. Today there are altogether 646 families.

Initially tobacco, wheat and maize were cultivated but better paying crops like citrus, grapes, cotton and vegetables are now grown.

DIMENSIONS

Type	Gravity
Height above lowest foundation	45 m
Length of crest	433 m
Volume content of dam	0,235 $m^3 \times 10^6$
Gross capacity of reservoir	207,644 $m^3 \times 10^6$
Purpose	Irrigation
Maximum discharge capacity of spillway	2 830 m^3/s
Type of spillway	Uncontrolled

CONSTRUCTION

The canal system commands an area of about 25 000 ha of which 16 248 ha are listed for supply from the Loskop Dam. This dam, on the Olifants River, 48 km north of the town Middelburg was completed in 1939; it is a mass concrete gravity overspill structure. A

Loskop Dam (D.W.A.)

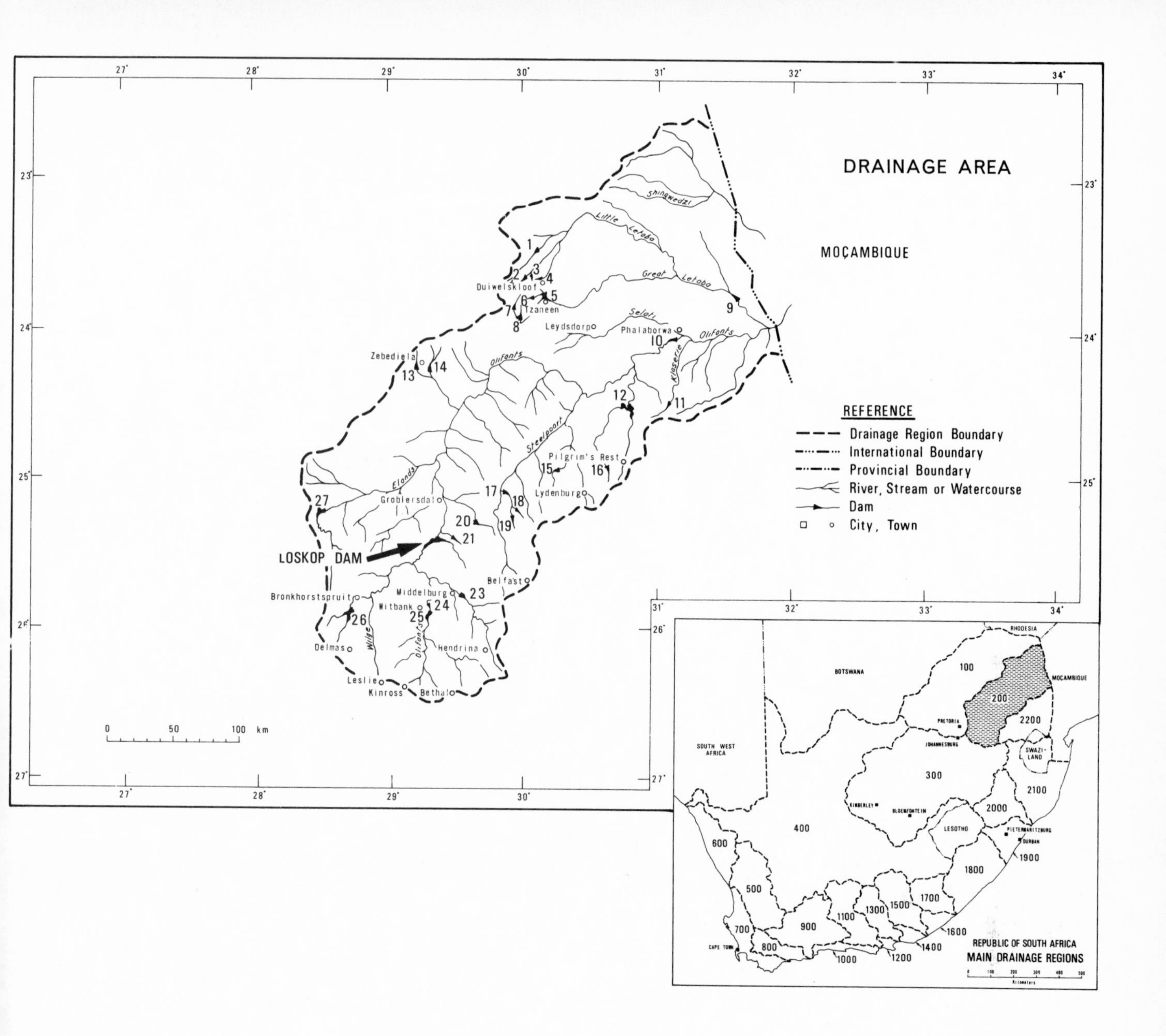

novel feature, and one which has since been incorporated in all overspill dams built in the Republic, was the method developed and designed by an engineer of the Department of Water Affairs for dissipating the energy of water discharging over the spillway during floods. This is achieved by the provision of a step on the downstream face about 6 m below crest, surmounted by a row of reinforced brackets or "splitters". The effect is to separate the solid nappe into a system of jets impinging one upon the other in such a way that the overspilling water is broken up into rain which descends upon the apron below the wall with much lessened destructive energy. Not only is this form of construction practically effective but it also creates a fine spectacle during floods.

A start was made recently on the raising of the dam by 9 m, which will very nearly double the capacity to 330 million m^3. The raising will be achieved by adding concrete to the downstream slope and crest, i.e. solid raising. It is owned by the Department of Water Affairs which carried out the engineering and construction.

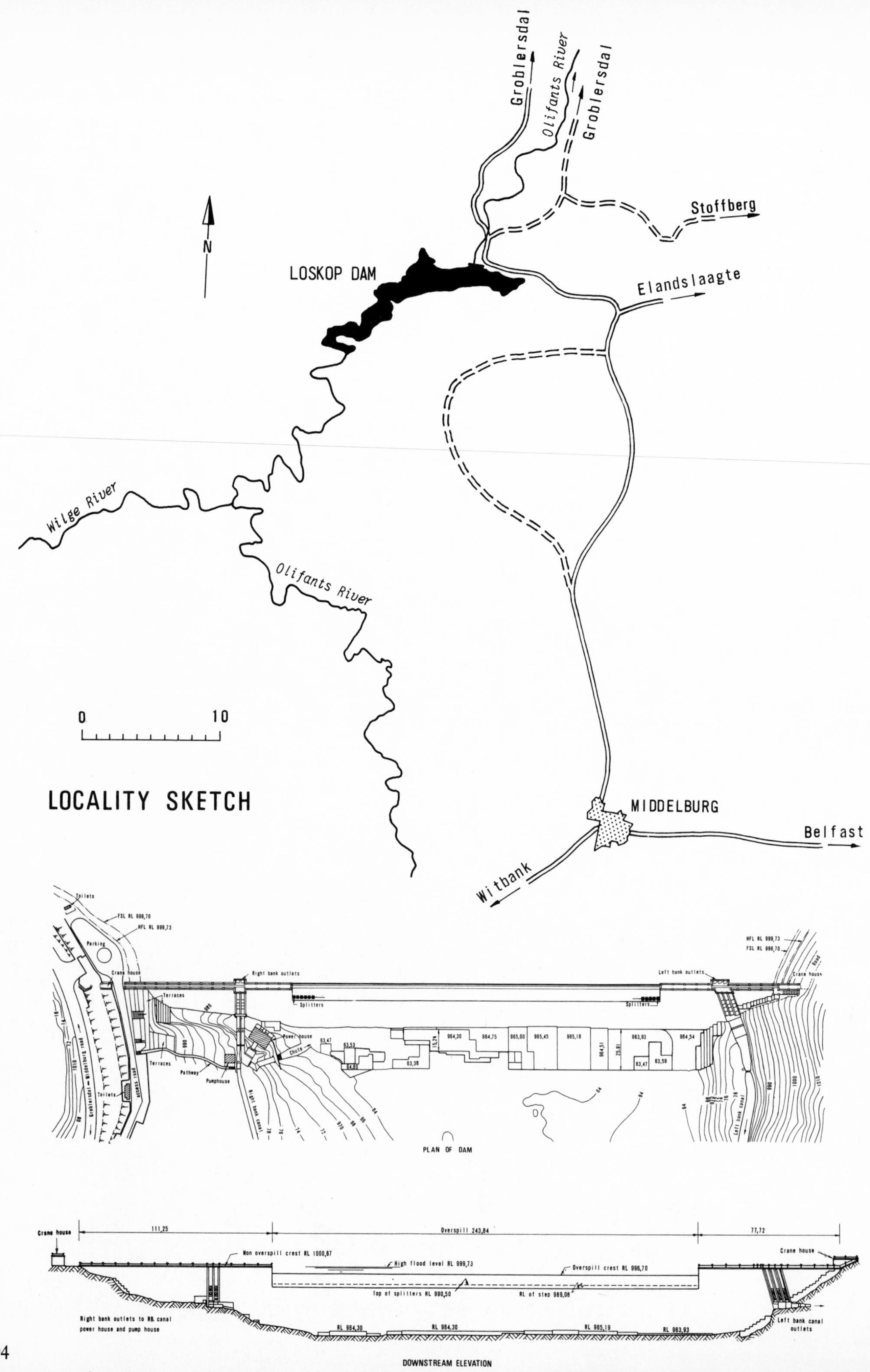

Groblersdal
Olifants River
Groblersdal
Stoffberg
N
LOSKOP DAM
Elandslaagte
Wilge River
Olifants River
0
10
LOCALITY SKETCH
MIDDELBURG
Belfast
Witbank
Toilets
FSL RL 996,70
HFL RL 999,73
Parking
Crane house
Right bank outlets
Left bank outlets
Splitters
Terraces
Power house
Chute
Pathway
Pumphouse
Right bank canal
Left bank canal
964,30
964,75
965,00
965,45
965,19
963,93
964,54
PLAN OF DAM
Crane house
111,25
Overspill 243,84
77,72
Non overspill crest RL 1000,67
High flood level RL 999,73
Overspill crest RL 996,70
Top of splitters RL 980,50
RL of step 989,08
Right bank outlets to RB. canal power house and pump house
RL 964,30
RL 964,30
RL 965,19
RL 963,93
Left bank canal outlets
DOWNSTREAM ELEVATION

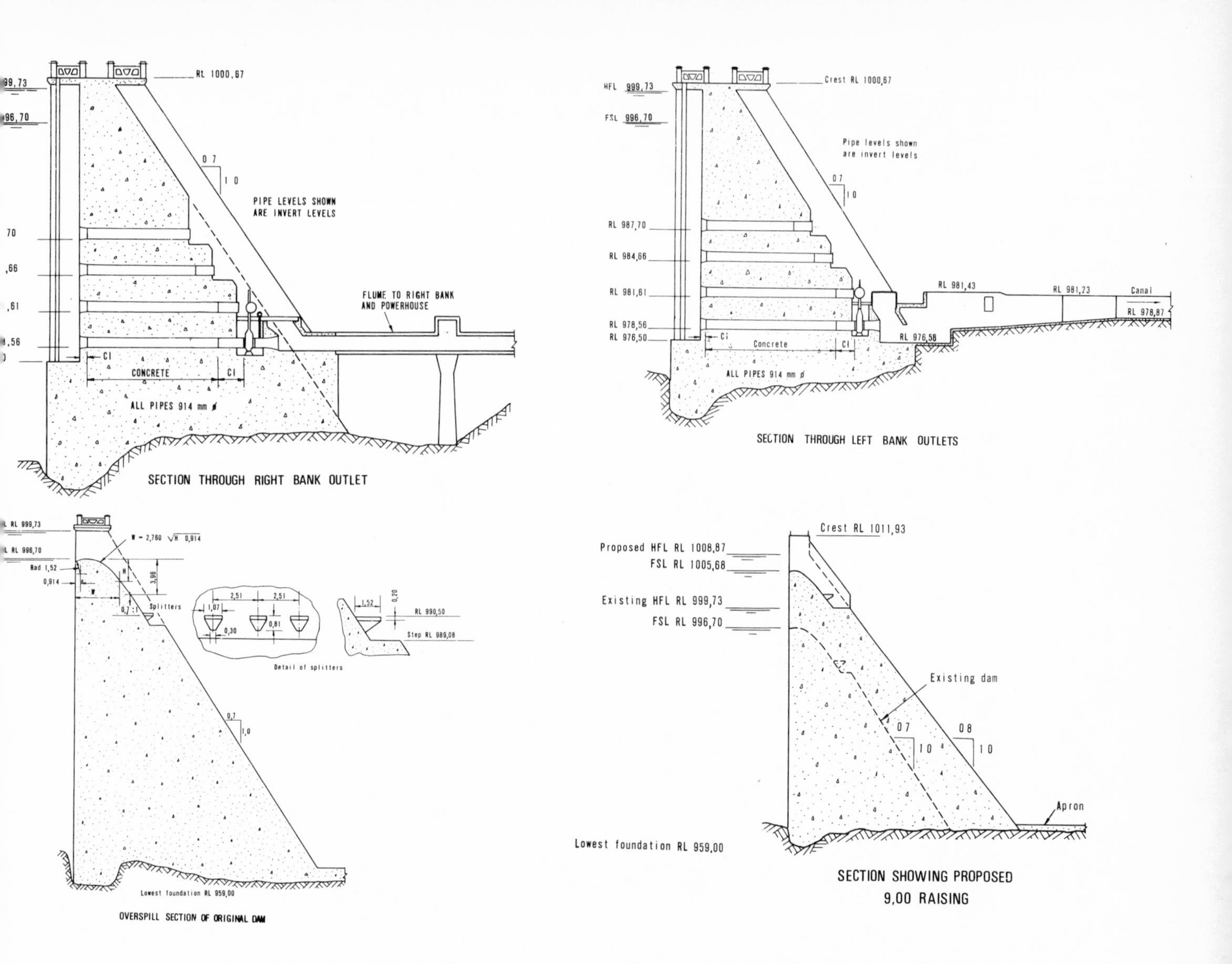

Loskop Dam (D.W.A.)

HARTEBEESPOORT DAM

Hartebeespoort Dam, completed in 1925, is situated on the Crocodile River about 16 km south-west of the town of Brits. Pretoria is 37 km due east of the dam. The dam is a variable radius arch structure of mass concrete. It has a trough spillway on its left flank and carries the Pretoria-Rustenburg main road on its crest. Two outlets, one for each bank, have been provided to supply water to a canal system that stretches 64 km down both sides of the Crocodile River valley. The canal system serves an area of 13 842 ha and the principal crops are wheat, tobacco, citrus and lucerne.

In 1971 the full supply level of the dam was raised by 2,44 m by means of crest gates on the spillway to increase the gross capacity of the dam from 168 077 million m³ to 211 990 million m³.

The rock formation at the site consists of compact quartzites with a northerly dip of about 30 degrees. A fault line on the right flank of the dam passes under the dam site at a depth of approximately 150 m; the safety of the dam is not affected by this fault. The quartzite on the right flank is compact with widely spaced joint planes and was found to be suitable for crushing. On the left flank however, the quartzite, although hard on the surface, showed up very poorly when excavations for the spillway were commenced, and as a result could not be used for crushing as was originally intended. The dam is owned and was engineered by the Department of Water Affairs.

DIMENSIONS

Height above lowest foundation	59 m
Length of crest	140 m
Volume content of dam	0,068 m³ × 10⁶
Gross capacity of reservoir	211,990 m³ × 10⁶
Purpose	Irrigation
Maximum discharge capacity of spillway	2 322 m³/s
Type of spillway	Controlled

Hartebeespoort Dam spillway

P.K. le Roux Dam nearing completion after closure of the river

Hartebeespoort Dam – the spillway

Wagendrift reservoir showing dam in centre distance

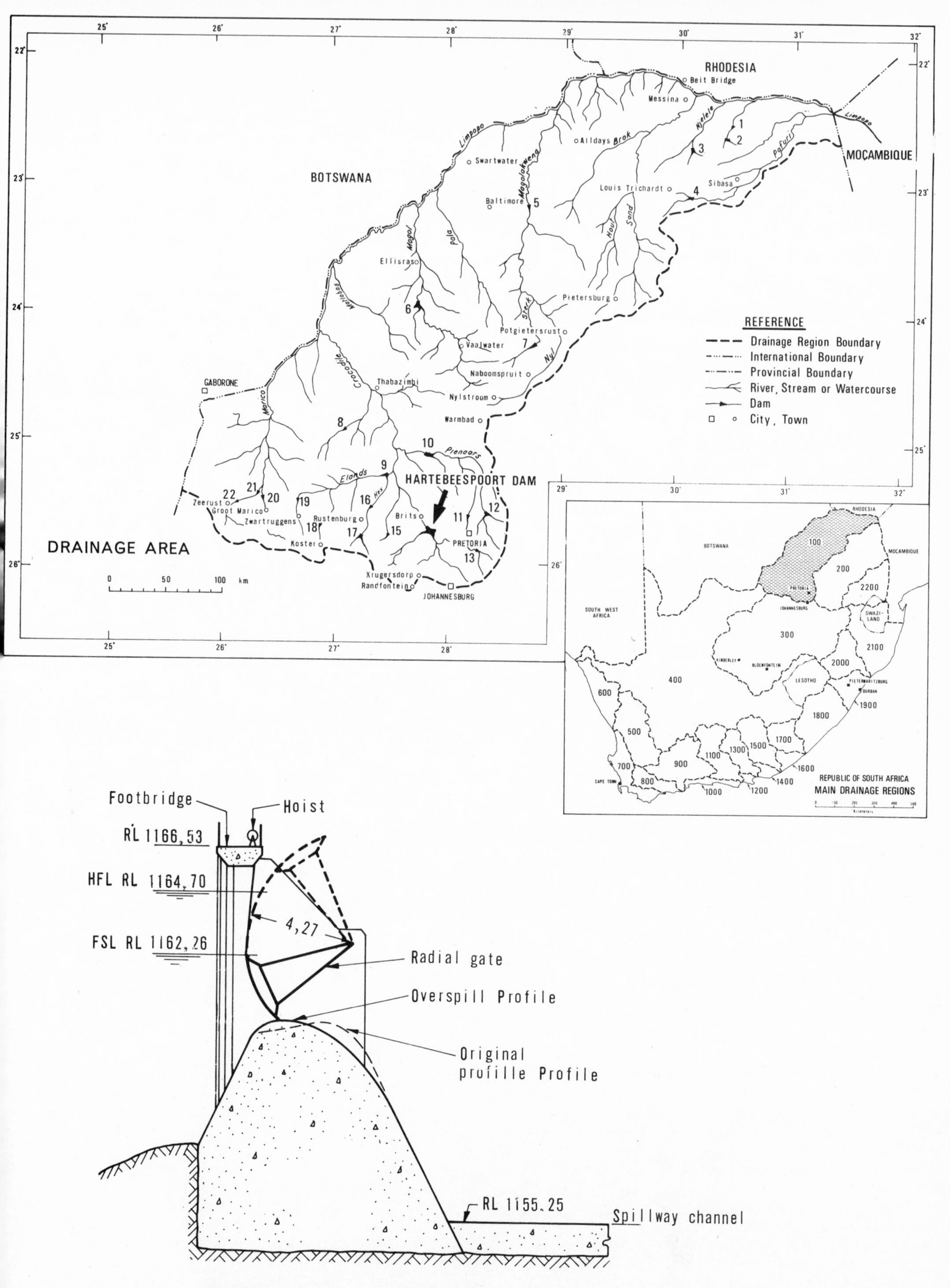
RHODESIA
BOTSWANA
MOÇAMBIQUE
HARTEBEESPOORT DAM
DRAINAGE AREA
REFERENCE
Drainage Region Boundary
International Boundary
Provincial Boundary
River, Stream or Watercourse
Dam
City, Town
Beit Bridge
Messina
Alldays
Swartwater
Baltimore
Louis Trichardt
Sibasa
Ellisras
Pietersburg
Potgietersrust
Vaalwater
Naboomspruit
Thabazimbi
Nylstroom
Warmbad
GABORONE
Zeerust
Groot Marico
Zwartruggens
Rustenburg
Koster
Brits
PRETORIA
Krugersdorp
Randfontein
JOHANNESBURG
Limpopo
Mogalakwena
Mogol
Pala
Brak
Nzhelele
Pafuri
Hout
Sand
Steck
Nyl
Crocodile
Marico
Elands
Hex
Pienaars
REPUBLIC OF SOUTH AFRICA
MAIN DRAINAGE REGIONS
Footbridge
Hoist
RL 1166,53
HFL RL 1164,70
FSL RL 1162,26
4,27
Radial gate
Overspill Profile
Original
profille Profile
RL 1155.25
Spillway channel
SPILLWAY SECTION

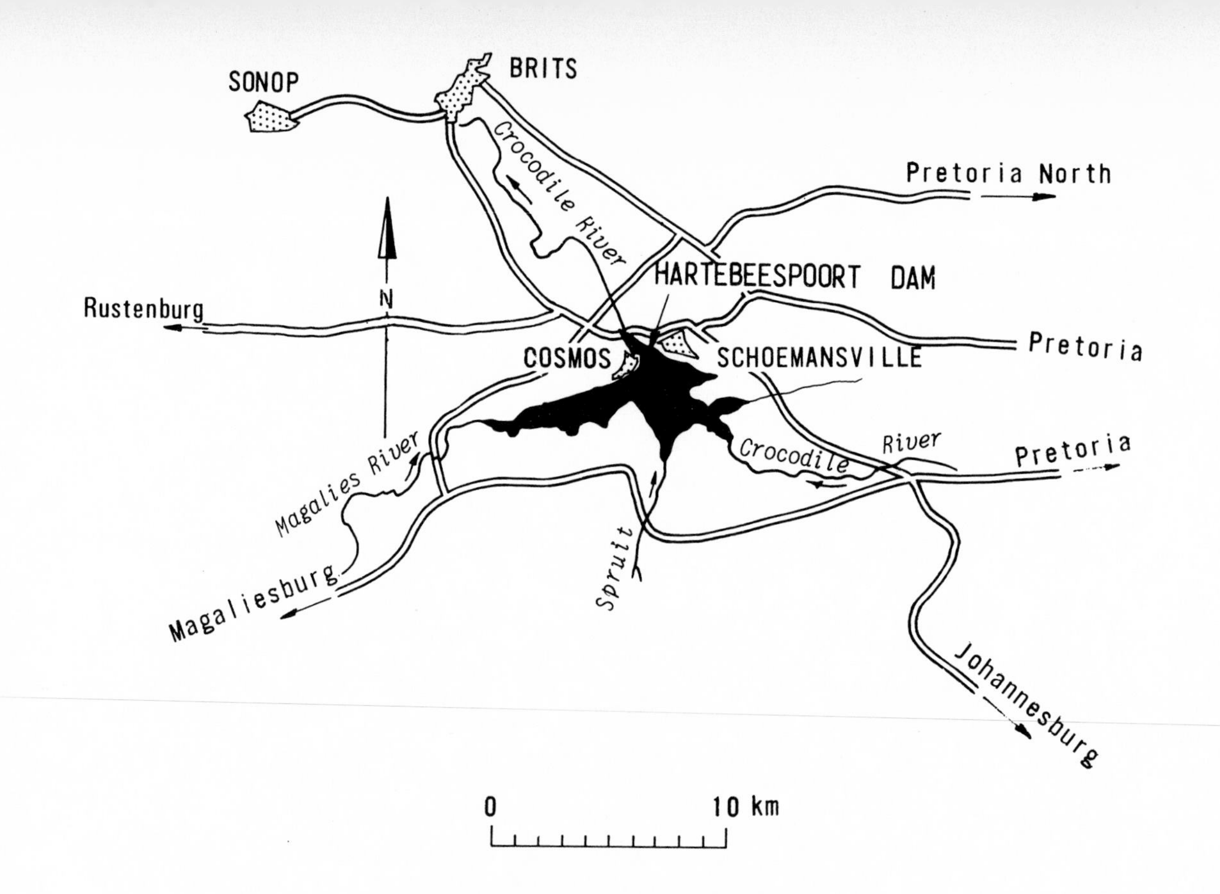

LOCALITY SKETCH

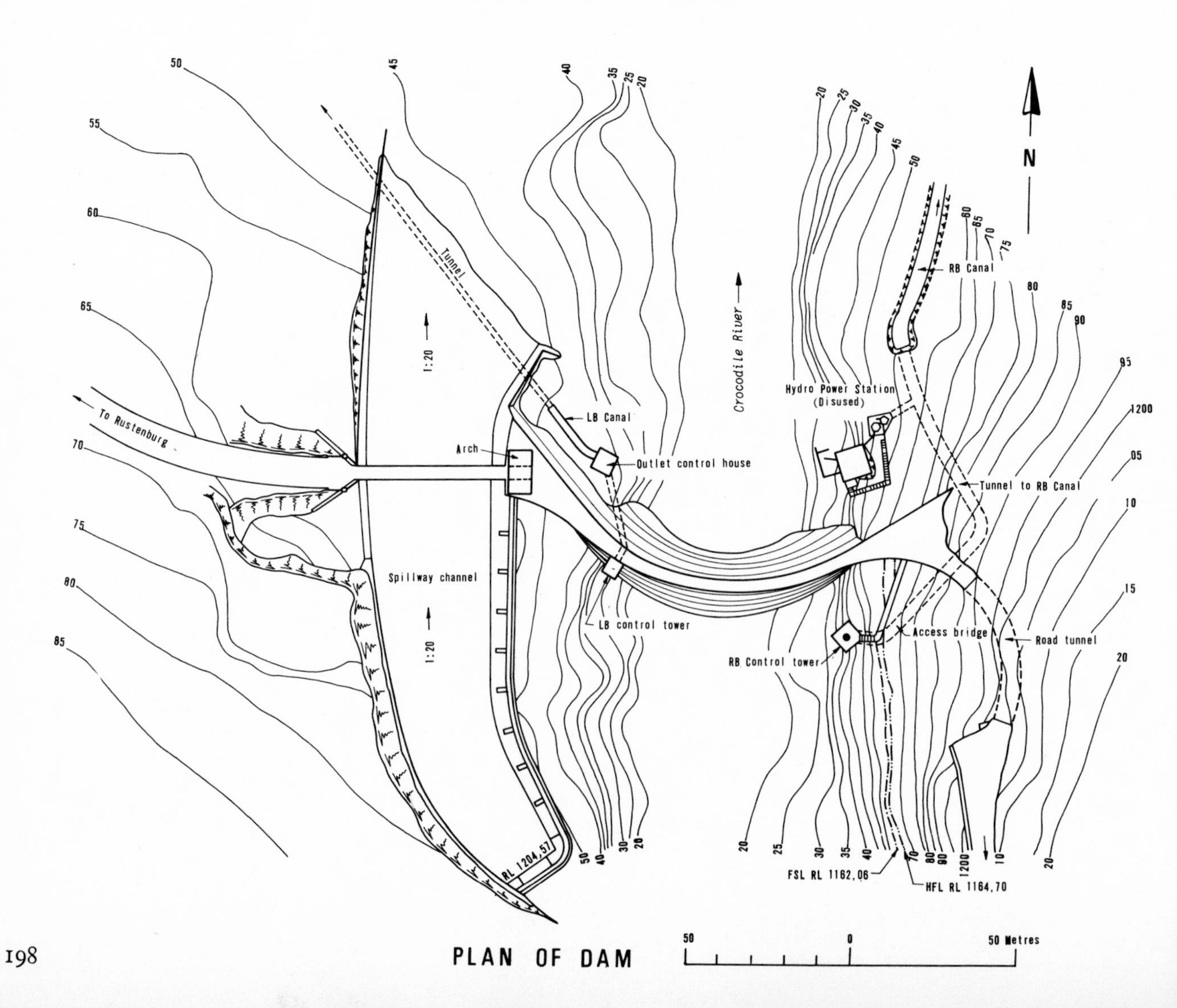

PLAN OF DAM

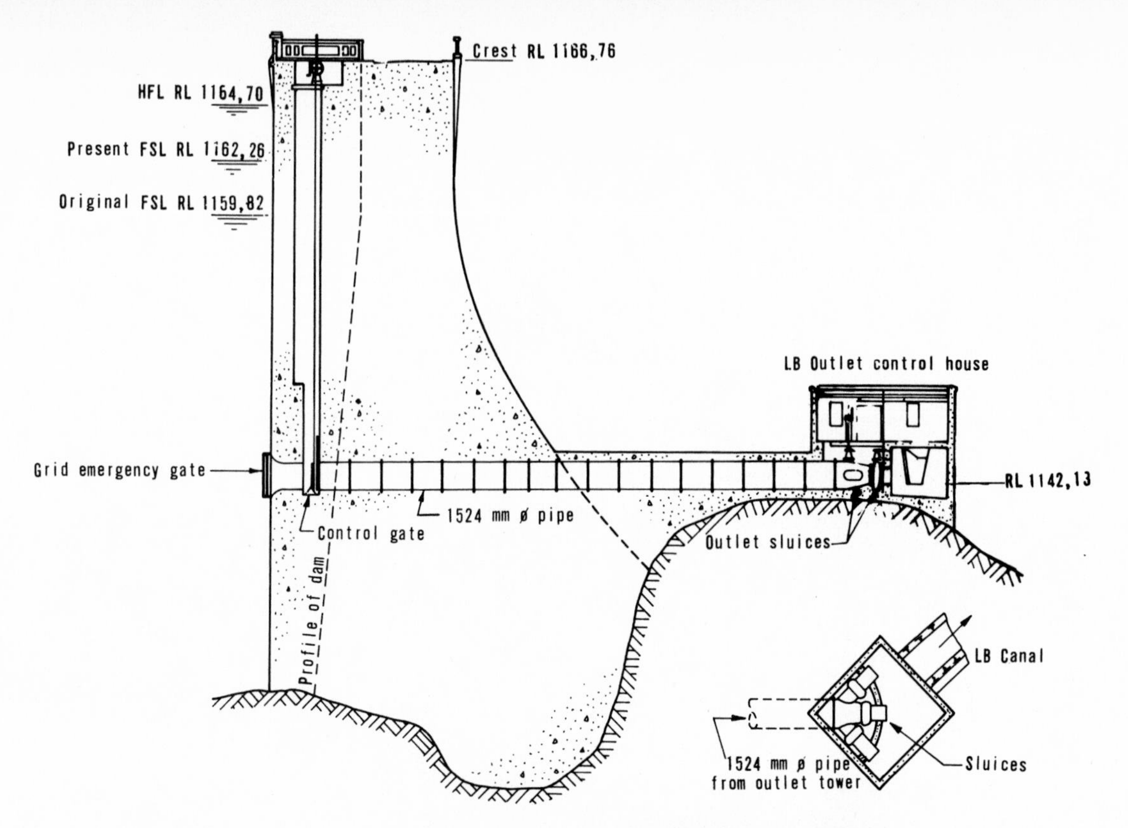

SECTION : L.B. OUTLET TOWER

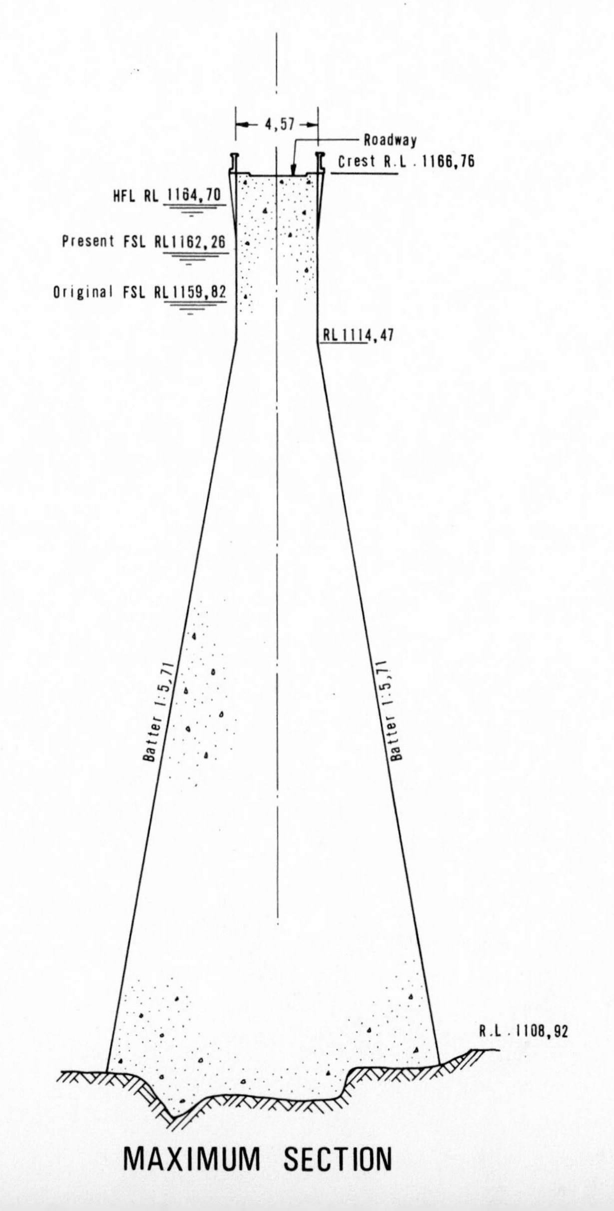

MAXIMUM SECTION

ERFENIS DAM

Erfenis Dam looking upstream, showing spillway

HISTORY

Erfenis Dam is situated about 23 kilometres due west of the town of Winburg in the Orange Free State and is one of the storage units of the Sand-Vet Government Water Scheme. The other dam in the scheme is the Allemanskraal Dam on the Sand River. Erfenis Dam is a mass gravity concrete structure with an overspill section 183 metres long, approximately in the middle of the structure.

The main canal from the dam is on the left bank of the Vet River and extends to well below its confluence with the Sand River, a distance of about 40 kilometres. Farms traversed by the canals are each supplied with water to irrigate about 17 hectares of ground. Farms on the right bank of the river are also served by means of siphons through the river and small distributary canals. Where the large patches of irrigable soil occur, the ground was bought by the State and given out as plots to settlers to be farmed on the intensive system. The two dams and their canal systems are being run as an integrated unit. The total scheduled area is 7 640 hectares.

Water from the Vet Canals is also supplied to the towns of Brandfort and Bultfontein to augment the towns' water supplies.

The engineering and construction was carried out by the owners, the Department of Water Affairs.

DIMENSIONS

Type	Gravity
Height above lowest foundation	46 m
Length of crest	489 m
Volume content of dam	$0{,}129\ m^3 \times 10^6$
Gross capacity of reservoir	$237{,}184\ m^3 \times 10^6$
Purpose	Irrigation
Maximum discharge capacity of spillway	$3\ 170\ m^3/s$
Type of spillway	Uncontrolled

CONSTRUCTION

This dam was constructed in a novel way, viz. the pre-packed system. Boxes to be concreted were first packed with stone of various sizes and a cement mortar was then pumped in to fill the voids and make the contents of the box a solid block of concrete. This method was necessary as the local sand was so fine that it was unsuitable for good concrete, and coarse sand would have had to be imported over a long distance. The local fine sand was suitable for the cement intrusion process.

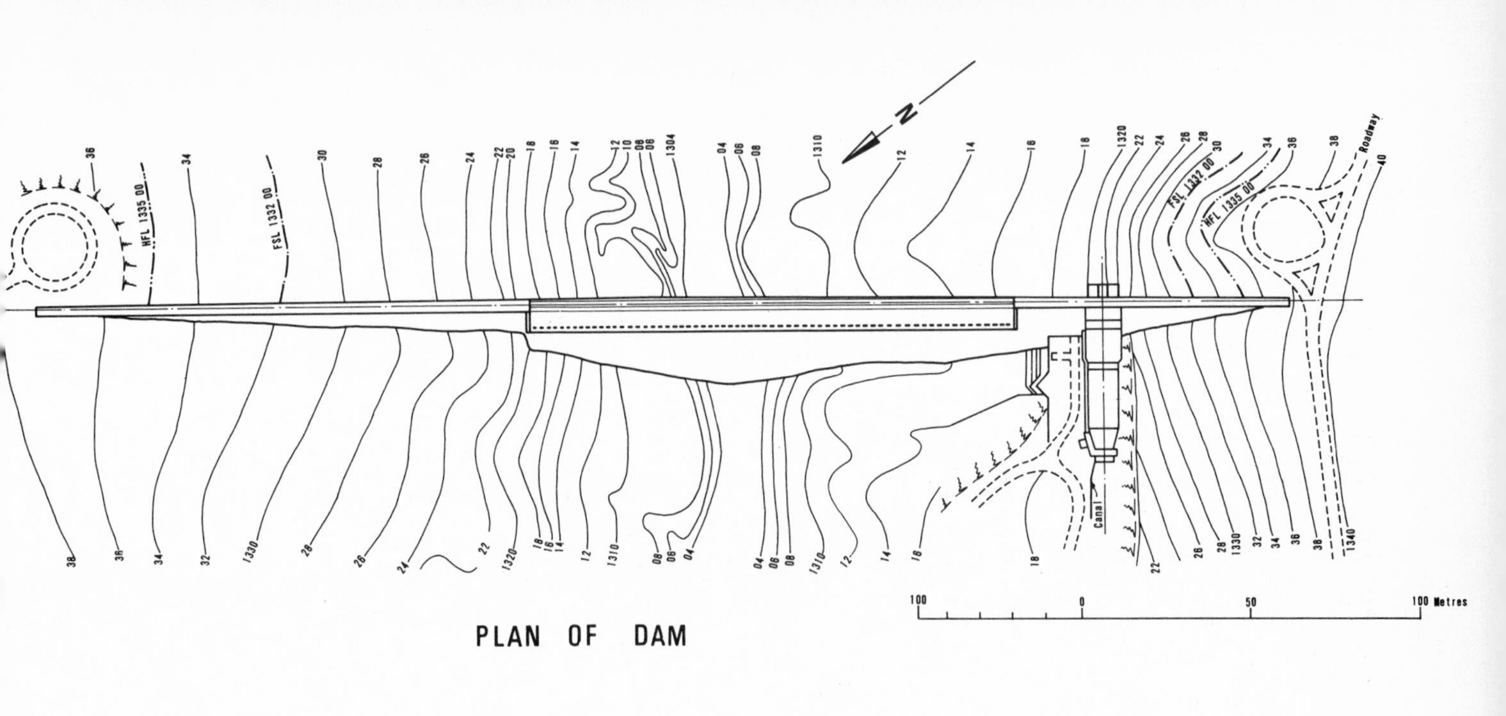
PLAN OF DAM
FSL 1332,00
WFL 1335,00
Roadway
Canal
100 Metres

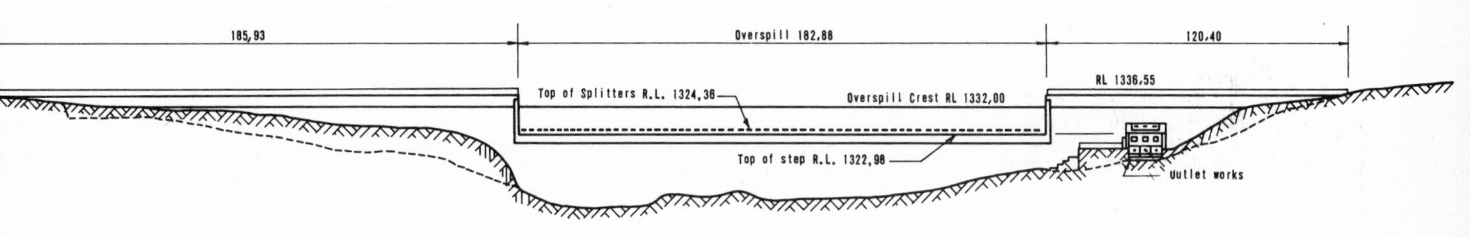
185,93
Overspill 182,88
120,40
RL 1336,55
Top of Splitters R.L. 1324,36
Overspill Crest RL 1332,00
Top of step R.L. 1322,98
Outlet works
DOWNSTREAM ELEVATION

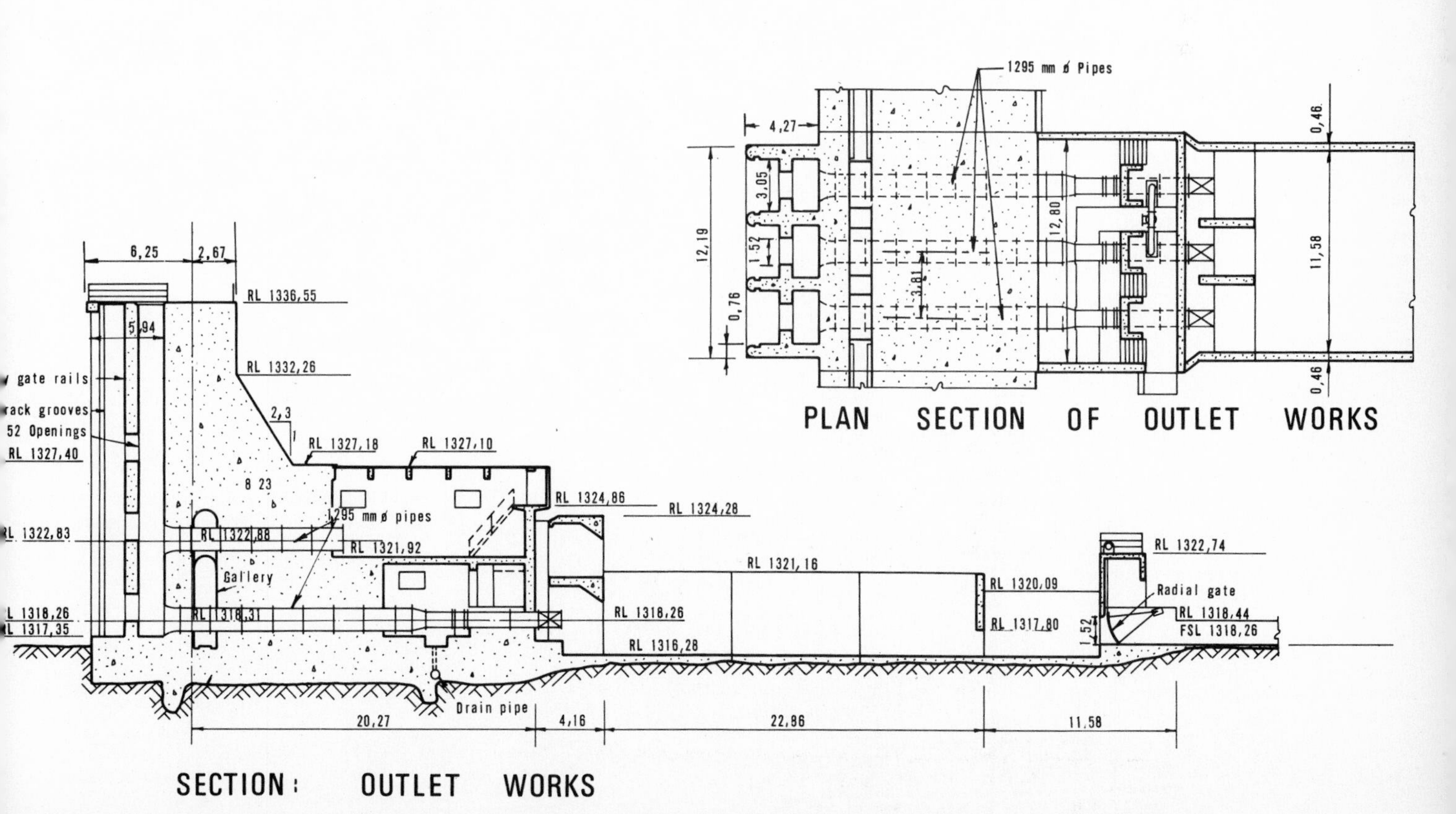
1295 mm ø Pipes
PLAN SECTION OF OUTLET WORKS
RL 1336,55
RL 1332,26
RL 1327,18
RL 1327,10
RL 1324,86
RL 1324,28
1295 mm ø pipes
RL 1322,88
RL 1321,92
Gallery
RL 1318,31
RL 1321,16
RL 1320,09
RL 1318,26
RL 1317,80
RL 1316,28
RL 1322,74
Radial gate
RL 1318,44
FSL 1318,26
Drain pipe
SECTION: OUTLET WORKS

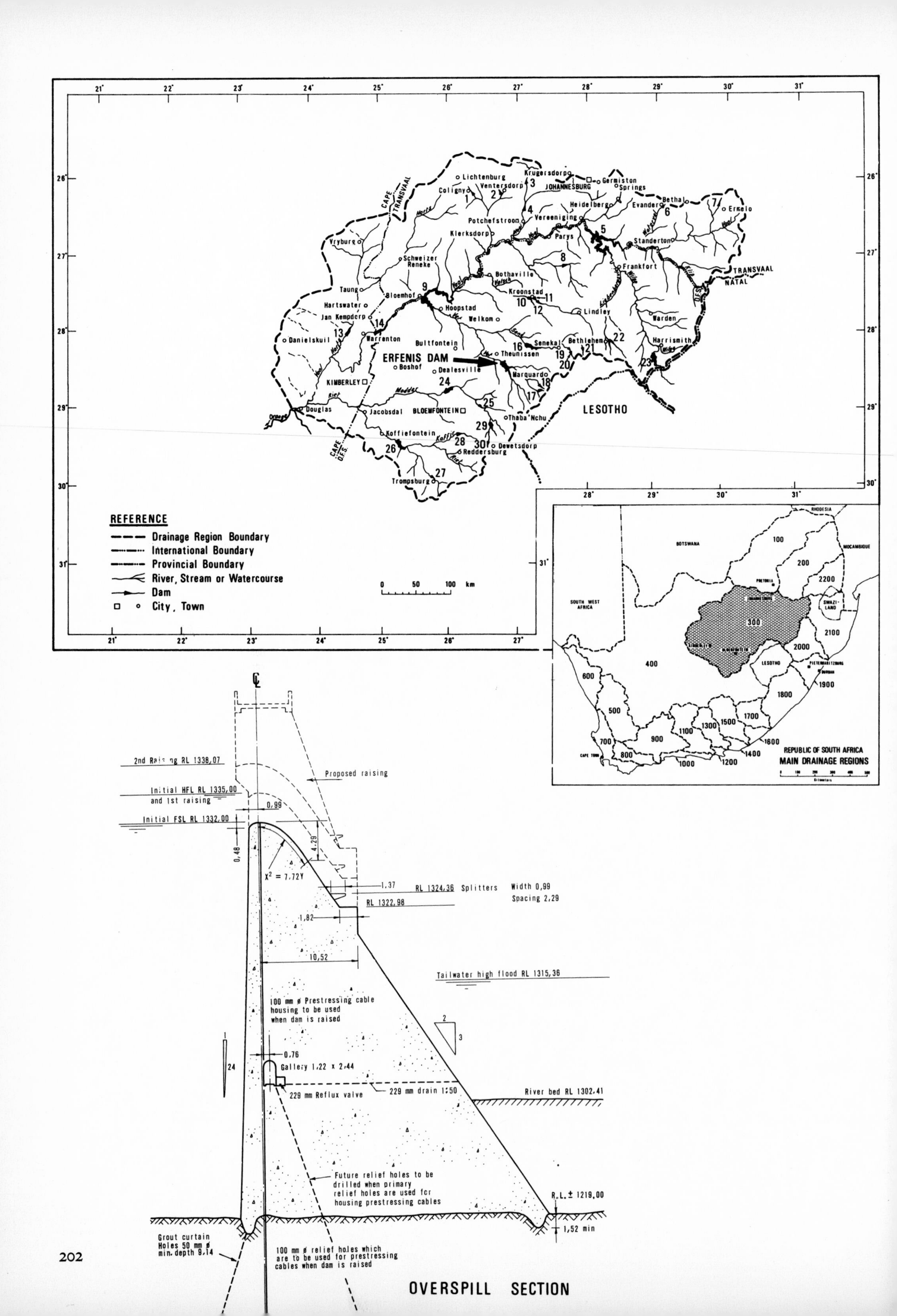
REFERENCE
Drainage Region Boundary
International Boundary
Provincial Boundary
River, Stream or Watercourse
Dam
City, Town
ERFENIS DAM
LESOTHO
REPUBLIC OF SOUTH AFRICA
MAIN DRAINAGE REGIONS
Proposed raising
2nd Raising RL 1338,07
Initial HFL RL 1335,00 and 1st raising
Initial FSL RL 1332,00
$X^2 = 7.72Y$
RL 1324,36 Splitters Width 0,99 Spacing 2,29
RL 1322,98
Tailwater high flood RL 1315,36
100 mm ø Prestressing cable housing to be used when dam is raised
Gallery 1,22 x 2,44
229 mm Reflux valve
229 mm drain 1:50
River bed RL 1302,41
Future relief holes to be drilled when primary relief holes are used for housing prestressing cables
R.L. ± 1219,00
1,52 min
Grout curtain Holes 50 mm ø min. depth 9,14
100 mm ø relief holes which are to be used for prestressing cables when dam is raised
OVERSPILL SECTION

CHURCHILL DAM

HISTORY

Churchill Dam, on the Kromme River about 110 km west of the city of Port Elizabeth is one of the water supply sources for that city. The dam, the first multiple arch structure built in South Africa, was designed by consulting engineers J. C. Hawkins, G. A. Stewart and G. Begg and built by contractors for the Port Elizabeth Municipality. It was completed in 1943.

DIMENSIONS

Type	Multi-Arch
Height above lowest foundation	40 m
Length of crest	213 m
Volume content of dam	0,075 m³ × 10⁶
Gross capacity of reservoir	35,467 m³ × 10⁶
Purpose	Water Supply
Maximum discharge capacity of spillway	1 980 m³/s
Type of spillway	Controlled

CONSTRUCTION

During the construction of the dam, 1940–43, provision was made for future raising by extending the buttress foundation in readiness for additional work. In 1957 some cracking of the buttresses was observed and investigation showed that this was due to expanding properties of the local aggregates. Though there was no threat to the stability of the dam, it was decided to strengthen the buttresses by constructing them to the section proposed for the raised dam. This was done by the "pre pack" or "colgrout" method, i.e. the river aggregates were placed dry and filled with pumped cement grout. The original capacity of 35 000 million m³ was not increased as other sources of water had become available.

CONSULTING ENGINEERS AND CONTRACTORS

Constructed by	Clark and Downie and Lewis Construction

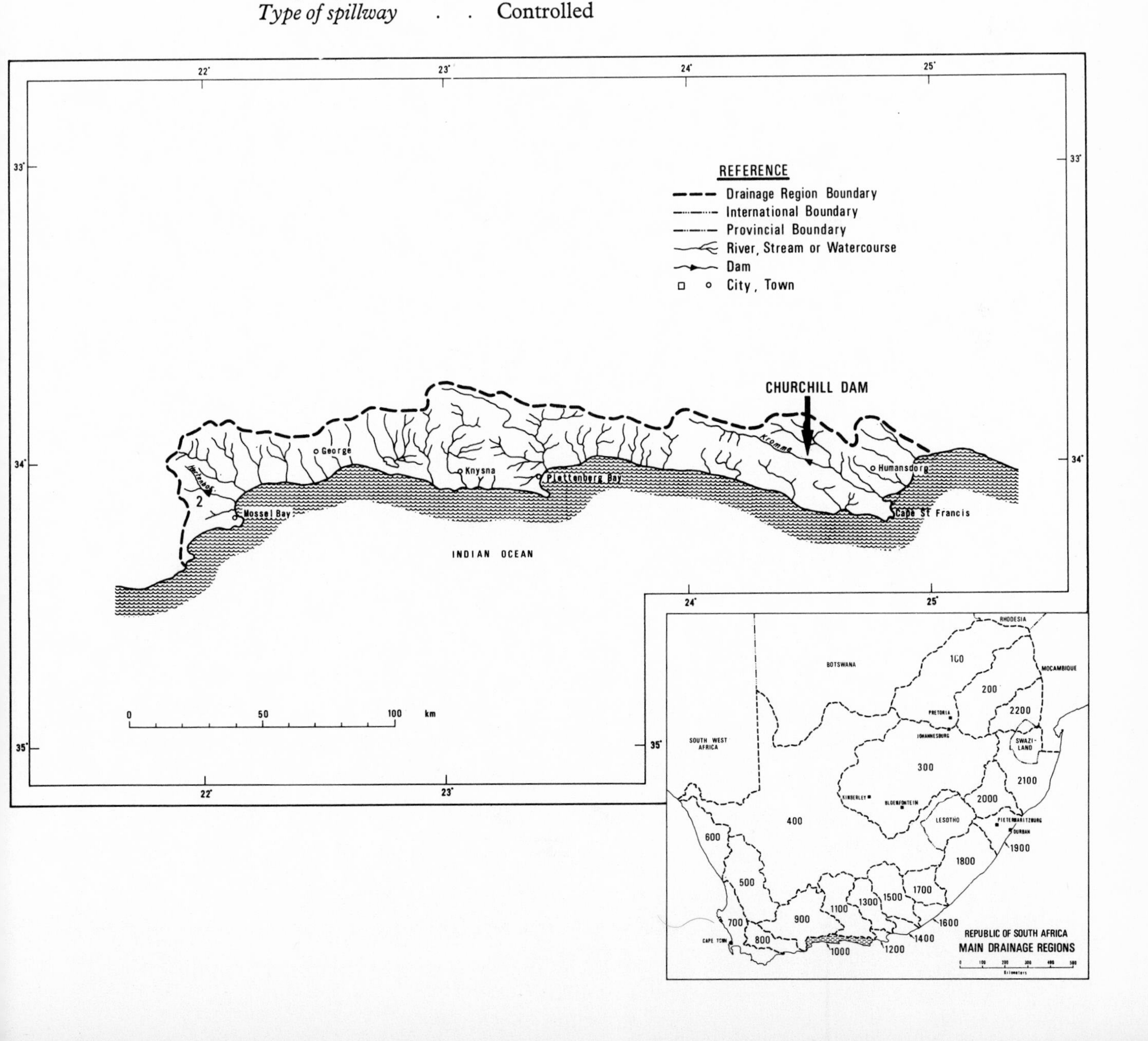

Above: Looking Upstream

CHURCHILL DAM

Below: Looking Downstream

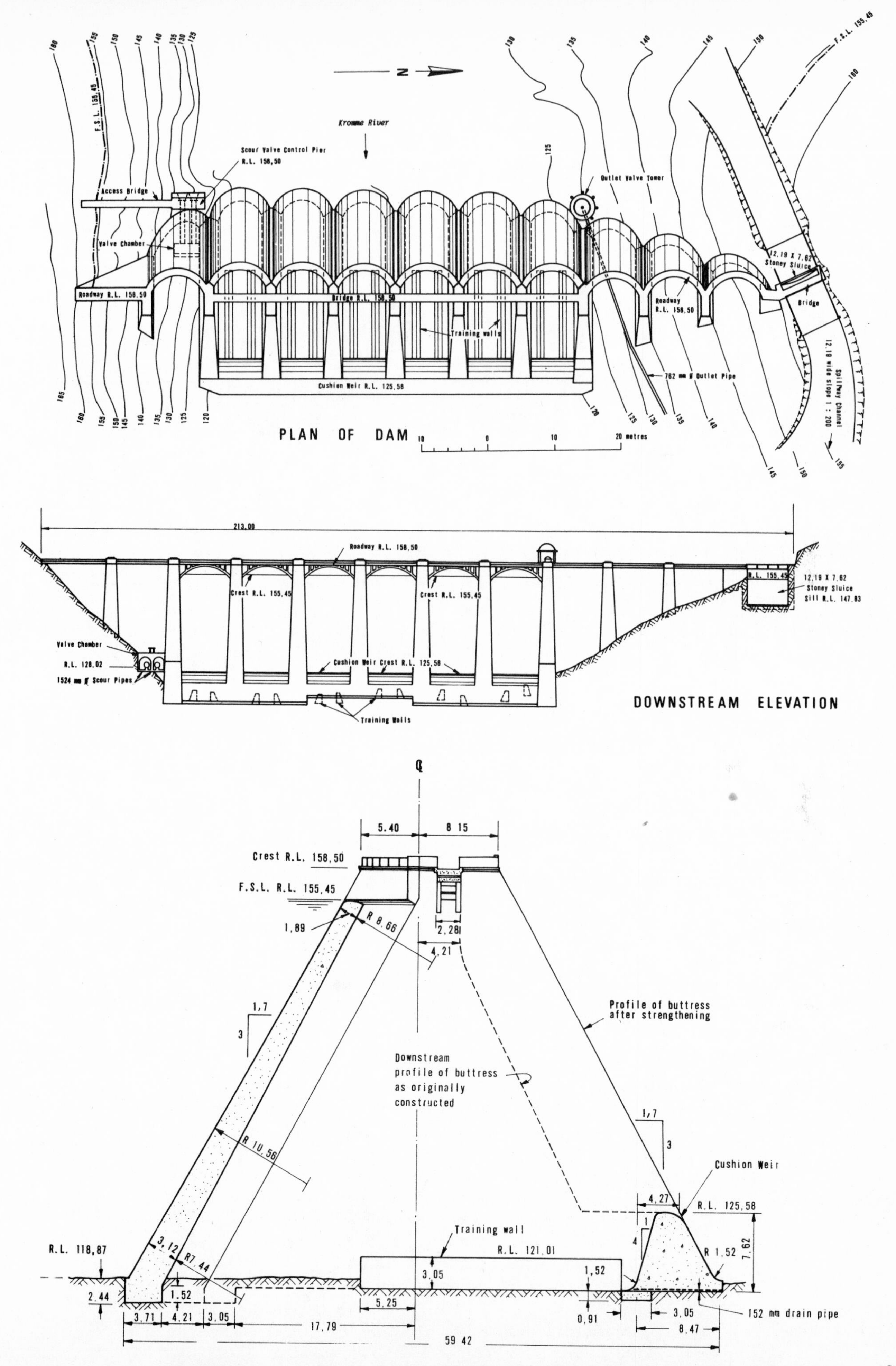
Kromme River
N
Scour Valve Control Pier R.L. 158,50
Access Bridge
Valve Chamber
Roadway R.L. 158,50
Bridge R.L. 158,50
Training walls
Cushion Weir R.L. 125,58
Outlet Valve Tower
762 mm ø Outlet Pipe
Roadway R.L. 158,50
12,19 X 7,62 Stoney Sluice
Bridge
12,19 wide slope 1 : 200 Spillway Channel
F.S.L. 155,45
PLAN OF DAM
20 metres
213,00
Roadway R.L. 158,50
Crest R.L. 155,45
R.L. 155,45
12,19 X 7,62 Stoney Sluice Sill R.L. 147,83
Valve Chamber
R.L. 128,02
1524 mm ø Scour Pipes
Cushion Weir Crest R.L. 125,58
Training Walls
DOWNSTREAM ELEVATION
Crest R.L. 158,50
F.S.L. R.L. 155,45
Profile of buttress after strengthening
Downstream profile of buttress as originally constructed
Cushion Weir
R.L. 125,58
Training wall
R.L. 121,01
R.L. 118,87
152 mm drain pipe
CENTRAL ARCH: SECTIONAL ELEVATION

CHURCHILL DAM

ALLEMANSKRAAL

*Allemanskraal Dam**

Allemanskraal Dam, completed in 1960, is one of the two storage units in the Sand-Vet Government Water Scheme. The other unit in the scheme is the Erfenis Dam on the Vet River. Allemanskraal is situated on the Sand River south-east of Virginia, about 50 kilometres upstream of the town. It is owned and was engineered and constructed by the Department of Water Affairs. The dam comprises a mass gravity concrete section and an earth embankment on the right flank of the dam. The overspill which is part of the concrete wall is 212 metres long.

Canals from the dam irrigate areas along both banks of the Sand River to its confluence with the Vet River, about 115 kilometres below the dam, where the Sand River canals meet the Vet River canal system that comes from the Erfenis Dam. The two canal systems are run as an integrated unit and the total scheduled area is 7 640 hectares. Farms traversed by the main canals are each supplied with water to irrigate 17 hectares of ground. Wherever large patches of good irrigable soil occurred, the ground was bought by the State and given out to settlers to be farmed intensively.

The ground bordering on the dam has been made a game reserve known as the Willem Pretorius Game Reserve which is stocked with a wide selection of game including Rhinoceros and Giraffe but excluding carnivore.

A pleasure resort has been established on the northern shore of the dam, near the wall, to cater for visitors to this reserve, anglers, yachtsmen and watersports enthusiasts.

DIMENSIONS

Type . . .	Concrete Gravity and Earth
Height above lowest foundation . .	38 m
Length of crest .	1 338 m
Volume content of dam . .	0,237 m³ × 10⁶ (Concrete: 0,092 m³ × 10⁶ (Earth: 0,145 m³ × 10⁶
Gross capacity of reservoir .	213,035 m³ × 10⁶
Maximum discharge capacity of spillway . . .	2 265 m³/s
Type of spillway	Uncontrolled

**Note: For location see reference 16 on page 202.*

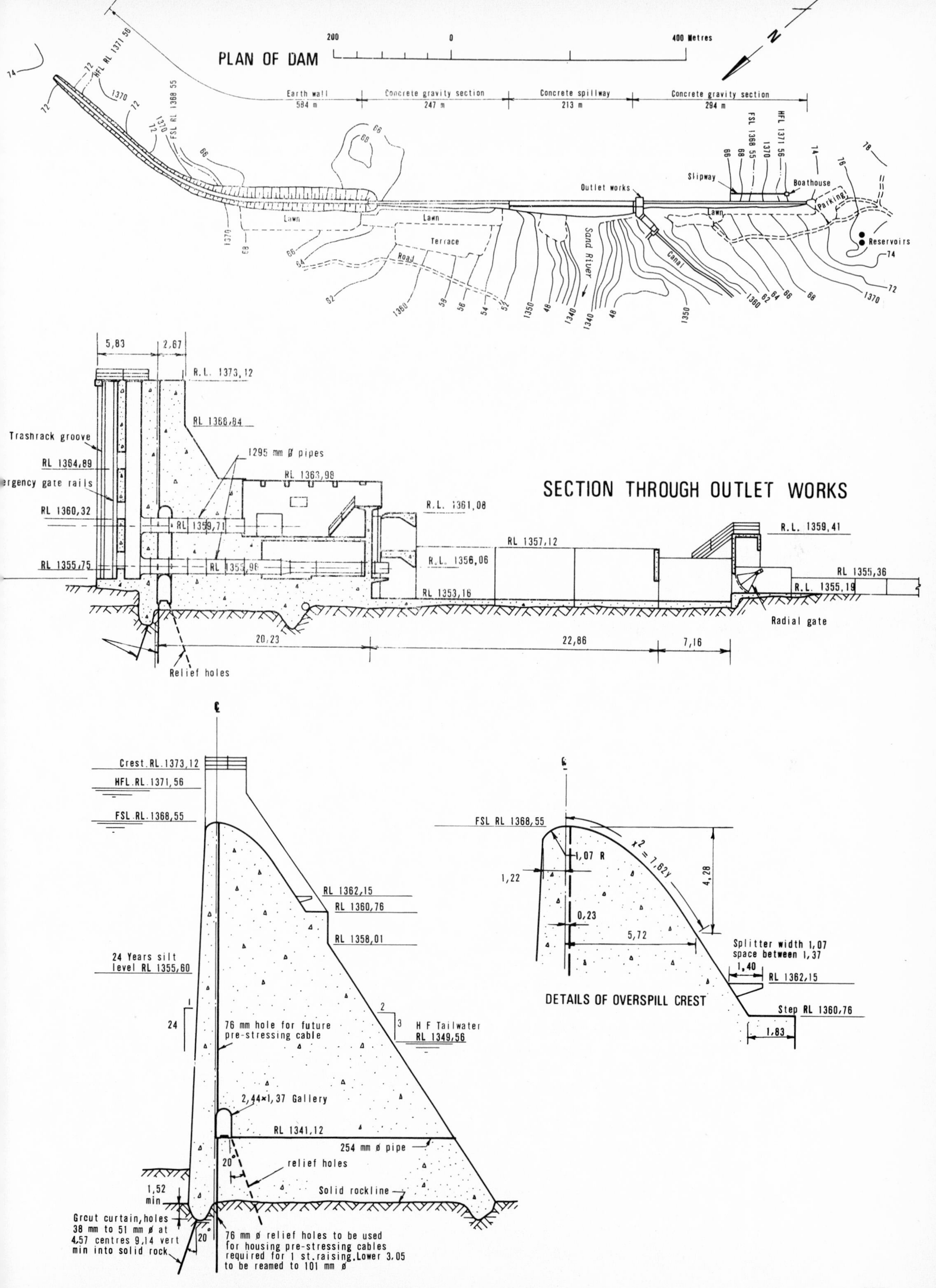
PLAN OF DAM
200
0
400 Metres
N
Earth wall
584 m
Concrete gravity section
247 m
Concrete spillway
213 m
Concrete gravity section
294 m
HFL RL 1371 56
FSL RL 1368 55
Slipway
Boathouse
Outlet works
Parking
Reservoirs
Lawn
Terrace
Road
Sand River
Canal
SECTION THROUGH OUTLET WORKS
5,83
2,67
R.L. 1373,12
RL 1368,84
Trashrack groove
RL 1364,89
Emergency gate rails
1295 mm Ø pipes
RL 1363,98
R.L. 1361,08
RL 1360,32
RL 1359,71
RL 1357,12
R.L. 1359,41
R.L. 1356,06
RL 1355,75
RL 1353,98
RL 1355,36
R.L. 1355,19
RL 1353,16
Radial gate
20,23
22,86
7,16
Relief holes
SECTION THROUGH OVERSPILL
Crest.RL.1373,12
HFL.RL.1371,56
FSL.RL.1368,55
RL 1362,15
RL 1360,76
RL 1358,01
24 Years silt level RL 1355,60
76 mm hole for future pre-stressing cable
H F Tailwater RL 1349,56
2,44×1,37 Gallery
RL 1341,12
254 mm ø pipe
relief holes
Solid rockline
1,52 min
Grout curtain, holes 38 mm to 51 mm ø at 4,57 centres 9,14 vert min into solid rock
76 mm ø relief holes to be used for housing pre-stressing cables required for 1 st.raising. Lower 3,05 to be reamed to 101 mm ø
DETAILS OF OVERSPILL CREST
FSL RL 1368,55
1,07 R
$x^2 = 7,62y$
4,28
1,22
0,23
5,72
Splitter width 1,07
space between 1,37
1,40
RL 1362,15
Step RL 1360,76
1,83

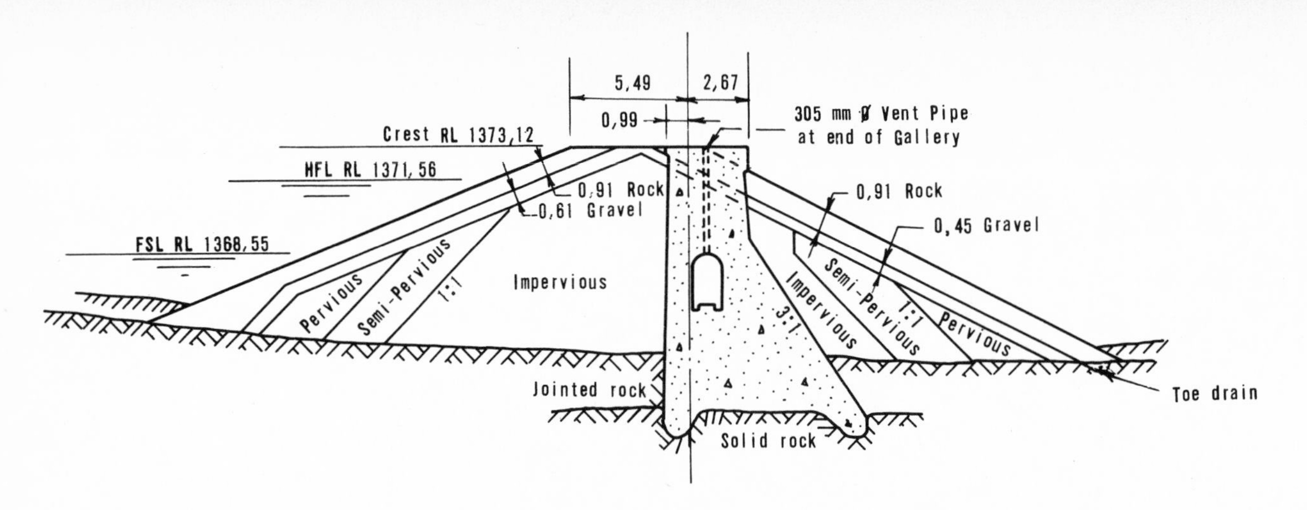

SECTION AT JUNCTION BETWEEN CONCRETE GRAVITY WALL AND EARTH EMBANKMENT

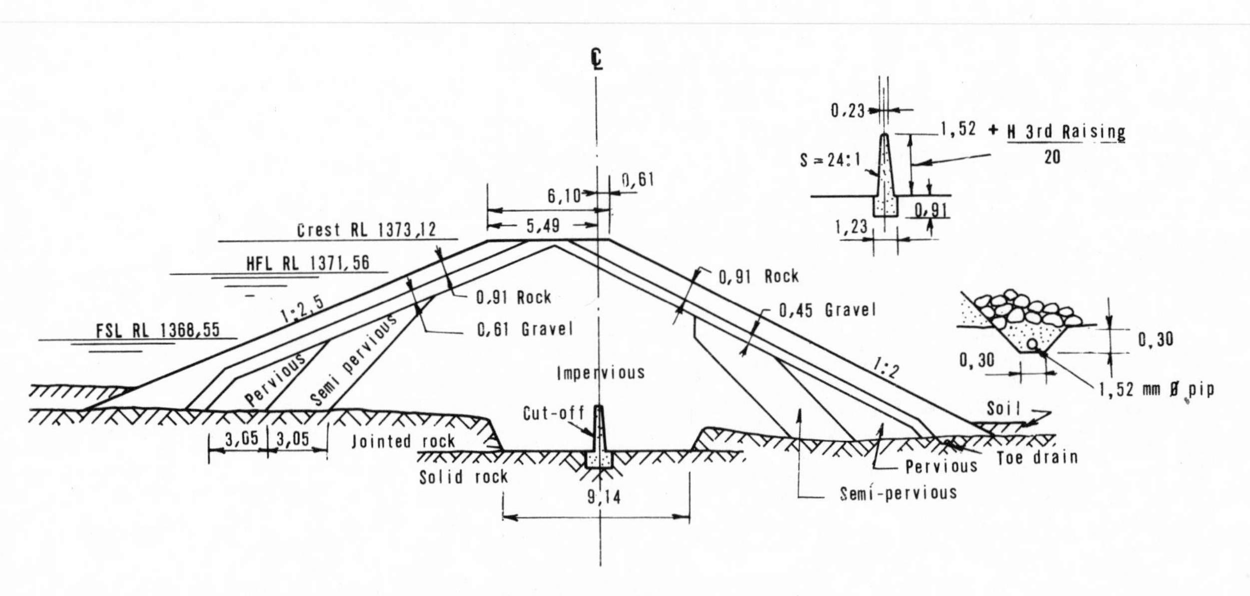

SECTION THROUGH EARTH EMBANKMENT

ALLEMANSKRAAL DAM

THE TUGELA – VAAL WATER SCHEME

As in the case of the Orange River Project, the Tugela-Vaal scheme is a major example of a multi-purpose, inter-regional project in Southern Africa: the application of inter-catchment engineering for the supply of water for all purposes and the generation of electricity by pumped storage. It involves dams, siphons, tunnels, pumps, generating plant etc.

Gold in the reefs and brown water in the Vaal River are the two resources of nature on which the greatest mining, industrial and metropolitan complex of the African continent has developed.

Because of its high degree of industrialisation, the area supplied from the Vaal River is particularly sensitive to a situation where the demand for water exceeds the supply during periods of subnormal rainfall.

In this highly developed region of South Africa, water is one of the major factors ruling development. The augmentation of the supply of water available from the Vaal River to the Pretoria-Witwatersrand-Vereeniging complex is therefore of vital importance to all.

The potential of the Vaal River as a reliable source of water supply is more highly developed than perhaps any other river in the country. Vaal Dam, the Vaal River Barrage, Bloemhof Dam and Vaalhartz weir are the main storage and diversion structures on the river, and, together with other minor pumping installations, provide a reliable yield estimated to be 4,02 million m³ per day.

The available supply from the Vaal River was used during 1973 as follows:

Jagers Rust pump station

Water Use	Average rate million m³d	Percentage of total
Irrigation . .	1,373	33,4
Municipal and Industrial .	1,841	44,8
Power station cooling . .	0,214	5,2
Losses along river . . .	0,682	16,6
	4,110	100,0

It is expected that the amount of Vaal River water used for irrigation and power generation will remain reasonably constant in future and that the transit losses, too, will not vary much from the present rate. The available evidence indicates however that municipal and industrial water demands in the region will grow at a rate of some 6,3 per cent a year. The total estimated future water demands are:

Year	Demand million m³d	Year	Demand million m³d
1973 . .	4,11	1993	6,23
1978 . .	4,56	1998	7,06
1983 . .	5,06	2000	8,36
1988 . .	5,62	2003	9,60

A wide variety of alternative sources of supply have been investigated and assessed. These include the Orange, Tugela, Caledon, Buffalo, Olifants, Crocodile, Komati and Usutu Rivers neighbouring the Vaal River basin. Many of these projects are attractive but those involving the Orange and the Tugela Rivers clearly offer the most beneficial solution to the problem under present circumstances.

Some of the benefits to be gained from linking neighbouring rivers are fairly obvious. They include the redistribution of scarce natural resources from areas of relative plenty to areas of relative scarcity; the ability to exploit periods of plenty in two separate systems; spreading the risk of shortages over a wider geographical area; and the stimulus of development in a less developed region. A less obvious but far more significant benefit accruing from linking the Tugela and Vaal Rivers is that the total system will make possible operating rules for the dams in the Vaal River that will spectacularly improve the utilisation of the natural run-off in the Vaal at the expense of evaporation loss and spillage down river during times of flood.

In developing the Tugela-Vaal project the concept of insurance storage was crystallised and exploited to considerable advantage. The idea is to create a very large reservoir of imported water in the upper reaches of the Vaal River basin where storage and evaporation characteristics are most advantageous.

With this reservoir of water available at short notice, withdrawal of water from storage units downstream can be increased far beyond the otherwise reliable yield in the knowledge that when shortages in natural run-off do occur, withdrawals can be made from insurance storage.

The essence of the concept is that the volume of insurance storage must be very large and the storage site must be efficient, with a minimum loss by evaporation of expensive imported water. Because water supplies can be withdrawn from the existing Vaal system at a rate higher than the natural reliable yield, the reservoirs will in the long run be less full for longer periods than has been the case in the past. This leads to a smaller evaporation loss from the main dams and more space to store and regulate incoming flood waters. The benefits gained in this way are initially very large in proportion to the importation rate, but decrease as the importation rate increases and the utilisation efficiency tends to a maximum possible level.

As is the case with all insurance policies, there is a premium to pay and the provisions of the insurance cover in the case of the Tugela-Vaal project change in sympathy with the changing economic climate and ever-increasing knowledge of the environment. The premium to be paid is the cost of building a system for transferring water from the Tugela to the Vaal basin and the cost of operating and maintaining the system.

In 1967 parliamentary authority was obtained to build Spioenkop Dam on the Tugela River about 16 km downstream from Bergville. The proposal was to pump water from below Spioenkop Dam via a rising main over the Drakensberg into a storage reservoir in the upper reaches of the Wilge River near Harrismith in the Orange Free State.

Further investigations showed that a site in the Nuwejaarspruit on the farm Sterkfontein, 15 km south-west of Harrismith, would be more appropriate for storing water transported over the Drakensberg. The site of Sterkfontein Dam is not attractive but the basin it commands is impressive. This site was finally adopted and the task of aligning the aqueduct proceeded. After careful examination and analysis of the hydrology and the capital and running costs it was decided to modify the original scheme to take advantage of the physiographic situation with the object of minimising capital investment and running costs, while at the same time meeting the target date for commissioning the project and allowing room for future development.

A system comprising diversions from the Upper Tugela at about 1 225 m above sea-level, a barrage with a full supply level 1 134 m above sea-level with a pump station giving a 90 m lift and a main pump station giving a 506 m lift at the foot of the Drakensberg, was developed with the necessary canal and siphon links. From the main pump station the system includes a short rising main, three tunnels and a canal which delivers the pumped water into Sterkfontein Dam.

The catchment area commanded by the Sterkfontein Dam is 19 300 ha and the surface area of the reservoir at ultimate capacity

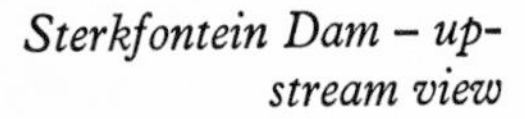
Sterkfontein Dam – upstream view

is 6 937 ha. With an ultimate capacity of about 2 656 million m^3 the run-off from the small catchment area becomes insignificant. This rather peculiar set of circumstances accounts for the fact that although the dam is an earth embankment it has no spillway, since all probable floods will either be absorbed in the storage basin or released through the outlet tunnel.

In the first phase Sterkfontein Dam is being built to a full supply level of 1 678 m above sea-level, giving a storage capacity of 1 203 million m^3. This phase requires the placing of about 6 million m^3 of earth in the embankment, excavations of about 600 000 m^3 and the placing of about 195 000 m^3 of stone pitching for protecting the embankment.

By increasing the full supply level by 24 m to 1 702 m above sea-level the capacity is increased to 2 656 million m^3. Approximately 11 million m^3 of earth will be required to raise the embankment, giving a total volume of earth placed of about 17 million m^3.

The outlet from the dam comprises a tunnel 4 m in diameter and 618 m long with a discharge capacity of 220 cumecs (cubic metres per second) – about 19 million m^{3}d, or twice the expected demand from the Vaal River in the year 2003.

When the supply in Vaal Dam and lower down the Vaal River reaches prescribed levels, water will be released from Sterkfontein at a rate which will minimise the proportion lost in transmission down the Wilge River. This rate will be set within the bankfull capacity of the river and will be such that damage to flood plain development is minimised.

Water pumped up the escarpment is conveyed to Sterkfontein Dam by means of the Tzamenkomst tunnel and the OFS aqueduct.

Placed at the level of a sandstone bed of the Beaufort Series in order to reduce tunnelling problems, the Tzamenkomst tunnel lies 1 705 m above sea-level, has a lined diameter of 4 m and is 1 040 m long. Associated with the tunnel is a surge chamber at the inlet end and a transition structure at the outlet end where the tunnel joins with a canal.

Separated by lengths of canal are two further concrete-lined tunnels between Tzamenkomst and Sterkfontein Dam. Grenshoek tunnel is 156 m long and was driven through a spur of the Drakensberg foothills near the historic Kerkenberg, in place of 3 500 m of canal that would have been required to pass round the spur.

Metz tunnel is the final link in the aqueduct from the Tugela River to Sterkfontein Dam and passes beneath the Bergville-Harrismith road at the top of Oliviershoek Pass. This tunnel is 510 m long and was the last to be holed through.

The main portion of the OFS aqueduct comprises 9,35 km of trapezoidal canal built on a gradient of 1:5 000. The canal has side slopes of $1\frac{1}{2}$:1 with a bed width of 3 m and was lined by hand. The canal was designed to carry 11 cumecs, i.e. 950 400 m^{3}d – sufficient for future extensions to the project.

Whereas all the civil engineering work on the Tugela-Vaal project was designed and built by the Department of Water Affairs, the pump stations and rising mains were engineered by the Rand Water Board, acting as consultants to the department, and built by contract. The main pump station is on the farm Jagers Rust and delivers water through a 3 730 m rising main to Tzamenkomst tunnel. At Driel barrage a pump station delivers water from storage via a 1 740 m rising main to the head of the main canal.

Jagers Rust pump station is situated at the foot of the Drakensberg on the banks of a balancing dam or forebay. The station houses four pump-motor sets each capable of lifting 110 000 m^{3}d through a lift of 506 m into the OFS aqueduct. Each of the four motors has a power of 7 730 kW. Associated with the pump station is a switchyard built by the Electricity Supply Commission.

Driel pump station is on the left bank of the Tugela River just downstream from Driel barrage.

This pump station houses three pump-motor sets each capable of delivering 110 000 m^{3}d through a lift of 90 m. The rising main from the station is 1 740 m of 1 700 mm steel pipeline. This main discharges into the headworks of the main canal. Each of the three motors requires 1 400 kW of power supplied by the Electricity Supply Commission.

The total power demand from the two pump stations is 35,2 MW.

Jagers Rust pump station is the heart of the whole project and houses the administrative and control staff, as well as the electronic equipment for receiving and logging data telemetered from various points on the project. This data will be automatically processed and the results converted to control instructions. These instructions will be telemetered to two canal gates, Driel barrage and the two pump stations where the necessary adjustments and settings will be made automatically.

Accommodation for operating and maintenance personnel is also provided at Jagers Rust.

The main Jagers Rust-Driel canal is 38,1

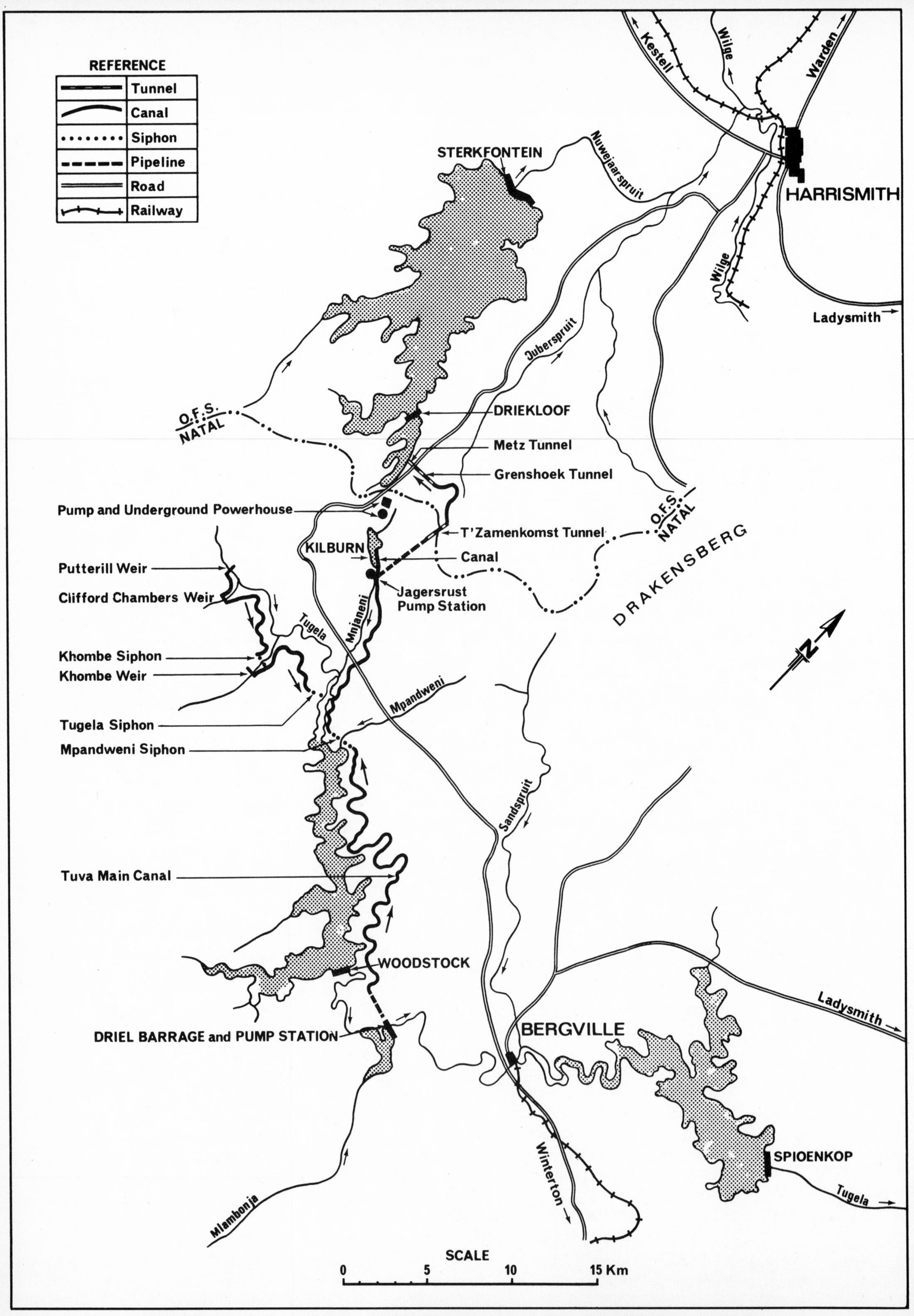

TUGELA-VAAL GOVERNMENT WATER WORK

km long and was designed to carry 11 cumecs on a gradient of 1:10 000. The canal is trapezoidal in shape with $1\frac{1}{2}$:1 side slopes and a 4,27 m bed width. At design discharge flow will be 2 m deep. A feature of this canal is that it is lined with 100 mm of unreinforced concrete, placed mechanically by a slip-form liner. Excavation was done by drag-line excavator and trimming by a mechanical trimmer fitted with a bucket chain running on guides set to the desired shape.

Associated with the main canal is the Mpandweni siphon which crosses the Mpandweni Spruit. This siphon is 1 740 m long and was built from 7 m-long prestressed concrete pipes with an internal diameter of 1 600 mm. The first stage makes provision for one barrel, and two more are to be built when the project is extended.

Driel barrage is the main storage unit in the first phase of the project. The barrage has a net storage capacity of 18,0 million m^3 and is situated in the Tugela River about 6 km upstream from Bergville, just below the confluence with the Mlambonja River. The structure comprises concrete gravity, outlet and spillway sections on the left bank, a retaining wall in the present river bed and an earth embankment on the right bank.

An interesting feature of the spillway section is that the crest level is 8,5 m above foundation level and that this portion of the structure is stressed on to the foundation by means of 120 cables each carrying a force of 108 tonnes. The cables are anchored at a depth of about 10 m below foundation level.

Three radial crest gates each 12,0 by 12,2 m surmount the spillway and impound water in storage to a full supply level of 1 134 m above sea-level. The maximum water depth is 22 m. The crest gates were designed to cope with a flood of 3 113 cumecs.

The whole structure required the excavation of about 46 000 m^3 of materials, the placing of 31 000 m^3 of concrete in sections varying from heavily reinforced thin walls and slabs to mass concrete in the foundations, and the compaction of 130 000 m^3 of earth in the embankment.

Situated in the headwaters of the Tugela River are three diversion weirs – Khombe weir in the Umkombi River, Clifford Chambers weir in the Tugela River and Putterill weir in the Putterill Spruit. These weirs are intended only as diversion and control structures and have insignificant storage capacity. They are built to the same basic design, with differences in size and flank conditions to suit the sites.

The weirs are interconnected by a system of canals which leads the water into the Jagers Rust-Driel canal via a siphon through the Tugela River.

All the canals in the Upper Tugela were built to a gradient of 1:2 500 except the Khombe which has a gradient of 1:1 000. They were hand-lined by the traditional method of screening between templates. Siphon spillways and long weirs are placed at strategic points along these canals to reject surplus flow in the event of emergencies.

The system of gates, valves and pumps will be operated automatically from Jagers Rust. Operating decisions will be made according to preset rules using water level, gate opening, pump delivery and flow data recorded and telemetered from strategic points in the system. When changes in gate openings or pump settings are required instructions to this effect will be relayed through the system and acted upon automatically.

The operating rules for the system aim at achieving maximum yield at lowest cost. As much water as the river flow will allow will be diverted from the Upper Tugela after providing for riparian consumption.

As the flow varies during the season the amount of water available from the Upper Tugela will fluctuate, and make-up water will be pumped from Driel at varying rates to keep the Jagers Rust pumps turning. In the event of no surplus water being available from the Upper Tugela, three pumps each will run at Driel and Jagers Rust.

Analysis of hydrological data for the Tugela River indicates that the system has an export potential of about 134 million m^3 a year, and that in the long run a utilisation efficiency of 97,3 per cent should be achieved, giving an average inflow into Sterkfontein Dam of 130 million m^3 a year. With this amount of water flowing each year into Sterkfontein Dam, with a capacity of 1 203 million m^3, the yield of the Vaal River system will be increased from 4,02 million m^{3}d to 5,03 million m^{3}d – an increase of 1,01 million m^{3}d or 370 million m^3 a year.

This then is the benefit gained from the first phase of the Tugela-Vaal project: an importation rate of 130 million m^3 a year giving an increase in yield of 370 million m^3 a year – thus making available 240 million m^3 a year of Vaal River water which at the moment is going to waste by evaporation and spillage.

It is estimated that this 25 per cent increase in available supplies will satisfy demands until 1983, by which time at least an extension to the Tugela-Vaal project should be in operation, or some other source of water should be harnessed. Since the first-phase pumping capacity will take about nine years to fill Sterkfontein Dam, these extensions will have to be commissioned not later than 1981. In view of the importance and strategic value of the undertaking the target date for completion is about 1978.

Early estimates of the financial implications of the cost of the scheme indicated that an additional 0,50 cents per m^3 would have to be paid by all users of Vaal River water below Vaal Dam, except irrigators, if the whole investment were to be repaid over the expected life of the various components and the operating and maintenance costs were to be covered.

This figure is merely an indication of the costs to be borne by users of the more secure water supply. The actual tariffs will be determined from time to time as the cost of money, interest rates and operating costs vary.

A key aspect of the Tugela-Vaal project is its phased development. The timing of additional phases and the priorities allocated to various attractive alternatives are continually reviewed. When the time for making a decision arrives, the most attractive course of action can be selected with confidence.

Sterkfontein Dam – downstream view

WAGENDRIFT DAM

HISTORY

Wagendrift Dam on the Bushmans River, Natal, about 6 km from the town of Estcourt was completed in 1963. It was built to regulate the river flow and stabilise the irrigation pattern for farming areas between the dam and the river's confluence with the Tugela River, mainly in the vicinity of Weenen. The town of Estcourt also derives benefit from the more stable river flow.

The dam, engineered and constructed by the owner, the Department of Water Affairs, is a multiple dome type with 4 thin walled cupolas supported by 5 buttresses with a radial gate control undersluice for scouring purposes. Its height is 40 m and net capacity 59,7 million m^3. The location and structure are aesthetically very pleasing; it is a popular spot for water sports.

DIMENSIONS

Type	Multi Arch
Height above lowest foundation	40 m
Length of crest	281 m
Volume content of dam	0,061 $m^3 \times 10^6$
Gross capacity of reservoir	59,681 $m^3 \times 10^6$
Purpose	Irrigation
Maximum discharge capacity of spillway	1 300 m^3/s
Type of spillway	Uncontrolled

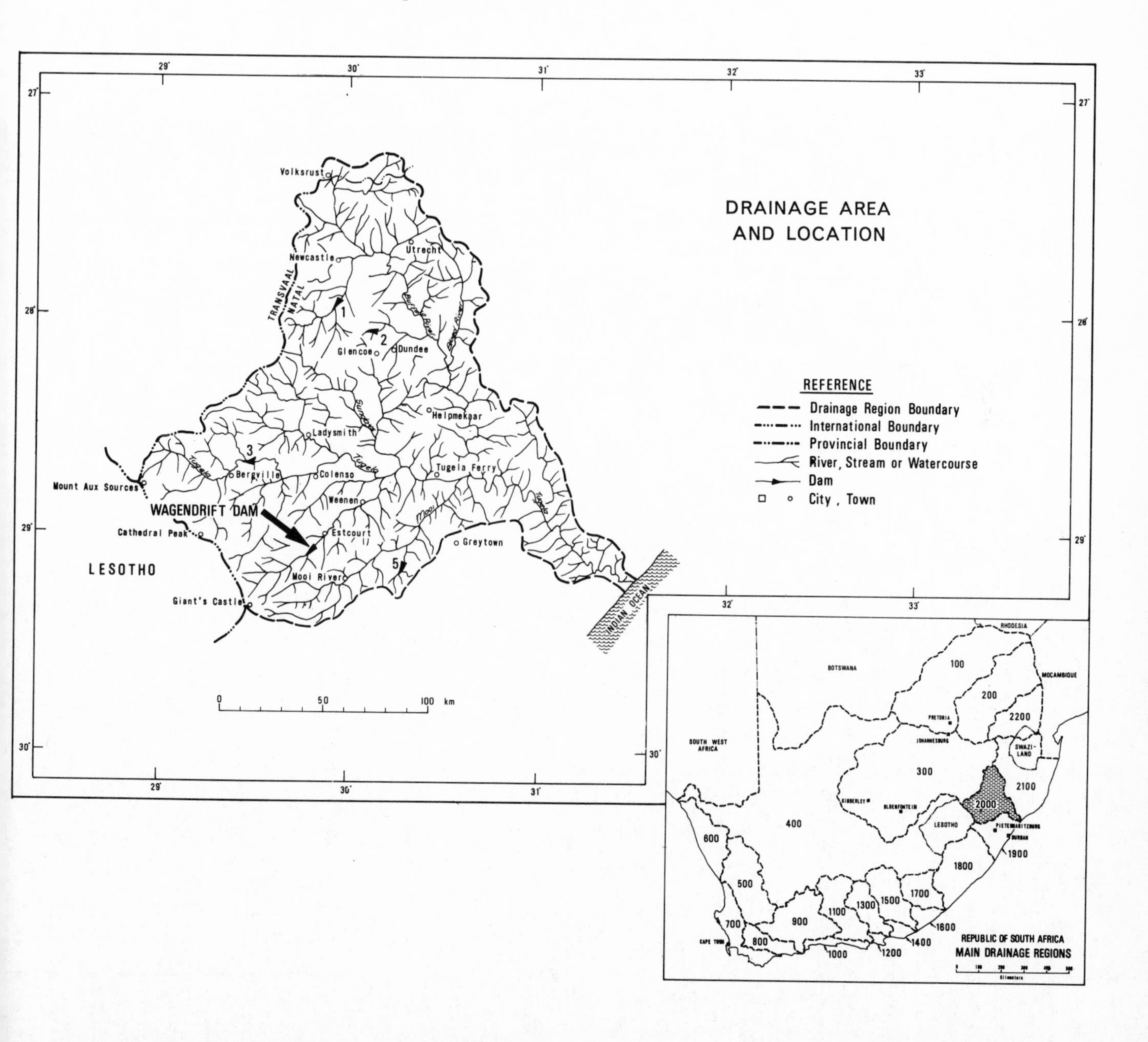

Wagendrift Dam

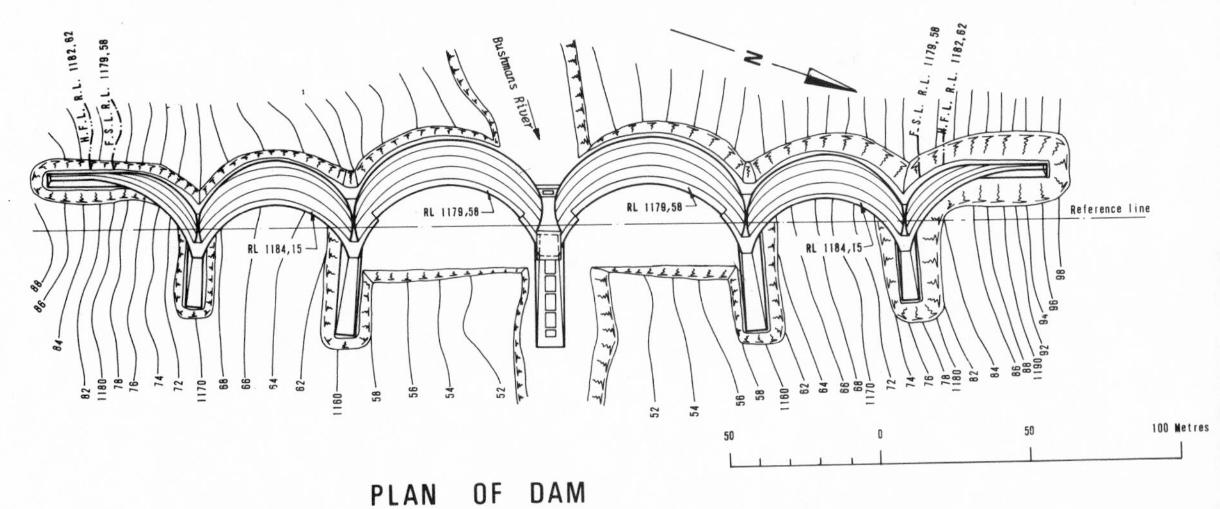

Wagendrift Dam showing dispersion valve in action (D.W.A.)

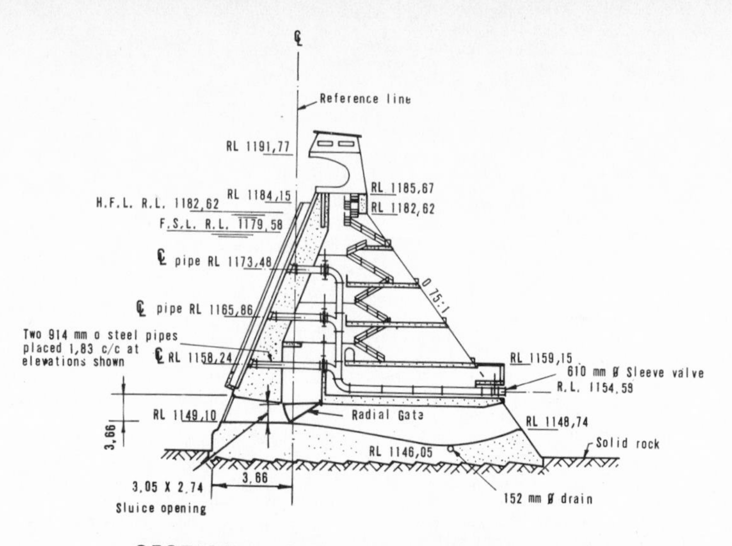

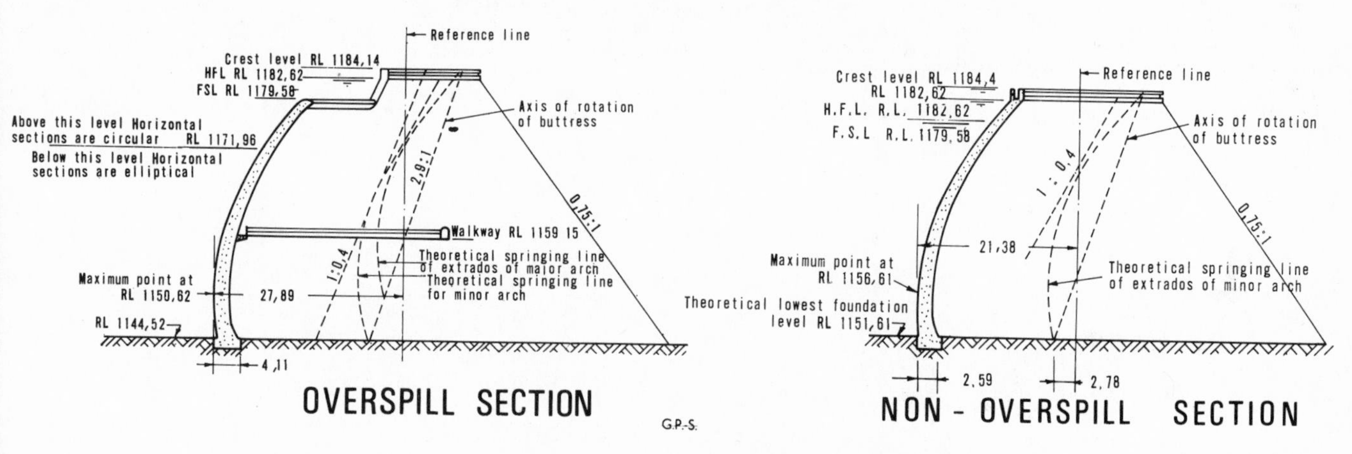

WAGENDRIFT DAM

CRAIGIE BURN DAM

HISTORY

The site of the dam is on the Mnyamvubu River about 32 km east of the town of Mooi River in the district Umvoti, Natal.

The prime reason for the construction was the need of stable irrigation of citrus, the main crop in this prosperous area.

The size of the catchment is 178 km² while the area irrigated by the scheme is 1 713 ha.

The dam, completed in 1963, consists of a double curvature concrete arch flanked by two earthfill embankments. It was engineered and constructed by its owner, the Department of Water Affairs.

The design of the dam made provision for future raising by 4,57 m but it was decided to build it up to its ultimate height of 39,3 m in one stage.

DIMENSIONS

Type . . .	Arch/Earth
Height above lowest foundation . . .	39,32 m
Length of crest	a) Concrete arch: 261,52 m b) Earth Emb: 259,08 m Total: 520,60 m
Volume content of dam . .	a) Concrete 19 880 m³ b) Earth Emb: 31 300 m³
Gross capacity of reservoir .	25 906 000 m³
Purpose . .	Regulation
Maximum discharge capacity of spillway .	685 m³/s
Type of spillway . . .	Uncontrolled

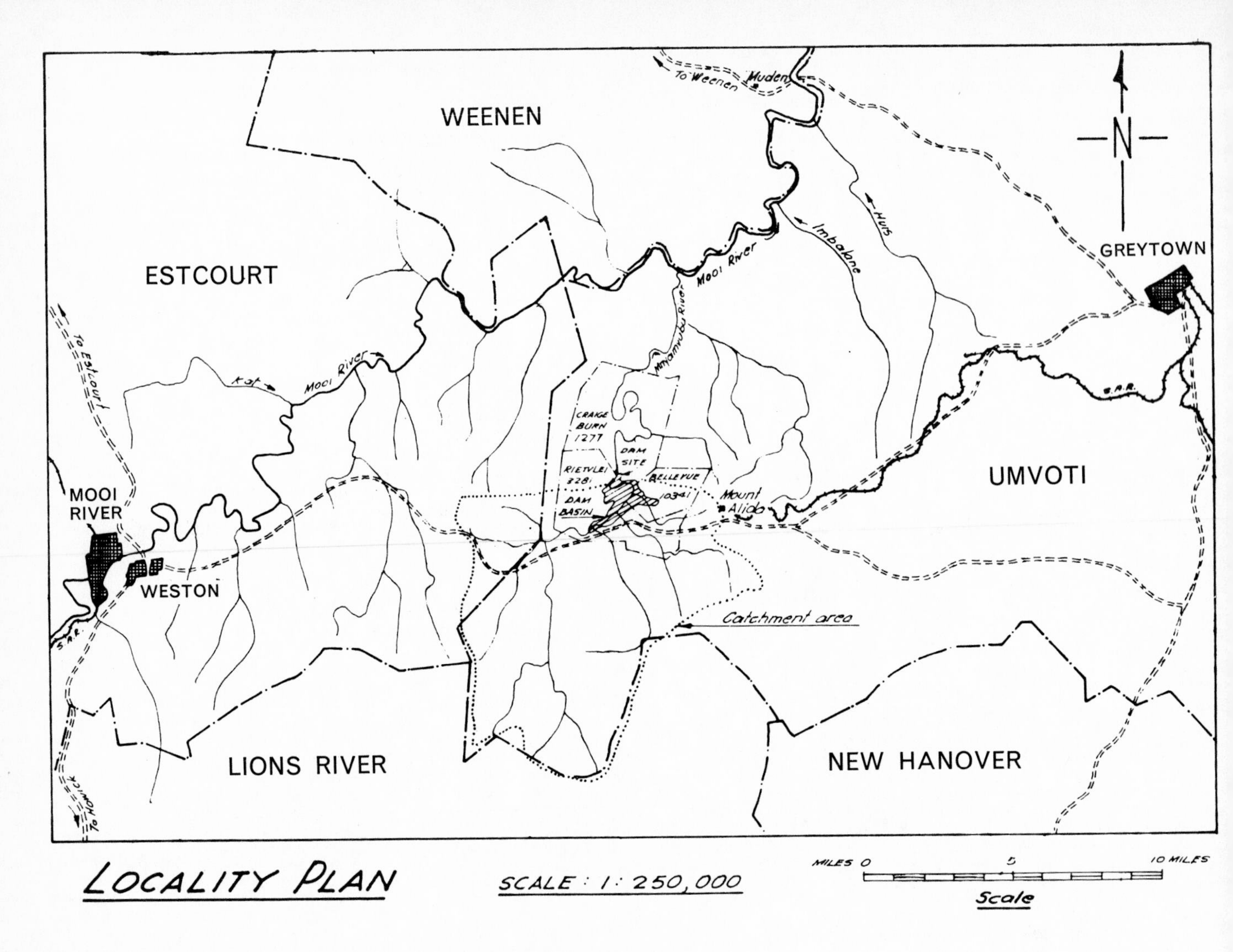

CRAIGIE BURN DAM

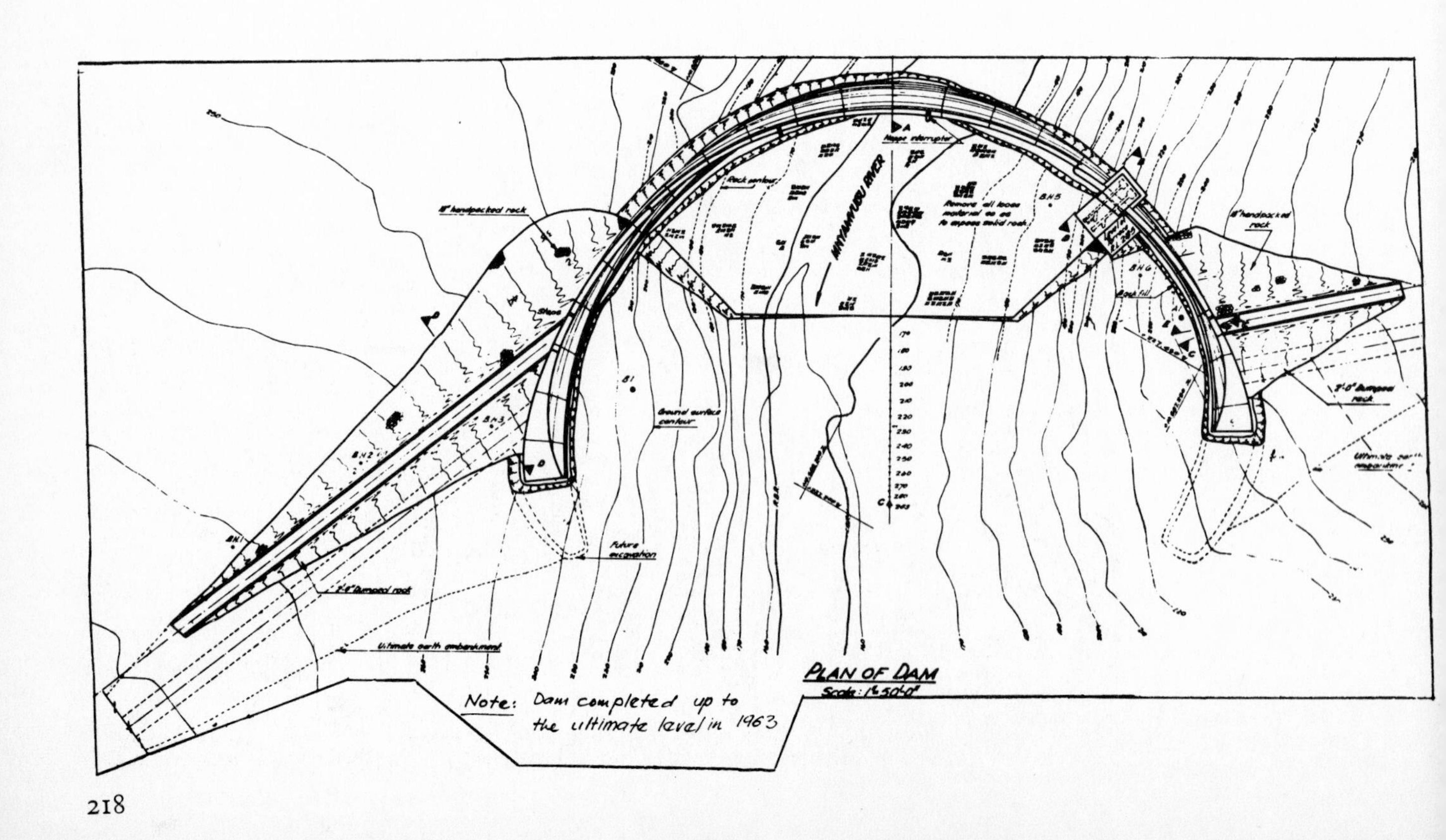

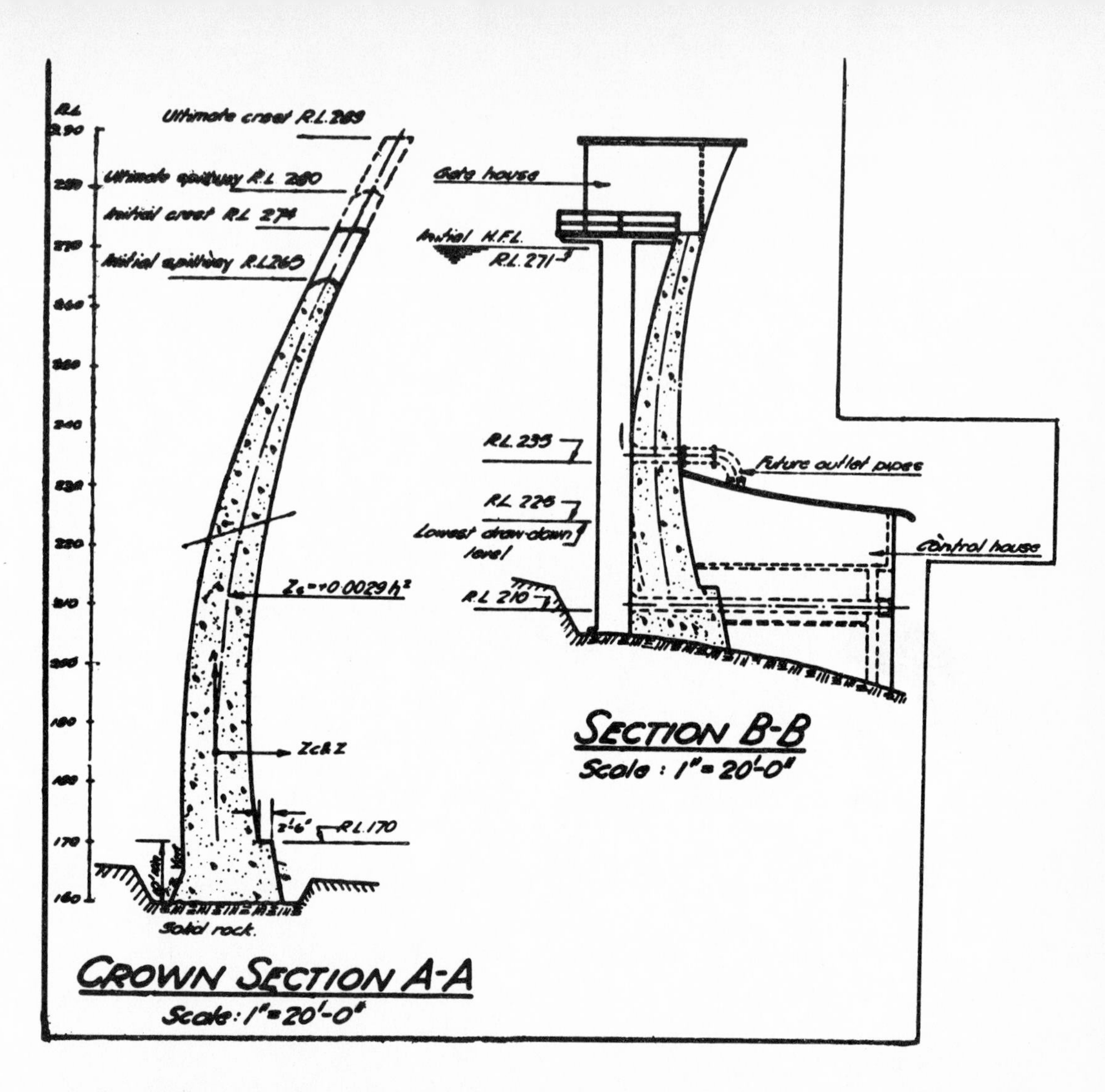

Craigie Burn Dam
Spillway and abutment
(D.W.A.)

NOOITGEDACHT

HISTORY

Nooitgedacht Dam on the Komati River in the Transvaal, supplies water to the Electricity Supply Commission Komati Power Station, which, when fully developed to its ultimate capacity of 1 000 megawatts, will require 90 920 m³ of water per day. It is an earth dam with a concrete overspill section on the left flank and is 44 m high, 1 021 m long and has a net capacity of 81 490 000 m. Water is pumped to the power station through a 68 km long pipeline. The dam, which was completed in 1962, was engineered and constructed by the owner, the Department of Water Affairs.

A unique structure that will decrease silting of the basin, by discharging silt-laden density currents instead of clear stored water over the spillway at times of flood, has been incorporated in the central section of the dam. This will increase the period before silt in the dam basin will necessitate raising the dam wall.

DIMENSIONS

Type	Earth
Height above lowest foundation	44 m
Length of crest	1 021 m
Volume content of dam	1 391 000 m³ (concrete 61 000 m³) (earth 1 330 000 m³)
Gross capacity of reservoir	84 146 000 m³
Purpose	Irrigation and water supply
Maximum discharge capacity of spillway	1 360 m³/s
Type of spillway	Uncontrolled

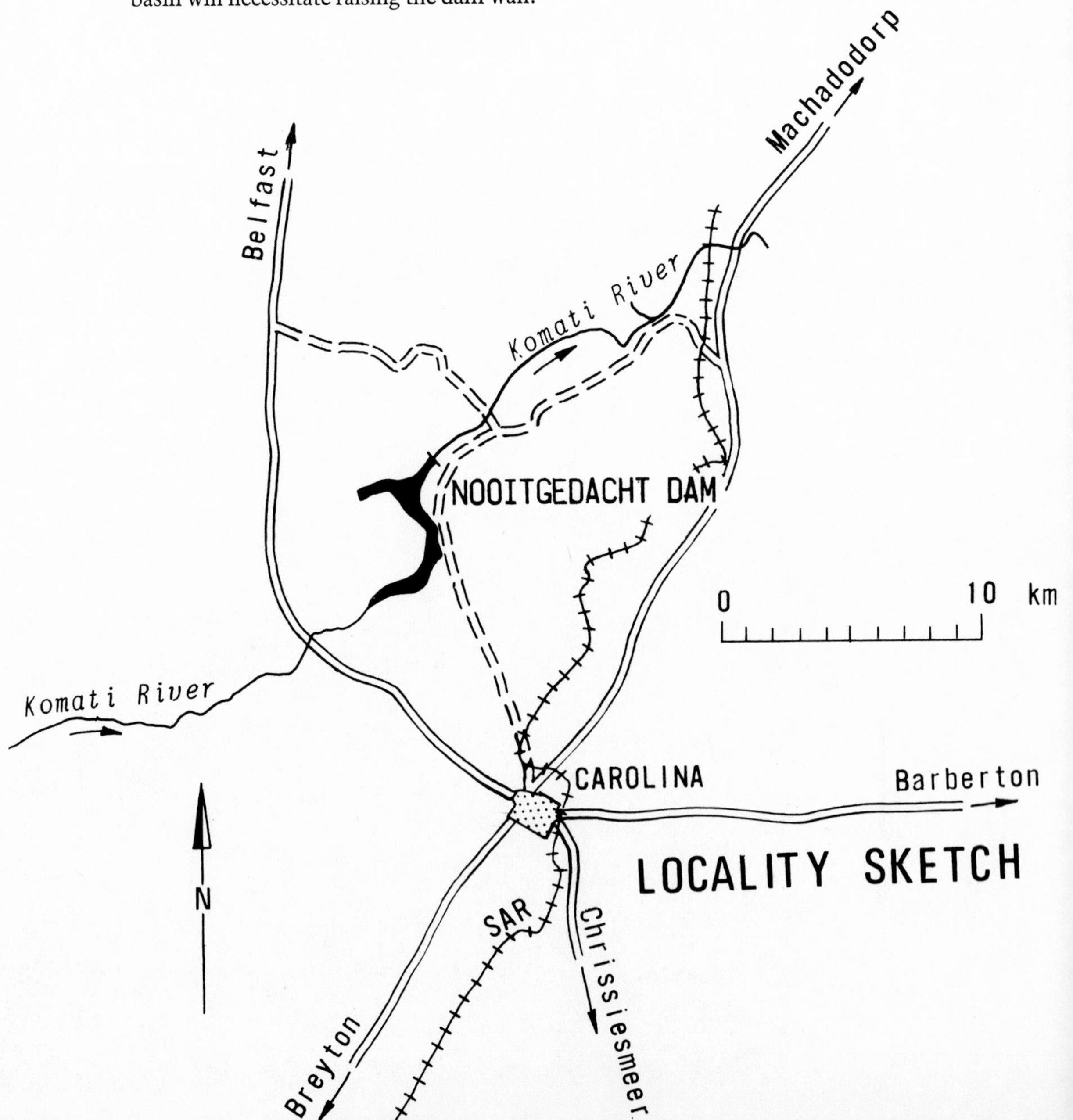

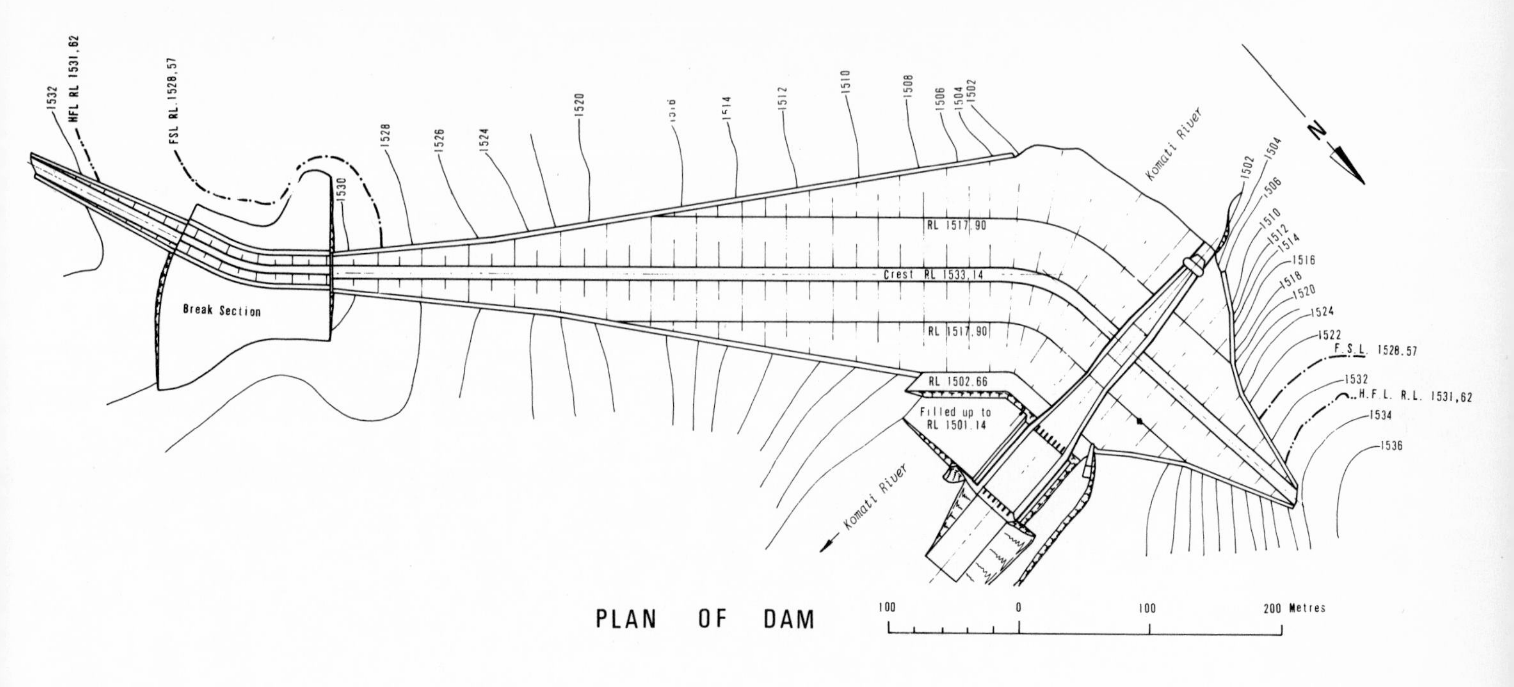

PLAN OF DAM

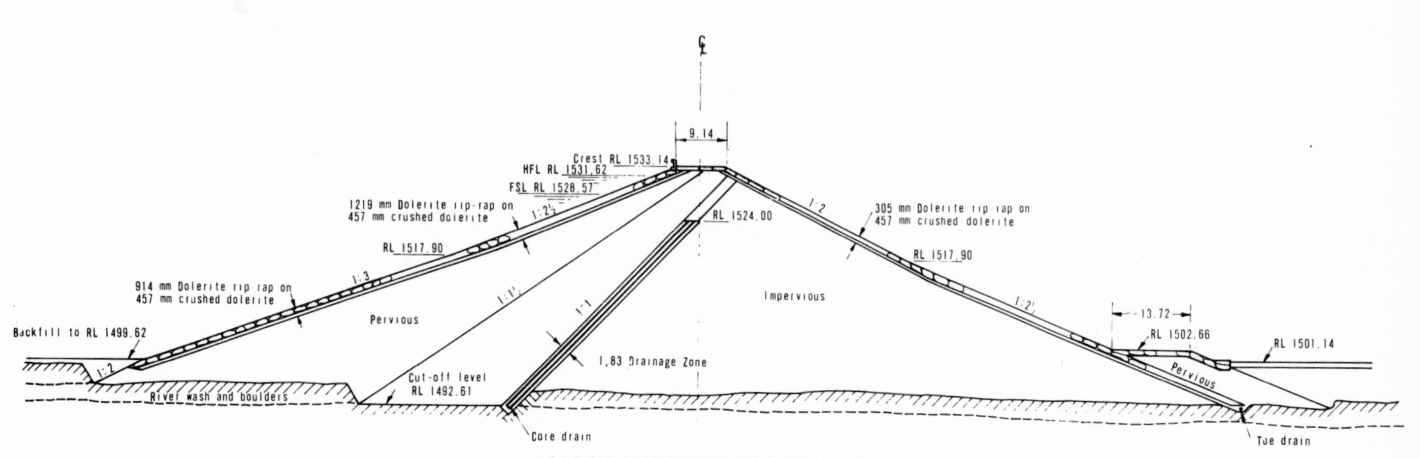

EARTH EMBANKMENT: SECTION

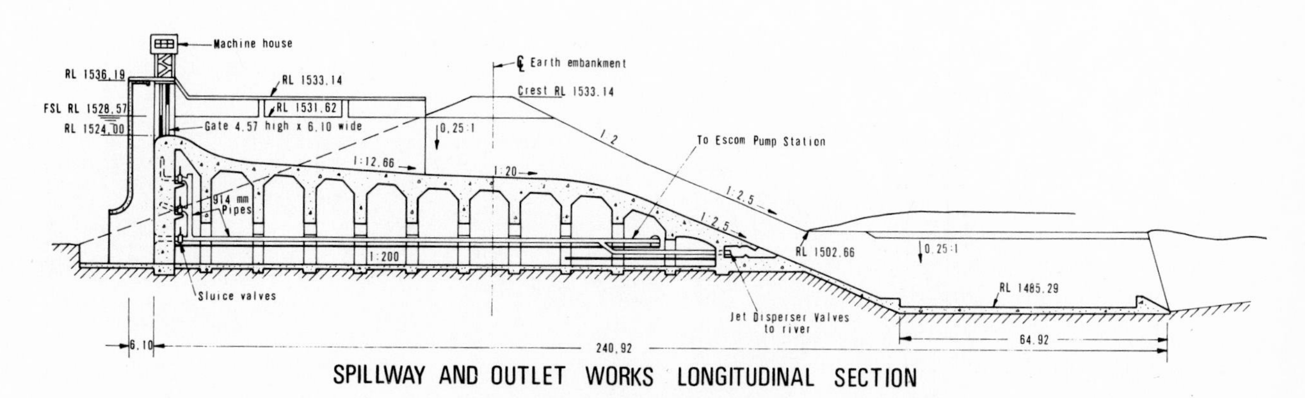

SPILLWAY AND OUTLET WORKS LONGITUDINAL SECTION

NOOITGEDACHT DAM

Machine-house looking upstream

Machine-house

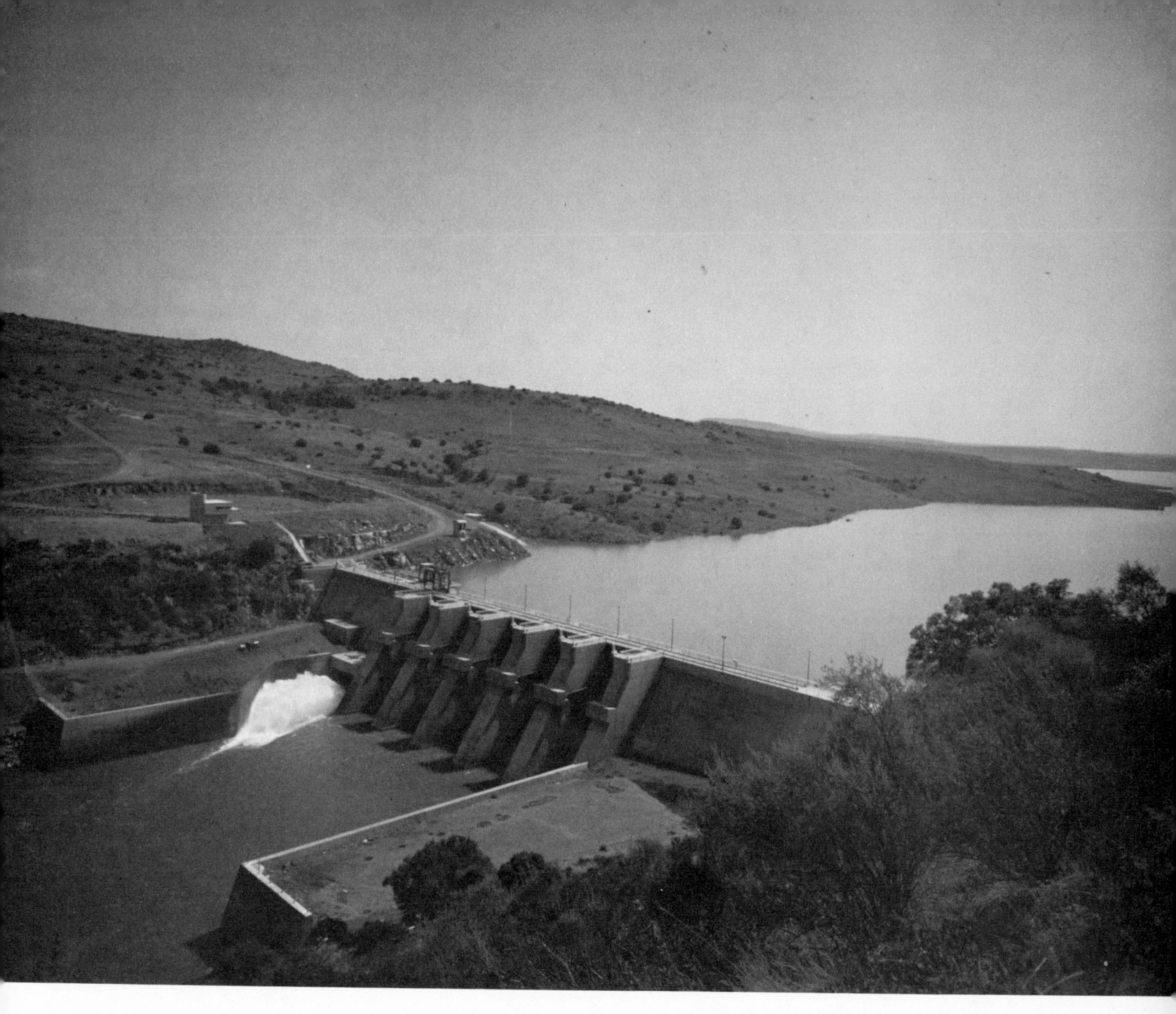

Welbedacht Dam looking upstream

WELBEDACHT DAM

HISTORY

Welbedacht Dam, completed in 1973, is on the Caledon River (a tributary of the Orange River) 25 km south-east of Wepener in the Orange Free State. It is mainly intended as an impounding reservoir for supply of water to the capital city, Bloemfontein, and other O.F.S. towns. The dam also functions in controlling floods and ensuring a more stable river flow for irrigation along the path of the river. Low-level pumps close to the dam pump water to a purification works, with an ultimate capacity 450 000 m³ per day. The clear water is pumped 6 km to a high level reservoir, from where the water flows by gravity pipeline 104 km to a terminal reservoir at Bloemfontein.

DIMENSIONS

Type	Gravity
Height above lowest foundation	32 m
Length of crest	192 m
Volume content of dam	100 000 m³
Gross capacity of reservoir	115 300 000 m³
Maximum discharge capacity of spillway	5 310 m³/s
Type of spillway	Controlled

CONSTRUCTION

The dam is a barrage type structure with five radial crest gates to control floods. The gates are 13,72 m high and 10,97 m wide, equipped with an automatic hydraulic lifting device. It was engineered and constructed by the owner, the Department of Water Affairs.

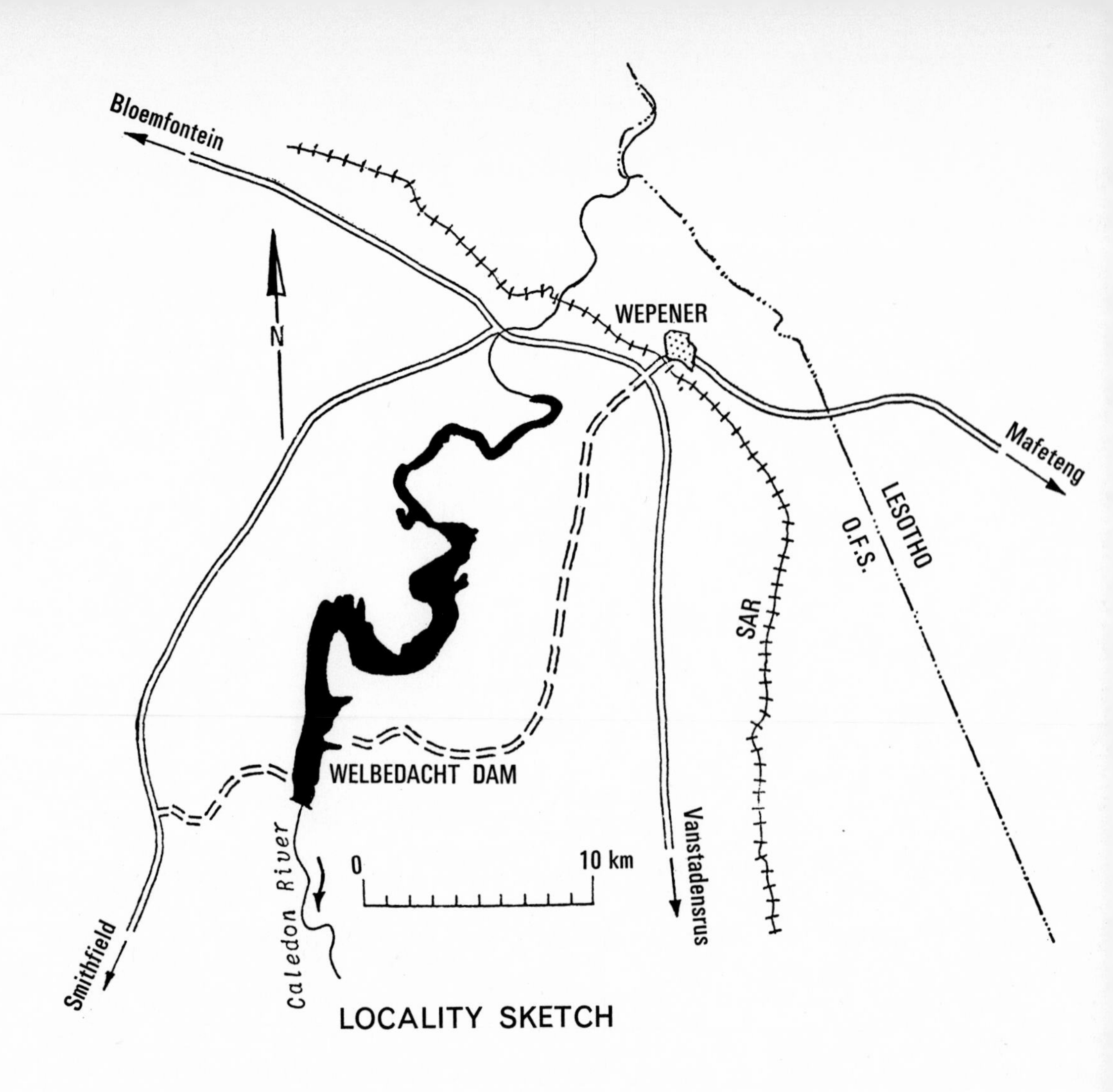

LOCALITY SKETCH

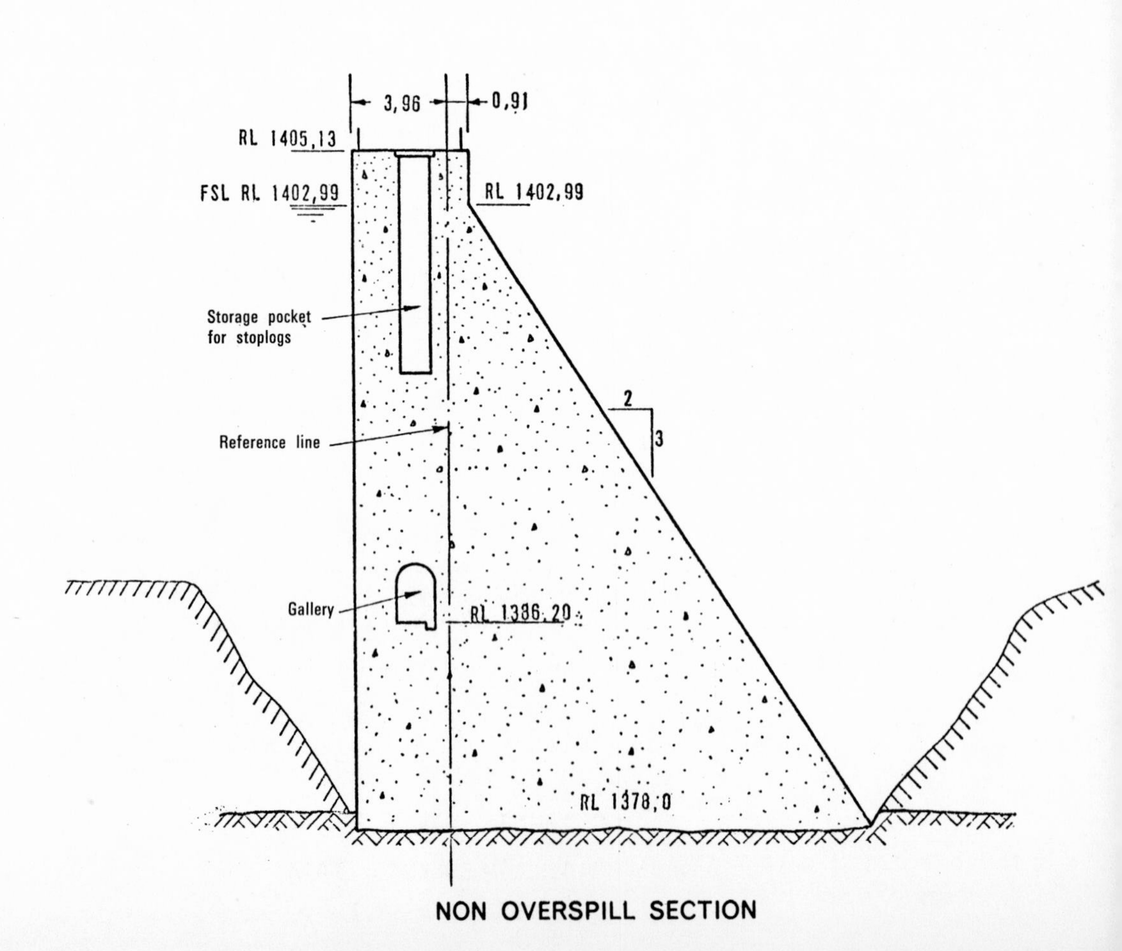

NON OVERSPILL SECTION

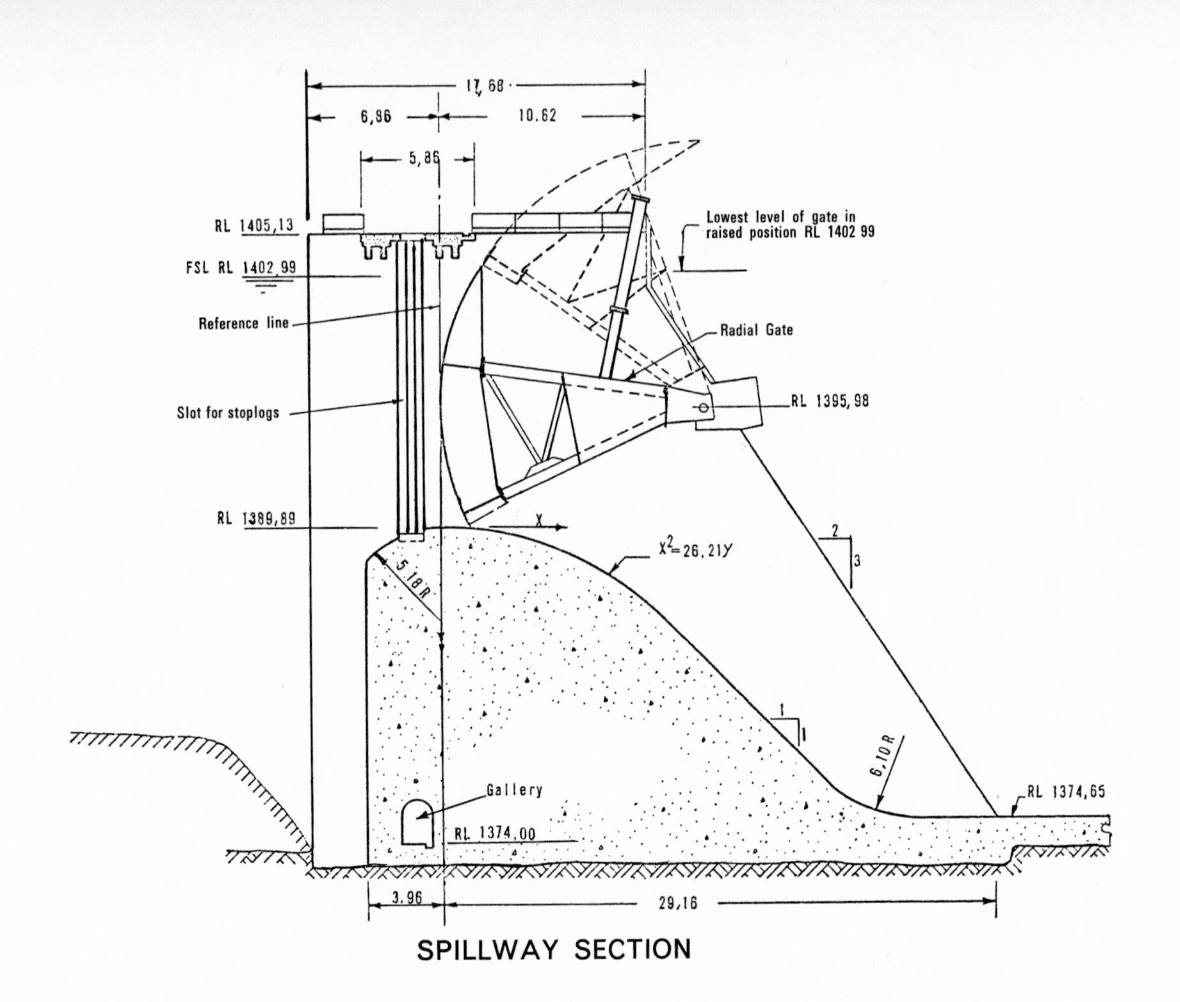

SPILLWAY SECTION

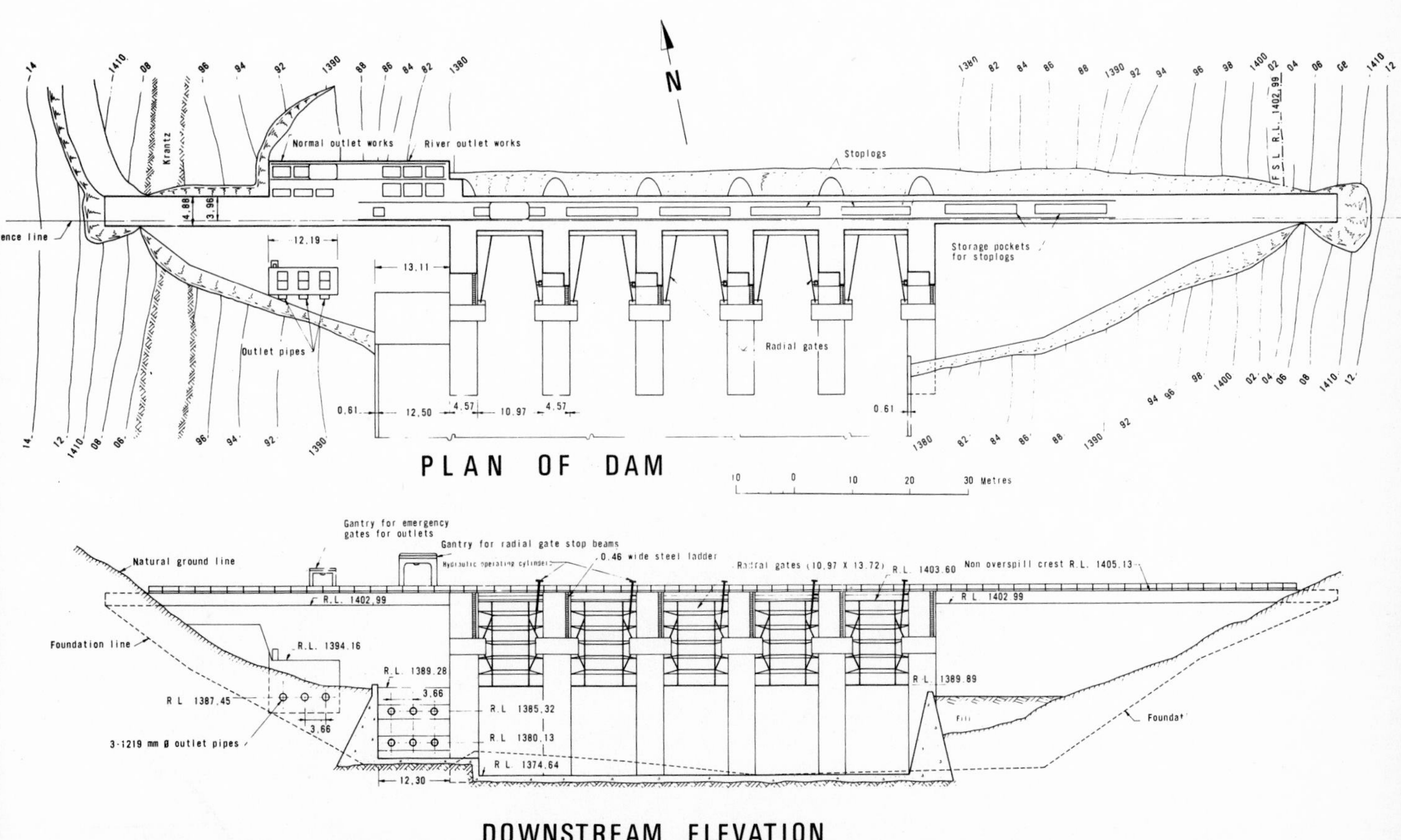

PLAN OF DAM

DOWNSTREAM ELEVATION

WELBEDACHT DAM

Glossary

It is realised that some countries have not yet converted to the metric system and others only recently. The following conversion tables and glossary of engineering terms are therefore included for those readers who may wish to familiarise themselves more fully with regard to the technical data quoted in the text:

Length:	1 mile	=	1,609 kilometres (km)
	1 foot	=	0,305 metres (m)
	1 inch	=	25,4 millimetres (mm)
Velocity:	1 foot per second	=	0,305 metres per second
	1 mile per hour	=	1,609 km per hour
Area:	1 sq. mile	=	2,59 square kilometres
	1 acre	=	0,4047 hectares
	1 sq. inch	=	645,2 square millimetres (mm^2)
	1 hectare	=	10 000 square metres (m^2)
Volume and Capacity:	1 acre ft.	=	1 233 cubic metres (m^3)
	1 cu. yard	=	0,765 cubic metres
	1 cu. ft.	=	28,32 litres
	1 m^3	=	1 000 litres = 1 kilolitre (kl)
	1 Imp. gallon	=	1,201 US gallons = 4,546 litres
Flow:	1 million gallons/day	=	4 546 m^3/day
		=	1,87 cusecs (cubic feet per second)
	1 cusec	=	0,028 m^3/s
	1 m^3/s	=	31,54 million m^3/annum
	mcm	=	million cubic metres
Force:	1 lb	=	0,45 kg
	1 tonne (t)	=	1 000 kg
	1 ton	=	1,02 tonne
Pressure:	1 pound per square inch	=	lb/in^2 or (p.s.i.) = 0,069 bar
Power and Energy	1 Horsepower (hp)	=	0,75 kilowatt (kW)
	1 Megawatt (MW)	=	1 000 kW
	1 Gigawatt hour (GWh)	=	1 million kilowatt hours (kWh)

Whereas the major part of the text is in metric units some of the drawings have been reproduced in foot-pound units as originally drawn.

References

Reference was also made to the following publications and documents:

1 *P. A. G. Bulletin, III* (3) 10–12.
2 *O.E.C.D. Observer*, No. 68, pp. 19–26, Paris 1974.
3 Ceres, F.A.O. *Review on Development*, No. 42, p. 6, 1974.
4 *Fertilizer Progress*, p. 19, May–June 1975.
5 Brandow, G. E., *Am. J. Agric. Econ.*, p. 1099, Dec. 1974.
6 *O.E.C.D. Observer*, No. 74, p. 20, 1975.
7 *Foreign Agricultural Economic Report No. 98*, U.S. Department of Agriculture, Washington, D.C.
8 Paarlberg, D. Paper delivered at Second General Assembly of World Future Society, June 3, 1975.
9 *Bioscience*, 25, 432, 1975.
10 *Africa Institute Bull.* XV(4), 131, 1975.
11 *Abstract of Agricultural Statistics*, 1975. Division of Agricultural Marketing Research, Pretoria.
12 Meadows, D. G. etal. *The Limits to Growth*, Universe Books, New York, 1972.
13 Nieuwoudt, A. D. *Immigration in the R.S.A.* (Unpublished report) June 1975.
14 *Agriculture Abroad*, XXX(3), 26, 1975.
15 Blakeslee, L. L. Heady, E. O. and Framingham, C. F. *World Food Production Demand and Trade*, Iowa State University, 1974.
16 *Homelands, the Role of the Corporations in the Republic of South Africa*, p. 61.
17 Raubenheimer, A. J. Speech at Congress of Soil Science Association of Southern Africa, July 2, 1975.
18 Kotze, J. M. *Africa Institute Bull.* XIV (8) 351, 1974.
19 *South West Africa Survey 1974*, pp. 3 and 34, Windhoek.
20 *Africa South of the Sahara 1975*, Europa Publications, London.
21 Fisher, A. C., *National Geographic*, May 1975. p. 662.
22 Verbeek, W. A., *Food and Agriculture in South and Southern Africa*, A paper delivered at a Conference entitled: "Resources of Southern Africa – Today and Tomorrow", held by the Associated Scientific and Technical Societies of South Africa at the Rand Afrikaans University, Johannesburg. September 22–26, 1975.
23 Faber, John. Oscar Faber and Partners, London. Private communication.
24 Reinius E. and Steby B. *Kidatu Powerplant, Tanzania.* International Water Power & Dam Construction. Vol. 28 No. 11, November 1976.
25 *The Economist* – September 25, 1976. Railways in Southern Africa.
26 Zambia Electric Supply Corporation Ltd. (ZESCO) *Kafue Hydro-electric Project.* Data Sheet.
27 Schmidt Carl Arne, *The Kafue Gorge Hydro-electric Power Project* Journal of the Engineering Institution of Zambia July 1969.
28 Sweco Stockholm *Kafue Gorge Hydro-electric Power Project* Technical Data Sheet, June 1970.
29 *Jane's Surface Skimmers Hovercraft and Hydrofoils* 1975–76.
30 Olivier, H. *Damit*, MacMillan South Africa (Publishers) (Pty.) Ltd. December 1975.
31 *The Naute State Water Scheme.* Brochure prepared for the opening of the scheme on 9th September 1972 by the Department of Water Affairs, South West Africa Branch.
32 *The Zambian Engineer*, Vol. 20 No. 2 April 1976. Article on Itezhitezhi dam by Mr. A. Mkandawire and Mr. S. Balasubrahmnayam.

GENERAL INDEX

Agricultural Statistics 32
Air Lubricated Barge (ALB) *Diagram* 36, 37
Algeria 19
Allemanskraal Dam 200, 206–207
Anglo American Corporation 17, 18, 93, 144
Angola 31, 32, 34, 76, 77, 123, 128, 143
'Apollo' Converter Station 78, 79, 81, 82, 83

Bafokeng Overflow Slimes Dam 18, 147
Bangala Dam 109–111
Bashee River 148
Beira 34, 39, 40, 41
Bloemhof Dam 175, 176, 208
Botletle River 144
Botswana 30, 32, 34, 35, 39, 77, 143–144
British Hydro-mechanics Research Assoc. 145
British Rail Seaspeed 35
Buffalo River 209
'Bunds' 144
Bushmans River 218

Cabora Bassa Hydro-Electric project 18, 22, 23, 27, 31, 32, 34, 35, 41, 54, 75, 77, 78, 79, 82, 85, 87, 147.
Caledon River 209, 223
Calueque 123, 125, 126, 128, 129
Caprivi Strip 143
Chikwawa 39, 63, 64
Chiredezi Canal 115, 117
Chobe River 30, 143
Churchill Dam 203–205
Churchill Falls, Canada 79
Churchill, Sir Winston 43
Craigie Burn Dam 217–219
Crocodile River 196, 209

Dar-es-Salaam 35, 39, 61
Darwendale Dam 112–113, 114
Driel Barrage 211, 213
Durban 39

Electrical Grid System, *Map* 21
Electricity Transmission Grid (S.A.) *Prologue* 33, 43, 60
Energy Crisis 19, 20, 25, 26
Erfenis Dam 200–202, 206
Esquilingwe Weir 100

Food Shortage 28
Frelimo 85, 86

Gamtoos Valley 185
Ghana 19
Gove Dam 123, 126, 128, 129
Grand Cooly, U.S.A. 79
Great Brak River 172
Great Dams in Southern Africa: *Map* 15
Great Fish River Valley 155, 156, 163, 165
Grootvlei Power Station 19
Gwenoro Dam 119–120

Hamilton Falls 39
Hartbeespoort Dam 196–199
Hendrik Verwoerd Dam 18, 25, 154, 155, 156, 157, 161, 163, 165, 169, 170, 175
'Herzov Furrow' 139, 140
Hex River Valley 181
Hippo Valley 102
Hovercraft 35, 37, 38, 40
Hunyani Poort Dam 114
Hunyani River 112, 114

ICOLD, *Preface* 144
Incomati River 32
Indwe River 150, 151
Inga, Grand Scheme, diagram 58, 60
Inga Project: Locality diagrams; Layout of Dams 17, 55
Inga Rapids 41

Ingwenzi Irrigation Project 120–122
Institution of Civil Engineers 17
Inyankuni Dam 108
Iran 20
Itezhitezhi Dam 17, 35, 65, 66, 68, 69

Jagers Rust Pump Station 209, 211, 213
Jane's Surface Skimmers 35

Kafue Gorge Dam 17, 35, 39, 65, 66, 68, 77, 92, 93
Kalahari Desert 30
Kalkfontein Scheme 192
Kariba 22, 23, 25, 26, 32, 35, 76, 77, 78, 79, 82, 89, 91, 97–101
Karoo System 166
Kenya 45, 48
Kidatu Hydro-Electric Scheme 61
Komati River 30, 32, 138, 140, 142, 209, 220
Kouga Dam (Paul Sauer Dam) 185
Krasnoyarsk, USSR 79
Kriel Thermal Station 19
Krokodil River 32
Kromme River 203
Kunene River 23, 123, 124, 125, 126, 127, 128, 129
Kyle Dam 102, 104, 109

Lake Albert 35, 45
Lake Cabora Bassa 34, 35, 76
Lake Kariba 76, 77, 89, 91, 93, 96
Lake MacDougall 115
Lake McIlwaine 112, 114
Lake Malawi 35, 39, 63
Lake Mentz 155
Lake Myoga 35
Lake Nyasa 39
Lake Tanganyika 35
Lake Victoria 35, 43, 44, 45, 46, 48
Lesapi Dam 105–107, 147
Lesotho 30, 32, 145–147
Letseng-Le-Terai Dam 18, 145
Limpopo River 30, 89, 143, 153
Liwonde Barrage 64
Lobito 34, 39
Loerie Dam 185
Lomati River 30, 138, 142
Loskop Dam 192–195
Löwen River 132, 133
Luanda 34
Lubisi Dam 18, 150
Lunsemfwa River 17, 71, 72
Lusaka 39, 41, 91
Lyra Da Silva, Dr. T. H. *Preface*

Malawi 26, 27, 31, 32, 34, 35
Manjirenji Dam 115–118
Maputo 34, 39
MaRobert 75
Matadi 35, 56
Mhlume Canal 18, 138, 142
Midmar Dam 189–191
Midshaft Township 165
Mita Hills Dams 17, 72
Mnyamvubu River 217
Moçambique 26, 27, 30, 31, 32, 34, 35, 40, 75, 76, 77, 78, 142
Mombasa 46, 53
Mopipi Dam 18, 143, 144
Mpatamanga 63
Mtilikwe River 102
Mulungushi Dam 71, 72
Murchison Cataracts 63, 64

Nagle Dam 189
Naute Dam 132–137
Ncala 34
Ncema Dam 108
Nile River 35, 43, 44, 45, 46, 48, 53
N'Kokolo Scheme 56

Nkula Falls Hydro-electric Scheme 64
Nooitgedacht Dam 220–222
Nova Lisboa 123, 126, 128
Nyamanyami, Kariba River god 91, 92

Odendaal Commission 123, 124–125
Okavango River 30, 123, 143
Olifants River 192, 209
Olushandja Balancing Dam 128
OPEC 19
Orange-Fish Tunnel 154, 156, 166
Orange River 23, 154, 155, 160, 161, 163, 170, 172, 175, 209, 223
Orange River Project 154, 155, 161, 208
Orapa Diamond Mine 144
Oviston Township 165, 167
Owen, Major E. R. 43
Owen Falls Hydro-electric Scheme 17, 43, 48
Oxbow Scheme 145

Paul Sauer Dam 18, 185–188
P. K. le Roux Dam 169–172
P. K. le Roux Power Station 18, 154
Putimolowane Pan 18, 143–144

Qamanco River 151
Qamata Irrigation Scheme 150
Queen Elizabeth II 44

Railways, Major Southern African *Map* 38
Rand Water Board 175, 176, 211
Rhodesia 25, 26, 27, 31, 32, 33, 34, 35, 39, 41, 77, 89
Richards Bay 34
Riet River 192
Ripon Falls 44, 45, 46, 53
Roode Elsberg Dam 181–184
Rovuma 23
Ruacana-Calueque Project 17, 125, 126–127, 128
Rufiji River 23

Sabi River 31, 32, 105
Saldanha Bay 34
Sanddrift River 181
Sand River Dam 139, 140, 142, 200, 206
Sand-Vet Government Water Scheme 200, 206
Shire River 23, 35, 39, 63, 64
Shongo Dam 56
South West Africa 32, 123, 130–137
Spioenkop Dam 210
Steenbras Dam 18, 179–180
Sterkfontein Dam 176, 210, 211, 213, 214
Suikerbos Pump Station 176
Sundays River 155, 156
Swakop River State Water Scheme 130
Swaziland 30, 32, 138–142
Swaziland Irrigation Scheme (SIS) 18, 30, 138, 140, 141
Sweden 19

Tanzania 26
Tedzani Falls 64
Teebus Spruit 165
Third World Countries *Prologue*, 27, 33
Transkei 24, 29, 148–151
Tsomo Dam 18, 151
Tugela River 23, 24, 153, 176, 209, 210, 211, 213, 215
Tugela-Vaal Water Scheme 208–214
Tumut 3, Australia 79
Tunnels, Table of Comparison 166
Tweerivieren Dam (Paul Sauer Dam) 185
Tzamenkomst Tunnel 211

Uganda Electricity Board 17, 44
Umgeni River 189
Umshagashe River 102
Umtata Hydro-electric Scheme 148–149
Umzimvubu River 148, 153
Umzingwane Dam 108
Union Corporation Ltd 147
Usutu River 30, 142, 209

Vaal Dam 18, 123, 154, 175–178, 208, 211, 214
Vaal-Hartz Irrigation Scheme 175, 192, 208
Vaal River 24, 154, 172, 176, 192, 208, 209, 211, 213, 214
'Vaal River Triangle' 176
Van Der Kloof Canal 170–172
Van Eck Power Station 125
Vet River 200, 206
Victoria Falls 91
Vila Fontes 35, 40

Wagendrift Dam 18, 215–217
Ward, James C. B. E. 94
'Water for Peace' Conference 22
Waterways 35–41
Welbedacht Dam 18, 223–225
Wemmershoek Dam 18, 173
White Kei River 150
Wilge River 176, 210, 211
Willem Pretorius Game Reserve 206

Zaire 26, 39, 77
Zaire River 23, 26, 35, 41, 54
Zambezi River 22, 23, 26, 27, 31, 32, 34, 35, 36, 37, 38, 39, 75, 76, 82, 85, 89, 91, 92, 96, 97, 147
Zambia 26, 27, 31, 32, 34, 35, 68, 77, 91

CONSULTING ENGINEERS AND CONTRACTORS INDEX

AEG 86, 160
African Batignolles Construction Ltd 159, 168
African Gate & Fenceworks (Rhod) Ltd 122
Airmec, Zambia 100
Air Survey Co Ltd 135
Allamanna Svenska-Elektriska Aktibolaget (ASEA) 99, 100, 129, 160
Alsthom, Paris 66, 86, 88
Amalgamated Power Engineering Ltd 160
Andrews & Kidd 108
Associated Electrical Industries Export Ltd 98

Babcock & Wilcox Ltd 99
Balfour Kilpatrick 100
Bateman, Edward L. Ltd 117, 159, 160
BECO, Stockholm 66
Begg, G 203
Bird & Robertson 159
Board of Engineers 173
Boart & Hard Metal Products 132, 168
Bonadei, A. Construction Ltd 137
Boving & Co Ltd 98
Boyles Bros. Drilling Co of USA 168
British General Electric Co of Central Africa 99
British Insulated Callender's Cables Ltd 98
Brown, Boveri & Cie Aktiengesellschaft 86, 98, 100
Bulawayo Blasting Co Ltd 119
Burton, A. G. Ltd 98, 159
Burton Construction Ltd, Lusaka 66, 72, 100
B.V.S. Ltd 129

Cabinete do Plano do Zambezi (CPZ) 76
Campos, A 123
Capco 17
Cape Town, Municipality of 179
Cementation (Africa Contracts) Ltd 159
Cementation Co (of Rhod) 72, 98, 107, 117, 121
Cementation Co (of SA) 145, 147, 159, 180
CGEE-COGELEX 88, 100
Chilanga Cement Ltd 99
Chloride Electrical Storage Co Ltd 160
Christiani and Nielson Ltd 53
Clifford Harris Ltd 114
Clough, Smith & Co (Central Africa) Ltd 98
Cohen & Co 100
Commonwealth Development Corporation 18, 30, 138, 140, 141, 142
Compagnie de Constructions Internationales 86, 168
Compagnie Industrielle de Travaux 86
Concor Construction Ltd 136, 137, 151
Connolly, O & Co Ltd 107
Consani Engineering Ltd 132
Consortium Zamco 17
Construzioni Interrazionali STA 129
Coopers & Lybrand 98, 100
Coyne, André & Bellier, J., Paris 93, 97, 102, 110, 158
C. R. C. Engineering Ltd 160
CSIR 154

Delkins Ltd, Lusaka 66
Department of Water Affairs 150, 157, 165, 168, 169, 170, 175, 181, 185, 189, 192, 193, 196, 200, 206, 211, 215, 217, 220, 223
Dessingtons 117
Dommisse & Durham 140
Dorman Long & Co Ltd 53, 72, 98, 120, 160
Dowson and Dobson 159

East Hunyani Earthmovers (Rhod) Ltd 122
Edmund Nuttall, Sons & Co (London) Ltd 53
Electrical Installations Ltd 53
Electricity Supply Commission (Malawi) 17
Electricity Supply Commission (SA) (ESCOM) 18, 19, 20, 22, 23, 24, 77, 86, 154, 159, 161, 169, 211
Electro-consult (Milan) 17
Energoprojekt (Belgrade) 66, 100
Engineering Design & Construction Co Ltd 159
English Electric Co Ltd 98, 160

Entreprises Campenon-Bernard 86
Etablissements Neyrpic 98

Fainsinger & Associates 137
Federal Hydro-electric Board 93
Federal Power Board 93, 94
Ferranti Ltd 99, 100
First Electric Corporation of SA Ltd 160
Fotogramensura Ltd 132
FOUGEROLLE 86, 88
Founders Steel Industries 121
Fuji Electric Co Ltd 159

Gabinete do Plano do Cunene (GPC) 123
G. F. T. Products Ltd 160
Gibb, Sir Alexander, and Partners 17, 18, 53, 97, 100, 138, 145
Gibb, Coyne, Sogei (Kariba) Ltd 97
Gibb, Hawkins & Partners 18, 145, 159
G. I. E., Milan 66
Grinaker Construction Ltd 122, 144, 150
Grist, J. B., Ltd 122
'Grupo de Trabalho para o Zambezi' (GTZ) 76
Gulliver, W. J. & R. L., Ltd 107, 112, 119, 121

Hawker Siddley Brush Ltd 117
Hawker Siddley, U.K. 100
Hawkins, J. C. 203
Herma Bros. Ltd 132
Hidro-Electrica do Cabora Bassa 86
Hidrotechnika, Belgrade 66
Hidrotecnica Cabora Bassa (H.C.B.) 17
Hidrotecnica Portuguesa 76, 78, 123
Hilton Barber, S. Ltd 140
HOCHTIEF (West Germany) 17, 86, 88
Hollandsche Beton Maatschappij N.V. 53
Hooper & Stopforth Ltd 117
Hoverlloyd 35
Hovermarine, U.K. 18, 37
Hubert Davies Ltd 137, 159, 160
Hume Pipe Co. Ltd 117, 120, 122, 132, 137, 140
Hydrauliques SARL 159
Hydroconsults 129, 132

Impregilo & Recchi 68
Impresa Ing. Di Penta of Italy 168
Impresit 72, 98, 120
Incledon, H & Co (SA) Ltd 122
Industrial Development Corporation (SA) 123
INGRA-LITOSTROJ, Yugoslavia 62
INGRA-METALNA, Yugoslavia 62
INGRA-RADE KONCAR, Yugoslavia 62
Internationale Gewapendbeton Bouw N.V. 53
International Orange River Consultants Co 158
International Power & Engineering Equipment Ltd 160
Iranian National Petroleum Company 19
Issels, F. & Sons Ltd 122
Italafrica/Kinshasa 17

James Thompson Manufacturing 159
Jeffares & Green, Salisbury 97
Jeffares, Mr 91
Jeffrey Manufacturing Co 132, 136
Jennings & Midgely 147
John Laing & Sons 139
Johnson & Fletcher Ltd 99

Kanthack, Dr F. E. 72, 180
Kariba Transport Ltd 99
Keir & Cawder Ltd 139
Kennedy & Donkin (Africa) 17, 53, 64
Kicon 62
Kier, J. L. & Co (London) Ltd 53
Kinnear Moody Group Ltd 100
Kvaerner Brug, Oslo 66

Lafrenz, E. Ltd 132
Leon Hurter, Chase and Holmes Ltd 137

Lewis Construction Co Ltd 98, 122, 203
Liebherr-Africa Ltd 140
Lilienthal 19
LTA Ltd, Johannesburg 17, 18, 86, 168
Lund, B. G. A. and Partner 18, 144

M & Z Motors & Engineering Ltd 132, 136
McAlpine, Sir Alfred & Son 98, 117
McAlpine & Concor Ltd 102, 110
McCullough & Naisbitt 122
Malawi Electricity Supply Commission 63, 64
Mann, J. & Co Ltd 98
Mariental Transport Ltd 135
Marryat & Scott Ltd 53
Mather & Platt, England 100, 132, 140
Merz & McLellan, London 98, 100
Metalna, Yugoslavia 100
Meyer, H. 132
Missao do Fomento e Povoamento do Zambeze (MFPZ) 76
Mitchell Construction 100
Modern Engineering Ltd 122
Morris Cranes 160
Morris, Dr S. S. 173
Mostofi, Dr Baghar 19
Mountbatten Hovercraft 36
Murphy, J. & Sons Ltd 132
Mwanandi Engineering & Contracting Co., Tanzania 62

National Trading Co Ltd 159
Nchanga Consolidated Copper Mines 17, 71, 72
Nederlandsche Aanneming Maatschappij N.V. 53
Nederlandsche Beton Maatschappij Nato N.V. 53
Ninham Shand & Partners 159
Norris Warming Co Ltd 53

O'Connell, Manthé & Partners 150, 151
Olivier, Dr H. 145, 150
Otis, South Africa 100
Owen Falls Construction Company 53

Parsons, C. A. & Co Ltd 99
Pirelli, Italy 100
Powerlines Ltd 86, 88

Rade Koncar, Yugoslavia 100
Rand Mines Ltd 71
Reunert & Lenz Ltd 160
Reyrolle & Co Ltd 99
Rheinstahl Union Bruckenbau A. G. 117
Rhodesian Broken Hill Development Co Ltd 71
Rhodesian Electricity Supply Commission 91
Rhodesia Power Lines Ltd 99
Richard Costain 98
Rio Tinto Co 145
Roberts Construction Co Ltd 98, 117, 119, 159
Rodio SA Ltd 135
Royal Netherland Harbour Works Co 62

SAE 86, 88
S.A. Liquid Meters Ltd
Savage, J. L. 173
Schindler Lifts Ltd 99
SENTAB, Sweden 62
Shaft Sinkers 86, 88
Shell Co (of Rhod) 99, 119, 122
Shell Co (SA) Ltd 160
SICAI (Rome) 17
Siemens Ltd (SA) 17, 86, 88, 129, 159
Simel 160
Skanska Cementgjuteriet, Sweden 62, 66
Smit, J. F. 160
Societa Nazionale Delle Officine Di Savigliano 160
Société B.V.S. 159, 161
Société d'Engineering pour l'Industrie et les Travaux Publics 158
Société des Grands Travaux de Marxeille 86
Société Francaise d'Entreprises de Dragages et de Travaux Publics 86
Société Générale de Exploitations Industrielles, Paris 97
Société Générales d'Entreprises 86
Société Grenobloise d'Etudes et d'Applications 158
SOREFAME 88, 102, 107, 110
South West Africa Water Electricity Corporation (SWAWEK) 17, 18, 123, 124, 125, 126, 129
South West Engineering Ltd 132
Sprecher & Schuh AG 160
Standard Telephones & Cables Ltd 98
Steel Metals Ltd 107, 112
Stein, J. W. & Partners 137
Stevens & Dawnays Ltd 119
Stewart, G. A. 203
Stewarts & Lloyds of Rhod. Ltd 117, 120, 159, 160
Stirling Astaldi, Italy 62, 139
Stuart, A. Ltd 159
Suidwes Konstruksie 132
Swart & Hancock 99
SWECO 62, 66, 68
Swemkor 132

Tanzania Electric Supply Company Ltd (TANESCO) 62
Taylor, Lourens & Haigh Ltd 122
Telewei Ltd 137
Thermal & Corrosion Contractors Ltd 117
Thompson-Cramond Ltd 140
Thornton's Transportation Rhodesia Ltd 99
Tokyo Shibaura Electric Co Ltd 160
Tunelogradnja, Belgrade 66

U. C.-Dumez-Borie Dams 159
United Transport Holdings 88
United Transport Overseas 100

Van N. E. P. Construction Ltd 136
Van Niekerk, Klein & Edwards 140, 159
Van Wyk & Louw 159
VBB, Stockholm 66
Voest, Austria 100, 129
VOITH 86, 88
Vorster, B. J., The Honourable *Foreword*, 160, 163
Vosper Thornycraft 36
V.P.C. 86, 88

Waring & Gillow Ltd 53
Watermeyer, Legge, Piesold and Uhlmann 17, 71, 72, 108, 159
Westinghouse, USA 129
Wimpey, Geo. & Co (London) 173
Wright Anderson Ltd 159

Zambian Electricity Supply Commission 17, 65, 68
Zamco Consortium 77, 78, 85
ZECCO, Lusaka 66